AF251956

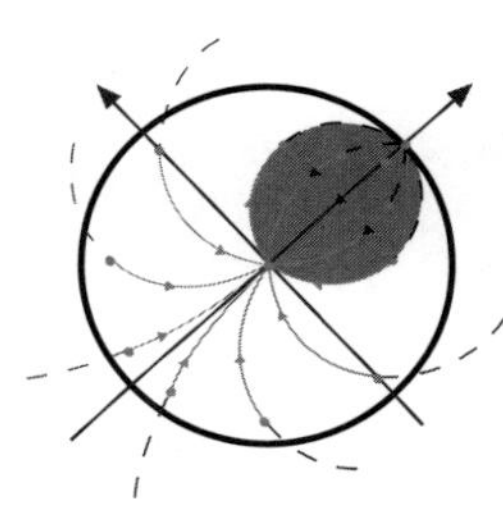

Nonlinear Semigroups, Fixed Points, and Geometry of Domains in Banach Spaces

NONLINEAR SEMIGROUPS, FIXED POINTS, AND GEOMETRY OF DOMAINS IN BANACH SPACES

Simeon Reich

The Technion - Israel Institute of Technology, Israel

David Shoikhet

ORT Braude College, Israel

Imperial College Press

Published by

Imperial College Press
57 Shelton Street
Covent Garden
London WC2H 9HE

Distributed by

World Scientific Publishing Co. Pte. Ltd.
5 Toh Tuck Link, Singapore 596224
USA office: 27 Warren Street, Suite 401-402, Hackensack, NJ 07601
UK office: 57 Shelton Street, Covent Garden, London WC2H 9HE

British Library Cataloguing-in-Publication Data
A catalogue record for this book is available from the British Library.

ISBN   1-86094-575-9

Printed in Singapore by World Scientific Printers (S) Pte Ltd

Preface

Nonlinear semigroup theory is not only of intrinsic interest, but is also important in the study of evolution problems. In recent years many developments have occurred, in particular, in the area of nonexpansive semigroups in Banach spaces. As a rule, such semigroups are generated by accretive operators and can be viewed as nonlinear analogs of the classical linear contraction semigroups.

In the last forty years the theory of monotone and accretive operators has been intensively developed by many mathematicians (see, for example, [Brézis (1973)] and [Barbu (1976)]) with many applications to nonlinear analysis and optimization. This theory is closely connected with the generation theory of nonlinear one-parameter semigroups of nonexpansive mappings and with nonlinear evolution problems.

In a parallel development (and even earlier) the generation theory of one-parameter semigroups of holomorphic mappings in $\mathbb{C}^n$ has been an object of interest in the theory of Markov stochastic processes and, in particular, in the theory of branching processes (see, for example, [Harris (1963)] and [Sevastyanov (1971)]). The central problem in the study of such processes is to locate the extinction probability which can be defined as the smallest common fixed point of a semigroup of holomorphic mappings or, equivalently, as the smallest null point of its generator.

Later such semigroups appeared in other fields: one-dimensional complex analysis [Löwner (1923); Kufarev (1943); Kufarev (1947); Lebedev (1975); Aleksandrov (1976)], finite-dimensional manifolds [Kaup and Vigué (1990); Abate (1992)], the geometry of complex Banach spaces [Arazy (1987); Isidro and Vigué (1984); Kaup (1983); Dineen (1989)], control theory and optimization [Helmke and Moore (1994)], and Krein spaces [Vesentini (1987a)–(1987b); (1991)]. For the finite-dimensional case, M. Abate proved in [Abate (1992)] that each continuous semigroup of holomorphic mappings is everywhere differentiable with respect to its parameter, i. e., is generated by a holomorphic mapping. In addition, he established a criterion for a holomorphic mapping to be a generator of a one-parameter semigroup. (Such a problem is equivalent to the global solvability of a complex dynamical system.) Earlier, for the one-dimensional case, similar facts were presented by E. Berkson and H. Porta in their study [Berkson and Porta (1978)] of linear continuous semigroups of composition operators in Hardy spaces. It seems that the first deep study of semigroups of holomorphic mappings in the infinite-dimensional case is due to E. Vesentini. In [Vesentini (1987a)] he investigates semigroups of those fractional-linear transformations on the open unit Hilbert ball $\mathbb{B}$ which are isometries with respect to the hyperbolic metric on $\mathbb{B}$. The approach used there is based on the correspondence between such nonlinear semigroups and the strongly continuous semigroups of linear operators which leave invariant the indefinite metric on a Pontryagin space of defect 1. In [Vesentini (1987b)] and [Vesentini (1991)] this approach has been developed for general Pontryagin spaces and also for Krein spaces. Note that, generally speaking, such semigroups are not everywhere differentiable, and the generator of the corresponding linear semigroup is only densely defined. As a matter of fact, it turns out that the everywhere differentiability of a semigroup of holomorphic mappings on a bounded domain is equivalent to its continuity in the topology of local uniform convergence. Since, in the finite-dimensional case, this topology is equivalent to the compact open topology, the study of complex dynamical systems generated by holomorphic mappings includes in this case the study of semigroups of holomorphic mappings which are pointwise continuous. On the other hand, holomorphic self-mappings of a domain D in a complex Banach space are nonexpansive with respect to any pseudometric ρ assigned to D by a Schwarz-Pick system [Harris (1979)]. Therefore it is natural to inquire whether a theory analogous to the theory of monotone and accretive operators can be developed in the setting of those mappings which are nonexpansive with respect to such pseudometrics. We note in passing that the class of ρ-nonexpansive mappings properly contains the class of holomorphic mappings.

It seems that the need to investigate holomorphic mappings in infinite-dimensional spaces arose for the first time in connection with the study of nonlinear integral equations with an analytic nonlinear part at the end of the 19th and the beginning of the 20th centuries by A. Liapunov, E. Schmidt, A. Nekrasov and others.

Later in the 20th century the interest in analytic methods diminished temporarily due to the rapid development of degree theory by J. Leray, J. Schauder, G. Birkhoff, M. Krasnoselskii, P. Zabreiko, Y. Borisovich and others; see the references in [Krasnoselskii and Zabreiko (1984)].

The traditional methods for solving nonlinear operator equations have been based on either the Banach fixed point principle for contractive maps or the Leray–Schauder principle for compact operators.

However, the application of these principles is not always possible, or else if the operator depends on a parameter, these methods (as well as the classical Lyapunov–Nekrasov method) give only local results.

Parallel with the achievements mentioned above, the first results regarding holomorphic mappings on infinite-dimensional spaces appeared in the works of H. Cartan, R. Phillips, L. Nachbin, L. Harris, T. Suffridge, M. Hervé, E. Vesentini, K. Goebel, T. Kuczumov, A. Stachura, S. Reich, J.-P. Vigué, P. Mazet and many others (see the references in [Franzoni and Vesentini (1980); Goebel and Reich (1984); Hervé (1989); Dineen (1989); Chae (1985)]). A bridge between nonlinear equations with noncompact analytic operators and the theory of holomorphic mappings has been built in the book [Khatskevich and Shoikhet (1994a)].

In the one-dimensional case, the classical Denjoy–Wolff theorem provides information on both the location of fixed points and the behavior of the iterates of a holomorphic self-mapping. Over the last twenty years this result has been developed in at least three directions. The first one concerns increasing the dimension of the underlying space. Finite-dimensional extensions can be found, for instance, in the papers by Kubota [Kubota (1983)], MacCluer [MacCluer (1983)], Chen [Chen (1984)], Abate [Abate (1989); (1998)], and Mercer [Mercer (1991)–(1993); (1997); (1999)]. Infinite-dimensional generalizations are due, for example, to Fan [Fan (1978); (1979); (1982); (1983); (1986); (1988)], Wlodarczyk [Wlodarczyk (1985)–(1987); (1995)], Goebel [Goebel (1981); (1982)], Vesentini [Vesentini (1983); (1985)], Sine [Sine (1989)] and Mellon [Mellon (1996)]. These authors used a variety of approaches and assumed diverse conditions on the mappings and the domains. The second direction is concerned with analogues of the Denjoy–Wolff theorem for continuous semigroups. This approach has been used by several authors to study the asymptotic behavior of solutions to Cauchy problems. The third direction yields extensions of this theorem to the wider class of those self-mappings which are nonexpansive with respect to Schwarz–Pick pseudometrics.

It turns out that the asymptotic behavior of solutions to evolution equations is applicable to the study of the geometry of certain domains in complex spaces. For example, it is a well known result, due to R. Nevanlinna (1921), that if f is holomorphic in $|z| < 1$ and satisfies $f(0) = 0$, $f'(0) \neq 0$, then f is univalent and maps the unit disk onto a starlike domain (with respect to 0) if and only if $Re[zf'(z)/f(z)] > 0$ everywhere. This result, as well as most of the work on starlike functions on the unit disk, can be obtained from the identity

$$\frac{\partial}{\partial\theta}\arg f(re^{i\theta}) = \operatorname{Re}\left\{\frac{re^{i\theta}f'(re^{i\theta})}{f(re^{i\theta})}\right\}.$$

This idea does not extend readily to a higher-dimensional space. Moreover, such an approach is crucially connected with the initial condition $f(0) = 0$. Much later, Wald [Wald (1978)] gave a characterization of those functions which are starlike with respect to another center. Observe that although the classes of starlike, spirallike and convex functions were studied very extensively, little was known about functions that are holomorphic on the unit disk Δ and starlike with respect to a boundary point. In fact, only in 1981 Robertson [Robertson (1981)] introduced two relevant classes of univalent functions and conjectured that they coincide. In 1984 his conjecture was proved by Lyzzaik [Lyzzaik (1984)]. Finally, in 1990 Silverman and Silvia [Silverman and Silvia (1990)], using a similar method, gave a full description of the class of univalent functions on Δ, the image of which is starlike with respect to a boundary point. However, the arguments used in their work have a crucially one-dimensional character (because of the Riemann mapping theorem, the de Branges theorem, and Carathéodory's theorem on kernel convergence). In addition, the conditions given by Robertson and by Silverman and Silvia, characterizing starlikeness with respect to a boundary point, essentially differ from Wald's and Nevanlinna's conditions of starlikeness with respect to an interior point. Hence, it is difficult to trace the connections between these two closely related geometric objects. Therefore, even in the one-dimensional case the following problem arises: to find a unified condition of starlikeness (and spirallikeness) with respect to an interior or a boundary point. It seems that the idea to use a dynamical approach was first suggested by Robertson [Robertson (1936)] and developed by Brickman [Brickman (1973)], who introduced the concept of Φ-like functions as a generalization of starlike and spirallike functions (with respect to the origin) of a single complex variable. Suffridge [Suffridge (1977); (1970); (1973)], Pfaltzgraff [Pfaltzgraff (1974); (1975)] and Gurganus [Gurganus (1975)] developed a similar approach in order to characterize starlike, spirallike (with respect to the origin), convex and close-to-convex mappings in higher dimensional cases. Since 1970 the list of papers on these subjects has become quite long. Nevertheless, it seems that there has been no extension of Wald's as well as Silvia and Silverman's results to higher dimensions.

The first chapter of this book is an introductory chapter which sets the stage for the remainder of the book by giving basic notions in functional analysis and operator theory on metric and normed spaces.

The second chapter defines differentiable and holomorphic (analytic) mappings and presents a generalization of classical function theory to Banach spaces.

The third chapter contains material that is not usually covered in basic graduate courses, but is needed in the study of fixed point theory in metric spaces and semigroups of nonexpansive mappings with respect to the so-called hyperbolic metric.

Chapter 4 contains some classical and modern fixed point principles while Chapter 5 demonstrates a special approach to fixed point theory of holomorphic mappings, which is based on the development of the classical Denjoy–Wolff Theorem in various settings.

Chapters 6–9 are devoted to nonlinear semigroup theory of those mappings which are nonexpansive with respect to some special metrics on domains in Banach spaces. The description is most complete in the case of nonlinear semigroups of holomorphic self-mappings of a convex domain (which are nonexpansive with respect to the hyperbolic metric).

The last chapter consists of some material devoted to less developed geometric function theory in infinite dimensional spaces. It demonstrates a dynamical approach to this theory which is based on the asymptotic behavior of semigroups of holomorphic mappings.

The latter topic is itself of intrinsic interest and is considered in more detail in Chapter 9.

Summing up, we hope that this book may be considered a first step in establishing bridges between nonlinear semigroup theory, fixed points, and the geometry of domains.

We are most grateful to Ms. Galya Khanin for her meticulous typing and for her devoted and careful work on all the technical aspects of this book. We also thank Ms. Tan Rok Ting of World Scientific for her patient encouragement.

Finally, we thank the Technion - Israel Institute of Technology and ORT Braude College for their support throughout this project.

Simeon Reich and David Shoikhet

Contents

Chapter 1

Mappings in Metric and Normed Spaces

1.1 Topological Spaces

1.1.1 *Topology*

Let X be a set. *A **topology** on X is a collection τ of subsets of the set X, satisfying three conditions:*

(a) the intersection of any two elements of τ is an element of τ;
(b) the union of the elements of any subfamily of the family τ belongs to τ;
(c) the set X and the empty set belong to τ.

The set X is called the space of the topology τ and the pair (X, τ) is called a **topological space**. When no confusion arises, we simply write "X is a topological space". The elements of the topology τ are called τ-**open** (*or simply open) subsets.*

Let τ_1 and τ_2 be two topologies on X; τ_1 is *weaker* (smaller, rougher) than τ_2, or τ_2 is *stronger* (greater, finer) than τ_1, if $\tau_1 \subseteq \tau_2$. It is possible that, for two given topologies τ_1 and τ_2 on X neither τ_1 is stronger than τ_2, nor τ_2 is stronger than τ_1; in this case τ_1 and τ_2 are said to be not comparable.

1.1.2 *Neighborhoods*

*A **neighborhood** (τ-neighborhood) of a point x in a topological space (X, τ) is any subset of this space which contains an element U of the topology τ with the property $x \in U$.* For example, in the case of the space $\mathbb{C}$ of complex numbers with the usual topology generated by the collection of open disks in $\mathbb{C}$, a neighborhood of a point is any subset of $\mathbb{C}$ containing an open disk which contains the point in question.

1

*A set $\mathcal{D} \subseteq X$ is **open** if and only if for every $x \in \mathcal{D}$, $\mathcal{D}$ is a neighborhood of x.*

The complement of an open set is a **closed** set. A set that is both closed and open is called a *clopen* set.

A set may be both open and closed, or it may be neither. In particular, both $\emptyset$ and X are both open and closed. The family of closed sets has the following properties, which are dual to the properties of the open sets.

- Both $\emptyset$ and X are closed.
- A finite union of closed sets is closed.
- An arbitrary intersection of closed sets is closed.

1.1.3 *Examples of topologies*

The following examples illustrate the variety of topological spaces:

Example 1.1   The trivial topology or the indiscrete topology on a set X consists of only X and $\emptyset$. These are also the only closed sets.

Example 1.2   The discrete topology on a set X consists of all subsets of X. Thus every set is both open and closed.

Example 1.3   The open intervals on the real line $\mathbb{R} = (-\infty, \infty)$ generate a topology on $X = \mathbb{R}$. The extended real line $\mathbb{R}^* = [-\infty, \infty] = \mathbb{R} \cup \{-\infty, \infty\}$ has a natural topology too. It consists of all subsets U such that for each $x \in U$:

(a)   If $x \in \mathbb{R}$, then there exists some $\varepsilon > 0$ with $(x - \varepsilon, x + \varepsilon) \subset U$;
(b)   If $x = \infty$, then there exists some $y \in \mathbb{R}$ with $(y, \infty] \subset U$; and
(c)   If $x = -\infty$, then there exists some $y \in \mathbb{R}$ such that $[-\infty, y) \subset U$.

Example 1.4   A different, and admittedly contrived, topology on $\mathbb{R}$ consists of all sets A such that for each x in A, there is a set of the form $U \backslash C \subset A$, where U is open in the usual topology, C is countable, and $x \in U \backslash C$.

Example 1.5   Let $N = \{1, 2 \ldots\}$. The collection of sets consisting of the empty set and all sets containing 1 is a topology on N. The closed sets are N and all sets not containing 1.

1.1.4 *Interiors and closures. Limit points*

Let (X, τ) be a topological space, and let A be any subset of X. The topology τ defines two sets intimately related to A.

The **interior** *of A, denoted by A°, is the largest (with respect to inclusion) open set included in A.* (It is the union of all open subsets of A.) The interior of a nonempty set may be empty.

The **closure** *of A, denoted by $\bar{A}$, is the smallest closed set including A; it is the intersection of all closed sets including A.*

It is not hard to verify that $A \subset B$ implies $A^\circ \subset B^\circ$ and $\bar{A} \subset \bar{B}$. Also, it is obvious that a set A is open if and only if $A = A^\circ$, and a set B is closed if and only if $B = \bar{B}$. Consequently, for any set $A, \overline{(\bar{A})} = \bar{A}$ and $(A^\circ)^\circ = A^\circ$. Thus, a neighborhood of a point x is any set V containing x in its interior.

The collection of all neighborhoods of a point x, called the **neighborhood base**, *or the neighborhood system, at x, is denoted by N_x.*

It is easy to verify that N_x has the following properties.

(a) $X \in N_x$.
(b)   For each $V \in N_x$, we have $x \in V$ (so $\emptyset \notin N_x$).
(c)   If $V, U \in N_x$, then $V \cap U \in N_x$.
(d)   If $V \in N_x$ and $V \subset W$, then $W \in N_x$.

*A topology on X is called **Hausdorff** (or **separated**) if any two distinct points can be separated by disjoint neighborhoods of the points. That is, for each pair $x, y \in X$ with $x \neq y$ there exist neighborhoods $U \in N_x$ and $V \in N_y$ such that $U \cap V = \emptyset$.*

*A point x is a point of closure or **closure point** of the set A if every neighborhood of x meets A.* Note that $\bar{A}$ coincides with the set of all closure points of A.

A point x is an **accumulation point** (or a **limit point**, or a **cluster point**) of A if for each neighborhood V of x we have $(V \backslash \{x\}) \cap A \neq \emptyset$.

To see the difference between closure points and limit points, let $A = [0, 1) \cup \{2\}$, a subset of $\mathbb{R}$. Then 2 is a closure point of A in $\mathbb{R}$, but not a limit point. The point 1 is both a closure point and a limit point of A.

Let A be any subset of a topological space X, and let A^c be its complement, *i.e.*, $A^c = X \backslash A$.

*A point x is a **boundary point** of A if each neighborhood V of x satisfies both $V \cap A \neq \emptyset$ and $V \cap A^c \neq \emptyset$.* Clearly, accumulation and boundary points of A belong to its closure $\bar{A}$. Let A' denote the set of all accumulation points of A (called the derived set of A) and let ∂A denote the **boundary** of A, the set of all boundary points of A. We have the following identities:

$$\bar{A} = A^\circ \cup \partial A \quad \text{and} \quad \partial A = \partial A^c = \bar{A} \cap \overline{A^c}. \tag{1.1}$$

From (1.1), we see that a set A is closed if and only if $A' \subset A$ (and also if and only if $\partial A \subset A$). That is,

- *A set is closed if and only if it contains all its limit points.*

To illustrate these definitions, again let $A = [0,1) \cup \{2\}$ viewed as a subset of $\mathbb{R}$. Then the boundary of A is $\{0,1,2\}$ and its derived set is $[0,1]$. The closure of A is $[0,1] \cup \{2\}$ and its interior is $(0,1)$. Also note that the boundary of the set of rationals in $\mathbb{R}$ is the entire real line.

A subset A of a topological space X is perfect (in X) if it is closed and every point in A is an accumulation point of A.

In particular, every neighborhood of a point x in A contains a point of A different from x. The space X is perfect if all of its points are accumulation points. A point $x \in A$ is an isolated point of A if there is a neighborhood V of x with $(V \backslash \{x\}) \cap A \neq \emptyset$. That is, if $\{x\}$ is a relatively open subset of A. A set is perfect if and only if it is closed and has no isolated points. Note that if A has no isolated points, then its closure, $\bar{A}$, is perfect in X. Also, note that the empty set is perfect.

1.1.5 *Dense subsets and separable spaces*

*A subset $\mathcal{D}$ of a topological space X is **dense** (in X) if $\bar{\mathcal{D}} = S$.*

In other words, a set $\mathcal{D}$ is dense if and only if every nonempty open subset of X contains a point in $\mathcal{D}$. In particular, if $\mathcal{D}$ is dense in X and x belongs to X, then every neighborhood of x contains a point in $\mathcal{D}$. This means that any point in X can be approximated arbitrarily well by a point in $\mathcal{D}$. A set N is nowhere dense if its closure has empty interior.

*A topological space is **separable** if it contains a countable dense subset.*

1.1.6 *Induced topology. Subspaces*

If Y is a subset of a topological space (X, τ), then an easy argument shows that the collection τ_Y of subsets of Y, defined by

$$\tau_Y = \{V \cap Y : V \in \tau\}, \tag{1.2}$$

is a topology on Y. *This topology is called the **relative topology** (or the topology induced by τ) on Y.*

When $Y \subset X$ is equipped with its relative topology, we call Y a (topological) subspace of X. A set in τ_Y is called (relatively) open in Y. For example, since $X \in \tau$ and $Y \cap X = Y$, then Y is relatively open in itself. Note that the relatively closed subsets of Y are of the form

$$Y \backslash (Y \cap V) = Y \backslash V = Y \cap (X \backslash V), \tag{1.3}$$

where $V \in \tau$. That is, the relatively closed subsets of Y are the restrictions of the closed subsets of X to Y.

1.1.7 *Continuous mappings*

*A mapping f from a topological space (X, α) into a topological space (Y, β) is called **continuous** if the preimage of any β-open set in Y is an α-open in X.*

It is easily seen that the composition of continuous mappings is continuous. Let $f : X \mapsto Y$ be a continuous mapping and $\mathcal{D} \subseteq X$. Then the restriction $f|_{\mathcal{D}}$ is also continuous. A mapping f for which $f|_{\mathcal{D}}$ is continuous is said to be continuous on $\mathcal{D}$.

The continuity of a mapping is characterized as follows.

Proposition 1.1 *Let X and Y be topological spaces and let $f : X \mapsto Y$ be a mapping. The following assertions are equivalent:*

(a) The mapping f is continuous.
(b) The preimage of any closed set is closed.
(c) For each point $x \in X$, the preimage of any neighborhood of $f(x)$ is a neighborhood of x.
(d) For each point $x \in X$ and any neighborhood U of the point $f(x)$, there exists a neighborhood V of the point x such that $f(V) \subseteq U$.

A mapping f from a topological space X into a topological space Y is called continuous at a point $c \in X$ if the preimage of any neighborhood of $f(x)$ is a neighborhood of x.

It is clear that f is continuous on a set if and only if it is continuous at each point of that set.

1.1.8 *Compactness*

We have already seen that the definition of a topology is sufficiently weak to allow some non-interesting topologies, such as, for example, the trivial topology. To obtain nontrivial results we need additional hypotheses regarding the topology.

*An open **cover** of a set K is a collection of open sets the union of which contains K.*

*A subset K of a topological space is **compact** if every open cover of K contains a finite subcover.*

That is, K is compact if every family $\{V_i : i \in I\}$ of open sets satisfying $K \subset \bigcup_{i \in I} V_i$ has a finite subfamily $V_{i_1} \ldots, V_{i_n}$ such that $K \subset \bigcup_{j=1}^{n} V_{i_j}$.

*A topological space is called a **compact space** if it is a compact set.*

*A subset of a topological space is called **relatively compact** (or precompact) if its closure is compact.*

For the trivial topology every set is compact; for the discrete topology only finite sets are compact. Compactness can also be characterized in terms of the finite intersection property.

Proposition 1.2   *For a topological space X, the following statements are equivalent*

(a) X is compact.
(b) Every family of closed subsets of X with the finite intersection property has a nonempty intersection.
(c) Every net in X has a limit point (or, equivalently, every net has a convergent subnet).

*A subset A of a topological space is **sequentially compact** if every sequence in A has a subsequence converging to an element of A. A topological space X is sequentially compact if X itself is a sequentially compact set.*

In many ways compactness can be viewed as a topological generalization of finiteness. There is an informal principle that compact sets behave like points in many instances. We now list a few elementary properties of compact sets.

- *Finite sets are compact.*
- *Finite unions of compact sets are compact.*
- *Closed subsets of compact sets are compact.*
- *If $K \subset Y \subset X$, then K is a compact subset of X if and only if K is a compact subset of Y (in the relative topology).*

We also note the following result, which we use frequently without any special mention. It is an instance of how compact sets behave like points: *single point sets are closed in Hausdorff spaces and so are compact sets.*

1.1.9 Ordered sets

*A set P is said to be **partially ordered** by a binary relation $\leq$ if:*

(i) $a \leq b$ and $b \leq c$ implies $a \leq c$.
(ii) $a \leq a$ for every $a \in P$.
(iii) $a \leq b$ and $b \leq a$ implies $a = b$.

*A subset Q of a partially ordered set P is said to be **totally ordered** if every pair $a, b \in Q$ satisfies either $a \leq b$ or $b \leq a$.*

Hausdorff's maximality theorem states:

- *Every nonempty partially ordered set P contains a totally ordered subset Q which is maximal with respect to the property of being totally ordered.*

1.1.10 Topological vector spaces

Let X be a set and let $\mathbb{K}$ be the field $\mathbb{R}$ of real numbers or the field $\mathbb{C}$ of complex numbers. *The set X is called a **vector space** provided that two operations, called addition of elements in X and multiplication by scalars, are given, satisfying the following axioms ($x, y, z \in X$, and $\alpha, \beta \in \mathbb{K}$):*

(1) Addition

    (a) commutativity: $x + y = y + x$;
    (b) associativity: $(x + y) + z = x + (y + z)$;
    (c) there exists a unique element $0 \in X$ such that $x + 0 = x$;
    (d) for any $x \in X$, there exists a unique element $(-x)$ such that $x + (-x) = 0$.

Instead of $x + (-y)$ we write $x - y$. The element 0 is called the zero element, or the zero, of the space X; the element $-x$ is called the inverse of x.

(2) Multiplication by scalars

    (a) associativity of multiplication: $\alpha(\beta x) = (\alpha\beta)x$;
    (b) distributivity: $(\alpha + \beta)x = \alpha x + \beta x$, $\alpha(x + y) = \alpha x + \alpha y$;
    (c) $1 \cdot x = x$.

The following facts are some consequences of the axioms defining a vector space:

- $0 \cdot x = 0$ (the first zero is the real number zero, while the second one is the zero vector);
- $(-1) \cdot x = -x$;
- if $\alpha x = \beta x$ and $x \neq 0$, then $\alpha = \beta$; if $\alpha x = \alpha y$ and $\alpha \neq 0$, then $x = y$.

Let τ be a topology on a vector space X. A natural assumption which enables an effective use of both the algebraic and topological structures is the continuity of linear operations on X with respect to the topology τ. If this is the case, then X is called a **topological vector space**.

Let X be a vector space over the field $\mathbb{K}$.

*A mapping $f : X \mapsto \mathbb{K}$ is said to be a **linear functional** on X if*

$$f(\alpha x + \beta y) = \alpha f(x) + \beta f(y) \tag{1.4}$$

for all x, y in X and for all α, β in $\mathbb{K}$. If, in addition, f is continuous, then it is called a continuous linear functional on X.

1.2 Metric Spaces

1.2.1 *Metrics and pseudometrics (semimetrics)*

An important class of topological spaces is the one consisting of those spaces where the topology is defined using a distance function (or a metric).

*A **metric** on a set X is a real-valued functional d defined on the Cartesian product $X \times X$ which satisfies (for $x, y, z \in X$) the following conditions (the axioms of a metric):*

(a) *Positivity*: $d(x, y) \geq 0$ and $d(x, x) = 0$ for all $x, y \in X$.
(b) *Positive-definiteness*: $d(x, y) = 0$ implies $x = y$.
(c) *Symmetry*: $d(x, y) = d(y, x)$ for all $x, y \in X$.
(d) *The Triangle Inequality*: $d(x, y) \leq d(x, z) + d(z, y)$ for all $x, y, z \in X$.

*The pair (X, d), where X is a space and d is a metric on X, is called a **metric space**.*

*A topological space X is **metrizable** if there exists a metric d on X generating the topology of X.*

The discrete metric, defined by $d(x, y) = 1$ if $x \neq y$ and $d(x, y) = 0$ if $x = y$, generates the discrete topology.

For $x, y \in X$, the number $d(x, y)$ is called the **distance** between x and y (or the d-distance).

Let $r > 0$. The set

$$B_r(X) = \{y : d(x, y) < r\} \tag{1.5}$$

is called the open r-ball centered at x.

The set

$$\overline{B_r(X)} = \{y : d(x, y) \leq r\} \tag{1.6}$$

is called the closed r-ball centered at x.

The family of all open balls provides a base for a topology τ on X, called the **metric topology**, *or the topology induced on X by the given metric.*

According to this definition any closed r-ball is closed in (X, τ).

A **pseudometric (or semimetric)** *on X is a function $d : X \times X \to \mathbb{R}$ satisfying (a), (c) and (d).*

A metric is a pseudometric that has the property that $d(x, y) = 0$ implies $x = y$.

Given a pseudometric d, let $B_\varepsilon(x) = \{y : d(x, y) < \varepsilon\}$ be the open ε-ball around x. A set U is open in the pseudometric (or semimetric) topology generated by d if for each point x in U there is an $\varepsilon > 0$ satisfying $B_\varepsilon(x) \subset U$. The triangle inequality guarantees that each open ball is an open set.

The zero pseudometric, defined by $d(x, y) = 0$ for all x and y, generates the trivial topology.

If d is a pseudometric, then the binary relation defined by $x \sim y$, if and only $d(x, y) = 0$, is an equivalence relation, and d defines a metric $\hat{d}$ on the set of equivalence classes by $\hat{d}([x], [y]) = d(x, y)$. For this reason one deals mostly with metric spaces.

For a nonempty subset A of a metric space (X, d), its **diameter** is defined by

$$\operatorname{diam} A = \sup\{d(x, y) : x, y \in A\}. \tag{1.7}$$

A set A is bounded if $\operatorname{diam} A < \infty$, while A is unbounded if $\operatorname{diam} A = \infty$. If $\operatorname{diam} X < \infty$, then X is bounded and d is called a bounded metric.

Similar terminology applies to pseudometrics.

In a semimetric space (X, d) the open ball centered at a point $x \in X$ with radius $r > 0$ is the subset $B_r(x)$ of X defined by

$$B_r(x) = \{y \in X : d(x, y) < r\}. \tag{1.8}$$

The closed ball centered at a point $x \in X$ with radius $r > 0$ is the subset $\overline{B_r(x)}$ of X defined by

$$\overline{B_r(x)} = \{y \in X : d(x,y) \le r\}. \tag{1.9}$$

Let (X, d) be a pseudometric space. A subset A of X is open if for each $a \in A$ there exists some $r > 0$ (depending on a) such that $B_r(a) \subset A$. We should verify that the collection of subsets

$$\tau_d = \{A \subset X : A \text{ is } d\text{–open}\} \tag{1.10}$$

is a topology on X called the topology generated or induced by d.

Once again, if d is a metric, we call τ_d the metric topology on (X, d). Two metrics generating the same topology are equivalent.

There are always several metrics on any given space that generate the same topology. Let (X, d) be a metric space. Then $2d$ is also a metric generating the same topology. More interesting is the metric $\hat{d}$ defined by $\hat{d}(x,y) = \min\{d(x,y), 1\}$. It too generates the same open sets as d, but is interesting because X is bounded under $\hat{d}$. That is, $\hat{d}$-diam $X < \infty$. A potential drawback of $\hat{d}$ is that the families of balls of radius r around x are different for d and $\hat{d}$. (For instance, $\{x \in \mathbb{R} : |x| < 3\}$ is a ball of radius 3 around 0 in the usual metric on $\mathbb{R}$, but in the truncated metric it is not a ball of any finite radius.)

Here is a list of some of the basic properties of metric and pseudometric topologies.

- The topology τ_d is Hausdorff if and only if d is a metric.
- A sequence $\{x_n\}$ in X satisfies $x_n \to x$ if and only if $d(x_n, x) \to 0$.
- Every open ball is an open set. To see this, let $B_r(x)$ be an open ball and let $y \in B_r(x)$. Put $\varepsilon = r - d(x,y) > 0$. Now if $z \in B_\varepsilon(y)$, then the triangle inequality implies $d(x,z) \le d(x,y) + d(y,z) < d(x,y) + \varepsilon = r$. So $B_\varepsilon(y) \subset B_r(x)$, which means that $B_r(x)$ is a d-open set.
- The topology τ_d is first countable, since for each $x \in X$ the countable family of open neighborhoods $\left\{B_{\frac{1}{n}}(x) : n \in N\right\}$ is a base for the neighborhood system at x. Thus x belongs to the closure $\bar{A}$ of a set A if and only if there is some sequence $\{x_n\}$ in A with $x_n \to x$.
- A closed ball is a closed set. Suppose $y \notin \overline{B_r(x)}$. Then $\varepsilon = d(x,y) - r > 0$, so by the triangle inequality, $B_\varepsilon(y)$ is an open neighborhood of y disjoint from $\overline{B_r(x)}$. This shows that the complement of $\overline{B_r(x)}$ is open.
- The closure of the open ball $B_r(x)$ is included in the closed ball $\overline{B_r(x)}$. But the inclusion may be strict–consider the open ball of radius one

under the discrete metric.

- A pseudometric $d : X \times X \to \mathbb{R}$ satisfies the inequality

$$\left| d(x,y) - d(u,v) \right| \leq d(x,u) + d(y,v). \tag{1.11}$$

This follows from the triangle inequality by observing that

$$\begin{aligned}
d(x,y) - d(u,v) &\leq \left[d(x,u) + d(u,v) + d(v,y) \right] - d(u,v) \\
&= d(x,u) + d(y,v).
\end{aligned} \tag{1.12}$$

In particular, d is a (uniformly) continuous function on $X \times X$. Although for general topological spaces the property of second countability is stronger than separability, for metrizable spaces these two properties coincide.

1.2.2 *Examples*

Example 1.6 The metric $d(x,y) = |x - y|$ defines a topology on the real line $\mathbb{R}$.

Unless we state otherwise, $\mathbb{R}$ is assumed to have this topology.

Every open interval (a,b) is an open set in this topology. Further, every open set is a countable union of disjoint open intervals (where the end points ∞ and $-\infty$ are allowed). To see this, note that every point in an open set must be contained in a maximal open interval, every open interval contains a rational number, and the set of rational numbers is countable.

Example 1.7 The Euclidean metric on $\mathbb{R}^n = \left\{ x = (x_1, x_2, \ldots, x_n) \right\}$,

$$d(x,y) = \left[\sum_{i=1}^{n} (x_i - y_i)^2 \right]^{\frac{1}{2}}, \tag{1.13}$$

defines its usual topology, also called the Euclidean topology.

Example 1.8 The metric $d(z,w) = |z - w|$ defines a topology on the complex place $\mathbb{C}$.

If $z = x_1 + ix_2$ and $w = y_1 + iy_2$ are algebraic representation of the complex number z and w respectively, then explicitly $d(z,w) = \left[\sum_{i=1}^{2} (x_i - y_i)^2 \right]^{\frac{1}{2}}$. In other words, topologically the complex plane $\mathbb{C}$ can be identified with the Euclidean topology on $\mathbb{R}^2$.

Example 1.9   Similarly, the Euclidean metric on $\mathbb{C}^n = \{z = (z_1, z_2, \ldots, z_n)\}$, is defined by

$$d(z, w) = \left[\sum_{i=1}^{n} |z_i - w_i|^2 \right]^{\frac{1}{2}} \tag{1.14}$$

for complex vectors $z = (z_1, z_2, \ldots, z_n)$ and $w = (w_1, w_2, \ldots, w_n)$ in $\mathbb{C}^n$. So the Euclidean complex metric space $\mathbb{C}^n$ can be identified with the Euclidean real metric space $\mathbb{R}^{2n}$.

1.2.3   *Completeness*

Let (X, d) be a metric space.

*A sequence $\{x_n : n \in \mathbb{N}\}$ is called **Cauchy** (or **fundamental**) if $d(x_n, x_m) \to 0$ for $n, m \to \infty$, i.e., for any $\varepsilon > 0$ there is $l \in N$ such that $d(x_n, x_m) < \varepsilon$ for all $n, m > l$.*

*A metric space is called **complete** if each one of its Cauchy sequences is convergent in the metric topology to an element of X.*

1.2.4   *Compactness and boundedness*

A subset U of a metric space is compact if every sequence in U contains a subsequence convergent to a point in U.

A subset U of a metric space X is called bounded if there exists a point $x \in X$ and a number $r > 0$ such that U is included in the r-ball centered at x. A compact subset of a metric space is bounded.

*A metric space is called a **Montel space** if all the closures of its bounded subsets are compact.*

A simple example of a Montel space is given by the space $\mathbb{C}$ of complex numbers, with the usual metric. The space $\mathbb{C}^n$, the Cartesian product of n copies of $\mathbb{C}$, is also a Montel space.

1.3   Normed and Banach Spaces

1.3.1   *Norms on a vector space*

Let X be a vector space over $\mathbb{K}$.

*A nonnegetive real valued function p on X is called a **seminorm** on X if the following conditions are satisfied:*

(a) $p(x + y) \leq p(x) + p(y)$ for all x and y in X;
(b) $p(rx) = |r|p(x)$ for all x in X and all r in $\mathbb{K}$.
  If, in addition,
(c) $p(x) = 0$ if and only if $x = 0$, then p is called a norm on X.

When p is a norm on X, it is customary to denote it by $\|\cdot\|$.

*A **normed space** is a pair $(X, \|\cdot\|)$, where X is a vector space and $\|\cdot\|$ is a norm defined on X.*

When confusion is unlikely we will denote the normed space $(X, \|\cdot\|)$ by X. There is a natural metric associated with a normed space X; we define the distance between any two points x and y of X by

$$d(x, y) = \|x - y\|. \tag{1.15}$$

Then d is a metric defined on X and d induces a metric topology on X that is called the norm topology of X.

*A normed space which is complete in the norm topology is called a **Banach space**.*

1.3.2 *Examples*

Example 1.10 The n-dimensional vector space $X = \mathbb{K}^n$ (real or complex) is a Banach space when equipped with any of the following norms:

$$(1) \qquad \|x\| = \sqrt{\sum_{i=1}^{n} |x_i|^2}; \tag{1.16}$$

$$(2) \qquad \|x\| = \max_{1 \leq i \leq n} |x_i|; \tag{1.17}$$

$$(3) \qquad \|x\| = \sum_{i=1}^{n} |x_i|. \tag{1.18}$$

Norm (1) is called the Euclidean norm on X (see examples (1.14)–(1.15)). Norm (2) is called the Chebyshev norm, while the norm (3) is called the hypercone norm.

Example 1.11 The space M of all bounded real or complex sequences $x = (x_1, \ldots, x_n, \ldots)$ with the norm $\|x\| = \sup |x_i|$ is an infinite dimensional Banach space.

Example 1.12 The space $L_p[a, b]$ of all (equivalence classes of) complex-valued functions $x(t)$ defined on an interval $[a, b] \subseteq \mathbb{R}$ and such that

$|x(t)|^p$ $(p \geq 1)$ is integrable on $[a, b]$ with the norm

$$\|x\| = \left(\int_a^b |x(t)|^p \, dt \right)^{\frac{1}{p}} \tag{1.19}$$

is (an infinite dimensional) Banach space.

Example 1.13   Consider now the infinite dimensional linear space of all functions $x(t)$ defined on an interval $[a, b]$, continuous on $[a, b]$, and having continuous derivatives up to an order k inclusively. On this space we define a norm by

$$\|x\| = \max \left\{ \max_{t \in [a,b]} |x(t)|, \ \max_{t \in [a,b]} |x'(t)|, \ldots, \ \max_{t \in [a,b]} \left| x^{(k)}(t) \right| \right\}. \tag{1.20}$$

Thus we obtain a Banach space denoted by $C^k[a, b]$.

- *A normed space is a Montel space if and only if it is finite dimensional.*

1.4   Hilbert Spaces

1.4.1   *Scalar product*

*A **scalar product** on a vector space X over $\mathbb{R}$ is a real valued function of two points x and y in X, denoted by (x, y), having the following properties:*

(i)   Bilinearity. For a fixed y the scaler product (x, y) is a linear function of x, i.e., $(ax_1 + bx_2, y) = a(x_1, y) + b(x_2, y), \quad a, b, \in \mathbb{R}$.
(ii) Symmetry. $(y, x) = (x, y)$.
(iii) Positivity. $(x, x) > 0$ for $x \neq 0$.

When the field of scalars is $\mathbb{C}$, (x, y) is complex valued and properties (i) and (ii) are altered as follows:

(i)   Sesquilinearity. For a fixed y, (x, y) is a linear function of x and for a fixed x, (x, y) is a skewlinear function of y, that is,

$$(ax, y) = a(x, y) \quad (x, ay) = \bar{a}(x, y), \quad a \in \mathbb{C}. \tag{1.21}$$

(ii) Skew symmetry.

$$(y, x) = \overline{(x, y)}. \tag{1.22}$$

Given a scalar product, we can define a norm, denoted by $\|\cdot\|$, as follows:

$$\|x\| = (x,x)^{\frac{1}{2}}. \tag{1.23}$$

The following are the main properties of the scalar product.

- **The Schwarz inequality.**
  A scalar product satisfying (i),(ii) and (iii) satisfies

$$|(x,y)| \leq \|x\| \cdot \|y\|, \tag{1.24}$$

where the norm is defined by (1.23). Equality holds for $x = ay$ or $y = 0$.

- **The parallelogram identity.**

$$\|x+y\|^2 + \|x-y\|^2 = 2\|x\|^2 + 2\|y\|^2. \tag{1.25}$$

- **Continuity of scalar product.**
  The scalar product depends continuously on its factors: that is, if $x_n \to x$, $y_n \to y$ in the sense of $\|x_n - x\| \to 0$, $\|y_n - y\| \to 0$, then (x_n, y_n) tends to (x,y).

- **Orthogonality.**
  *Two vectors x and y are called **orthogonal** if $(x,y) = 0$.*

*A linear space with a scalar product that is complete with respect to the induced norm is called a **Hilbert space**.*

Given a linear space with a scalar product, it can be completed with respect to the norm derived from the scalar product. It follows from the Schwarz inequality that the scalar product is a uniformly continuous function of its factors; therefore it can be extended to the completed space. Thus the completion is a Hilbert space.

1.4.2 *Examples*

Example 1.14 The real Euclidean space $\mathbb{R}^n$ of vectors with n real components $x = (x_1, x_2, \ldots, x_n)$, $y = (y_1, y_2, \ldots, y_n)$ is a Hilbert space with the scalar product defined by

$$(x,y) = \sum_{j=1}^{n} x_j y_j. \tag{1.26}$$

Example 1.15   Similarly, the complex Euclidean space $\mathbb{C}^n$ of vectors with n complex components $z = (z_1, z_2, \ldots, z_n)$, $\ w = (w_1, w_2, \ldots, w_n)$ is a (complex) Hilbert space with the scalar product defined by

$$(z, w) = \sum_{j=1}^{n} z_j \bar{w}_j. \tag{1.27}$$

In both cases (real and complex) the norm of a vector $x = (x_1, x_2, \ldots, x_n) \in \mathbb{K}^n$ ($\mathbb{K}$ is either $\mathbb{R}$ or $\mathbb{C}$) is given by

$$\|x\| = (x, x)^{\frac{1}{2}} = \sqrt{\sum_{i=1}^{n} |x_i|^2}. \tag{1.28}$$

Topologically, the space $\mathbb{C}^n$ is equivalent to $\mathbb{R}^{2n}$.

Example 1.16   The space l^2 of complex vectors with countably many components

$$x = (x_1, x_2, \ldots), \quad y = (y_1 y_2, \ldots) \tag{1.29}$$

subject to the restriction

$$\sum |x_j|^2 < \infty \quad \sum |y_j|^2 < \infty. \tag{1.30}$$

We define the scalar product as

$$(x, y) = \sum a_j \bar{b}_j. \tag{1.31}$$

In this case the norm of l^2 is defined by

$$\|x\| = (x, x)^{\frac{1}{2}} = \sqrt{\sum |x_i|^2}. \tag{1.32}$$

Example 1.17   The space of complex valued continuous functions $x(t)$ on the interval $[0, 1]$ with

$$(x, y) = \int_0^1 x(t)\bar{y}(t)dt \tag{1.33}$$

$$\|x\| = \left(\int_a^b |x(t)|^2 \, dt \right)^{\frac{1}{2}} \tag{1.34}$$

is incomplete.

1.5 Locally Convex Spaces

1.5.1 *Convex sets and convex hulls*

Definition 1.1 Let X be a vector (linear) space. A subset K of X is called **convex** if, whenever x and y belong to K, the whole segment with endpoints x and y, meaning all points of the form

$$ax + (1-a)y, \quad 0 \le a \le 1, \tag{1.35}$$

also belong to K.

The simplest examples of convex sets in the plane are the disk, triangle, and parallelogram. The following property of convex sets is an immediate consequence of the definition

Proposition 1.3 *Let K be a convex subset of a vector space X. Suppose that $x_1, \ldots, x_n$ belong to K. Then so does every x of the form*

$$x = \sum_1^n \lambda_j x_j, \quad \lambda_j \ge 0, \tag{1.36}$$

$$\sum_1^n \lambda_j = 1. \tag{1.37}$$

An element x of the form (1.36) is called a convex combination of $x_1, x_2, \ldots, x_n$.

Let X be a vector space. The following are elementary properties of convex sets.

- *The empty set is convex.*
- *A subset consisting of a single point is convex.*
- *Every linear subspace of X is convex.*
- *The sum of two convex subsets is convex.*
- *If K is convex, so is $-K = \{-x, \ x \in K\}$.*
- *The intersection of an arbitrary collection of convex sets is convex.*
- *Let $\{K_j\}$ be a collection of convex subsets that is totally ordered by inclusion. Then their union $\cup K_j$ is convex.*

Definition 1.2 Let S be any subset of a vector space X. The **convex hull** of S, denoted by $\operatorname{conv} S$, is defined as the **intersection** of all convex sets containing S, *i.e.*,

$$\operatorname{conv} S = \cap\{K \subset X : K \supset S, \;\; K \text{ is convex}\}. \tag{1.38}$$

Obviously, $\operatorname{conv} A$ is the smallest convex subset of X which contains S. Moreover, $x \in \operatorname{conv} S$ if and only if $x = \sum_{i=1}^{n} \lambda_i x_i$, where $x_i \in S$, $\lambda_i \geq 0$ and $\sum_{i=1}^{n} \lambda_i = 1$.

The closure of $\operatorname{conv} S$ is denoted by $\overline{\operatorname{conv}} S$ and called the **convex closure** of S.

As above,

$$\overline{\operatorname{conv}} S = \cap\{K \subset X : K \supset S, \;\; K \text{ is closed and convex}\}. \tag{1.39}$$

A fundamental property of the convex operation is given by **Mazur's theorem**:

- *If $\bar{S}$ is compact, then so is $\overline{\operatorname{conv}} S$.*

1.5.2 *Extreme points*

Definition 1.3 A subset E of a convex set K is called a (real) **extreme subset** of K if:

(i) E is convex and nonempty;
(ii) whenever a point x of E is expressed as

$$x = \frac{y + z}{2}, \quad y, z \text{ in } K, \tag{1.40}$$

both y and z belong to E.

An extreme subset consisting of a single point is called a **(real) extreme point** of K.

Proposition 1.4 *Let K be a convex set, E an extreme subset of K, and F an extreme subset of E. Then F is an extreme subset of K.*

Definition 1.4 A set A in a vector space X is:

- **absorbing** (or **radial**) if for any x some multiple of A includes the line segment joining x and zero. That is, if there is an $\alpha_0 > 0$ satisfying $\alpha x \in A$ for every $0 \leq \alpha \leq \alpha_0$.
- **circled** (or **balanced**) if for any point $x \in A$ and any $|\alpha| \leq 1$, the point αx is also in A.

- **symmetric** if $x \in A$ implies $-x \in A$.
- **star-shaped about zero** if it includes the line segment joining each of its points with zero. That is, if for any $x \in A$ and any $0 \le \alpha \le 1$, $\alpha x \in A$.

Note that an absorbing set must contain zero, and any set containing an absorbing set is itself absorbing. For any absorbing set A, the set $A \cap (-A)$ is nonempty, absorbing and symmetric. Every circled set is symmetric. Every circled set is star-shaped about zero, as is every convex set containing zero.

1.5.3　*Examples*

Example 1.18　K is the interval $0 \le x \le 1$; the two endpoints are extreme points.

Example 1.19　K is the closed disk

$$x^2 + y^2 \le 1. \tag{1.41}$$

Every point on the circle $x^2 + y^2 = 1$ is a (real) extreme point.

Example 1.20　The open disk

$$x^2 + y^2 < 1 \tag{1.42}$$

has no extreme points.

1.5.4　*The topology induced by seminorms*

Recall the definition of a seminorm.

A real-valued function p on a vector space X is called a seminorm on X if

(1) $p(x + y) \le p(x) + p(y)$ for al x, y in X;
(2) $p(\lambda x) = |\lambda| p(x)$ for all x in X and for all scalars λ.

For example, if f is a linear functional on X, then we can define a seminorm p_f on X by setting

$$p_f(x) = |f(x)| \tag{1.43}$$

for all x in X. Hence, on any vector space we can define a family of seminorms.

It follows from (2) that $p(0) = 0$. Conversely, if $p(x) = 0$ implies $x = 0$ then p is a norm. For any seminorm p on X, we have

$$|p(x) - p(y)| \leq p(x - y) \tag{1.44}$$

for all x and y in X.

Proposition 1.5   *If p is a seminorm on X, then:*

(1) $p(0) = 0$.
(2) *For all x, we have $-p(x) \leq p(-x)$. Consequently, p is linear if and only if $p(-x) = -p(x)$ for all $x \in X$.*
(3) *The function g defined by $g(x) = \max\{p(x),\ p(-x)\}$ is a seminorm.*
(4) *The function $p : X \mapsto \mathbb{R}$ is nonnegative, i.e., $p(x) \geq 0$ for all $x \in X$.*
(5) *The set $\{x : p(x) = 0\}$ is a linear subspace of X.*

A **norm** p is a seminorm satisfying $p(x) = 0$ and only if $x = 0$. A seminorm p defines a pseudometric (semimetric) d by $d(x, y) = p(x - y)$. If p is a norm, then the pseudometric is actually a metric.

1.5.5   *The Minkowski functional*

Definition 1.5   The **gauge**, or **Minkowski functional**, p_A of a set A is defined by

$$p_A(x) = \inf\{\alpha > 0 : x \in \alpha A\}, \tag{1.45}$$

where, by convention, $\inf \emptyset = \infty$.

If A is absorbing, then p_A is finite valued and positively homogeneous. These sets are important because any positively homogeneous function is completely determined by its values on any absorbing set.

Proposition 1.6   *A real function p on a vector space is a seminorm if and only if it is the gauge of a symmetric convex absorbing set.*

The next assertion collects some elementary properties of gauges.

Proposition 1.7   *For nonempty sets B and C:*

(1) $p_{-C}(x) = p_C(-x)$ *for all x.*
(2) $B \subset C$ *implies* $p_C \leq p_B$.
(3) *If C contains a subspace M, then $p_C(x) = 0$ for all $x \in M$.*

(4) If C is star-shaped about zero, then

$$\{x : p_C(x) < 1\} \subset C \subset \{x : p_C(x) \leq 1\}. \qquad (1.46)$$

Recall that a real function $f : \mathcal{D} \mapsto \mathbb{R}$ on a subset $\mathcal{D}$ of a topological vector space is uniformly continuous on $\mathcal{D}$ if for every $\varepsilon > 0$, there is a neighborhood V of zero such that $|f(x) - f(y)| < \varepsilon$ whenever $x, y \in \mathcal{D}$ satisfy $x - y \in V$.

Proposition 1.8 *A seminorm on a topological vector space is (uniformly) continuous if and only if it is bounded on some neighborhood of zero.*

Proposition 1.9 (Semicontinuity) *A seminorm is:*

(i) *<u>lower semicontinuous</u> if and only if it is the gauge of a closed convex set containing zero.*

(ii) *<u>continuous</u> if and only if it is the gauge of a closed convex neighborhood of zero.*

1.5.6 *Locally convex spaces and seminorms*

Definition 1.6 A topological vector space is locally convex, or is a **locally convex space**, if every neighborhood of zero contains a convex neighborhood of zero.

Since in a topological vector space the closure of a convex set is convex, in a locally convex space the closed convex circled neighborhoods of zero form a neighborhood base at the origin.

It turns out that the locally convex topologies are precisely the ones derived from families of seminorms.

Let X be a vector space. For a seminorm $p : X \mapsto \mathbb{R}$ and $\varepsilon > 0$, let us consider

$$V_p(\varepsilon) = \{x \in X : p(x) \leq \varepsilon\}, \qquad (1.47)$$

the closed ε-ball of p centered at zero. Now let $\{p_i\}_{i \in I}$ be a family of seminorms on X. Then the collection β of all sets of the form

$$V_{p_1}(\varepsilon) \cap \ldots \cap V_{p_n}(\varepsilon), \quad \varepsilon > 0, \qquad (1.48)$$

is a neighborhood base of convex sets at zero.

Consequently, β induces a locally convex **topology generated by the family of seminorms** $\{p_i\}_{i \in I}$.

In the converse direction, let τ be a locally convex topology on a vector space X, and let β denote the neighborhood base at zero consisting of all circled convex closed neighborhoods of zero. Then, for each $V \in \beta$, the gauge p_V is a seminorm on X. An easy argument shows that the family of seminorms $\{p_V\}_{V \in \beta}$ generates τ. Thus we have the following important characterization of locally convex topologies.

Proposition 1.10 (Seminorms and local convexity) *A linear topology on a vector space is locally convex if and only if it is generated by a family of seminorms. In particular, a locally convex topology is generated by the family of gauges of the convex circled closed neighborhoods of zero.*

Since the topology of a locally convex space is generated by a family of seminorms, it is easy to work with locally convex spaces. Most important topological vector spaces are indeed locally convex.

1.6 Linear and Multilinear Mappings in Banach Spaces

1.6.1 *Linear operators*

Definition 1.7 Let X and Y be normed spaces over the same field $\mathbb{K}$. A mapping $T : X \mapsto Y$ is said to be a linear operator if

$$T(\alpha x + \beta y) = \alpha T(x) + \beta T(y) \tag{1.49}$$

for all x, y in X and for all α, β in $\mathbb{K}$.

If $Y = \mathbb{K}$ then T is called a **linear functional**. If $T : X \mapsto Y$ is not a linear operator, then T is often referred to as a nonlinear operator (or just a mapping).

The following assertion characterizes continuity of linear operators.

Proposition 1.11 (Continuity at zero) *A linear operator $T : X \mapsto Y$ between topological vector spaces is continuous if and only if it is continuous at zero (in which case it is uniformly continuous).*

If $T : X \mapsto Y$ is a linear operator between normed spaces, then

$$\sup_{\|x\| \leq 1} \|Tx\| = \min\{M \geq 0 : \|Tx\| \leq M\|x\| \ \text{for all} \ x \in X\}, \tag{1.50}$$

where we adhere to the convention $\min \emptyset = \infty$. If the normed space X is nontrivial (that is, $X \neq \{0\}$), then we also have

$$\sup_{\|x\| \leq 1} \|Tx\| = \sup_{\|x\|=1} \|Tx\|. \tag{1.51}$$

Definition 1.8 The **norm of a linear operator** $T : X \mapsto Y$ between normed spaces is the nonnegative extended real number $\|T\|$ defined by

$$\|T\| = \sup_{\|x\| \leq 1} \|Tx\| = \min\{M \geq 0 : \|Tx\| \leq M\|x\| \text{ for all } x \in X\}.$$

$$\tag{1.52}$$

If $\|T\| = \infty$, we say that T is an **unbounded operator**, while in case $\|T\| < \infty$, we say that T is a **bounded operator**.

Consequently, a linear operator $T : X \mapsto Y$ between normed spaces is bounded if and only if there exists some positive real number $M > 0$ satisfying the inequality $\|Tx\| \leq M\|x\|$ for all $x \in X$. Another way of stating the boundedness of an operator is the following:

- *A linear operator $T : X \mapsto Y$ is bounded if and only if it carries the closed (or open) unit ball of X onto a norm bounded subset of Y.*

So we summarize:

Proposition 1.12 *Given normed spaces X and Y and a linear mapping $T : X \mapsto Y$, the following are equivalent.*

(a) T is continuous;
(b) T is continuous at 0;
(c) There exists M such that $\|Tx\| \leq M\|x\|$ for all $x \in T$;
(d) T maps bounded sets onto bounded sets.

In general, if A is a subset of X and $f : A \mapsto Y$ is a mapping (not necessarily linear) we let $\|f\|_A = \sup_{x \in A} \|f(x)\|$.

Definition 1.9 A mapping $f : A \mapsto Y$ is said to be **bounded** on A if $\|f\|_A < \infty$.

If $A = X$ is a Banach space we sometimes let X_0 denote the open unit ball of X and we write $\|f\|$ in place of $\|f\|_{X_0}$. Thus, *a linear operator on X is continuous if and only if it is bounded on a ball in X.*

Definition 1.10 If X and Y are Banach spaces and $T : X \mapsto Y$ is a linear mapping, then we say T is an **isometry** if $T(X) = Y$ and $\|T(x)\| = \|x\|$ for all $x \in X$.

If X_0 and Y_0 are the open unit balls of X and Y respectively, then it is easily seen that an injective linear mapping $T : X \mapsto Y$ is an isometry if and only if $T(X_0) = Y_0$.

1.6.2 *Examples*

Example 1.21   Consider the space $X = \mathbb{C}^n$ with any norm, and a rectangular matrix (a_{ik}) of order $m \times n$, $i \in \{1, \ldots, m\}$, $k \in \{1, \ldots, n\}$, $a_{uk} \in \mathbb{C}$. The equalities

$$y_i = \sum_{k=1}^{n} a_{ik} x_k, \quad i = 1, \ldots, m, \tag{1.53}$$

define a linear continuous operator $T : \mathbb{C}^n \mapsto \mathbb{C}^m$ ($y = Tx$ for $x = (x_1, x_2, \ldots, x_n) \in \mathbb{C}^n$, $y = (y_1, \ldots, y_m) \in \mathbb{C}^m$).

Note that in a finite dimensional normed space any linear operator is continuous, hence bounded.

Example 1.22   On the space $X = L_p[a, b]$ with $p = 1$, let us consider the integral operator given by

$$y(t) = \int_a^b K(t, s) x(s) ds, \tag{1.54}$$

where $K(t, s)$ is a continuous function on the square $a \leq t, s \leq b$. This is a linear and bounded operator from $L_p[a, b]$ into itself.

Example 1.23   Let $X = C(I; \mathbb{K})$ be the normed space of all continuous $\mathbb{K}$-valued functions on the unit interval $I = [0, 1]$ with the sup-norm, and let $Y = C^1(I; \mathbb{K})$ be the normed subspace of X consisting of those functions f which have continuous derivative df. Then the linear mapping $f \mapsto df$ of X into Y is not continuous. In fact, if we set $f_n(x) = (\sin nx)/n$, the sequence (f_n) converges to 0, whereas the sequence (df_n) does not converge to 0.

Example 1.24   Consider the space $C^k[a, b]$ and define a differential operator by

$$y(t) = \frac{d^{k+1}}{dt^{k+1}} x(t). \tag{1.55}$$

This operator is defined on the dense linear subspace of $C^k[a, b]$ consisting of all functions which have continuous derivatives on $[a, b]$ up to the order

$k + 1$ inclusively and takes values in $C^0[a, b]$; it is linear, but it is not bounded.

1.6.3 *The space of bounded linear operators*

The set $L(X, Y)$ of all continuous linear operators taking the whole space X into Y, where X and Y are normed spaces, becomes a normed space with the following linear operations:

(1) $(A + B)x = Ax + Bx$,
(2) $(\alpha A)x = \alpha(Ax)$, (for $A, B \in L(X, Y)$, $x \in X$, $\alpha \in \mathbb{K}$) and the norm
$$\|A\| = \sup_{\|x\|=1} \|Ax\|.$$

- *If Y is a Banach (i.e., complete) space, then $L(X, Y)$ is a Banach space, too.*

If $X = Y$, we write $L(X)$ instead of $L(X, X)$.

The convergence of a sequence of bounded linear operators in the norm of the space $L(X, Y)$ is called **uniform convergence**, *and the corresponding topology is called the* **uniform operator topology**.

1.6.4 *Multilinear mappings and polynomials*

Let $X_1, X_2, \ldots, X_m$ be a finite family of normed spaces over the same field $\mathbb{K}$, and let $E = X_1 \times X_2 \times \cdots \times X_m$ be its Cartesian product with coordinatewise addition and multiplication by scalars, *i.e.*,

$$E = \{x : x = (x_1, x_2, \ldots, x_m)\}. \tag{1.56}$$

The norms

$$\|x\|_\infty = \sup\{\|x_k\|, \quad k = 1, \ldots, m\} \tag{1.57}$$

$$\|x\|_p = \left(\sum_{k=1}^m \|x_k\|^p\right)^{\frac{1}{p}}, \quad 1 \le p < \infty, \tag{1.58}$$

are equivalent on E and induce the norm (product) topology on E.

Definition 1.11 Let $X_1, \ldots, X_m$ and Y be normed space over $\mathbb{K}$. A mapping $A : X_1 \times X_2 \times \cdots \times X_m \mapsto Y$ is said to be **multilinear** (or m**-linear**) if it is linear in each variable separately.

As in the case of linear operators the following assertion characterizes the continuity property of multilinear mappings.

Proposition 1.13   *Let $A : E = X_1 \times X_2 \times \cdots \times X_m \mapsto Y$ be a multilinear mapping. Then the following are equivalent:*

(i) A is continuous on E;

(ii) A is continuous on some neighborhood of the origin $(0, \ldots, 0)$;

(iii) A is bounded in the sense: there is $M > 0$ such that for $x = (x_1, \ldots, x_m) \in E$,

$$\|A(x_1, \ldots x_m)\| \leq M \|x_1\| \ldots \|x_m\|. \tag{1.59}$$

If now $X_1 = X_2 = \cdots = X_m = X$ and Y are normed spaces we denote by $L^m(X, Y)$ the space of all continuous m-linear mappings from $X^m = X \times \cdots \times X$ into Y. $L^m(X, Y)$ is a normed space with respect to the norm

$$\|A\| = \sup\{\|A(x_1, \ldots, x_m)\|, \ x_k \in X, \ \|x_k\| \leq 1\}. \tag{1.60}$$

It is a Banach space if Y is Banach. For $m = 1$, $L^1(X, Y) = L(X, Y)$ is the space of linear operators from X into Y.

Notice that for $m > 1$, $L^m(X, Y) \neq L(X^m, Y)$, where $X^m = X \times \cdots \times X$.

Let $L_S^m(X, Y)$ be the vector subspace of $L^m(X, Y)$ consisting of all $A \in L^m(X, Y)$ which are symmetric, *i.e.*,

$$A\big(x_{\sigma(1)}, \ldots, x_{\sigma(m)}\big) = A(x_1, \ldots, x_m) \tag{1.61}$$

for all $(x_1, \ldots, x_m)$ in X^m and for every permutation σ of $\{1, \ldots, m\}$. Clearly, $L_S^m(X, Y)$ is a Banach space if Y is complete.

Given any m-linear mapping $A : X^m \mapsto Y$, one can define a symmetric mapping by

$$A_S(x_1, \ldots, x_m) = \frac{1}{m!} \sum_{\sigma} A\big(x_{\sigma(1)}, \ldots, x_{\sigma(m)}\big), \tag{1.62}$$

where the summation is taken over all permutations σ of $(1, \ldots, m)$.

It is obvious that $\|A_S\| \leq \|A\|$. If A is symmetric, then $A_S = A$.

Given a multilinear mapping $A \in L^m(X, Y)$, define the polar form $\hat{A} : X \mapsto X$ by using the restriction of A to the diagonal:

$$\hat{A}(x) = A(x, \ldots, x). \tag{1.63}$$

Note that $\hat{A}_S = \hat{A}$.

Definition 1.12   A mapping $P : X \mapsto Y$ is called a **bounded homogeneous (m-homogeneous) polynomial** (or a **homogeneous polynomial of order** m) if there exists a polar form $\hat{A} : X \mapsto Y$ such that

$$P = \hat{A}. \tag{1.64}$$

It is clear that

- A bounded homogeneous polynomial from X into Y is a continuous mapping on X.
- The space $\mathcal{P}^m(X,Y)$ of all bounded m-homogeneous polynomials is a normed space with respect to the norm

$$\|P\| = \sup\{\|P(x)\| : x \in X, \quad \|x\| \le 1\}. \tag{1.65}$$

- $\mathcal{P}^m(X,Y)$ is a Banach space if Y is complete. Sometimes we will also write

$$P = P_m \quad \text{for} \quad P \in \mathcal{P}^m(X,Y). \tag{1.66}$$

- For all $x \in X$

$$\|P_m(x)\| \le \|P_m\| \cdot \|x\|^m. \tag{1.67}$$

Proposition 1.14   *The mapping $A \mapsto \hat{A}$ is a continuous isomorphism of $L_S^m(X,Y)$ onto $\mathcal{P}^m(X,Y)$. In addition,*

$$\|\hat{A}\| \le \|A\| \le \frac{m^m}{m!}\|\hat{A}\|. \tag{1.68}$$

1.6.5   *Banach algebra of linear operators*

Definition 1.13   A complex algebra is a vector space E over the field $\mathbb{K}$ in which multiplication is defined in such a manner that for all x, y and z in E,

(a) $x(yz) = (xy)z$,
(b) $(z + y)z = xz + yz$,
(c) $x(y + z) = xy + xz$,
(d) $\alpha(xy) = (\alpha x)y = x(\alpha y), \quad \alpha \in \mathbb{K}$.

If, in addition, E is a Banach space with respect to a norm that satisfies the multiplicative inequality

$$\|xy\| \le \|x\| \cdot \|y\|, \quad x, y \in E, \tag{1.69}$$

and E contains a unit element I such that

$$xI = Ix = x, \quad x \in E, \tag{1.70}$$

and

$$\|I\| = 1, \tag{1.71}$$

then E is called a **Banach algebra**. If $xy = yx$ for all $x, y \in E$, then E is called a **commutative algebra**.

If X is a Banach space over a field $\mathbb{K}$, then the space $L(X)$ of bounded linear operators on X (into itself) becomes a Banach algebra (noncommutative) if we define multiplication on E as the **composition operation** and let the unit I be the identity operator on X.

Therefore sometimes it is natural to study polynomials on $L(X)$ of the form

$$p(A) = \alpha_0 + \alpha_1 A + \alpha_2 A^2 + \cdots + \alpha_n A^n \tag{1.72}$$

as functions on a Banach algebra.

We will return to this matter later.

Definition 1.14   An operator $A \in L(X)$ is said to be **continuously invertible** (or just **invertible**) if there exists an operator $A^{-1} \in L(X)$ (the inverse of A) such that

$$A^{-1}A = AA^{-1} = I. \tag{1.73}$$

The set of (continuously) invertible operators on $L(X)$ endowed with the uniform operator topology is open in $L(X)$.

This fact is based on the following lemma which is important in itself.

Lemma 1.1   *Let $T \in L(X)$ have a norm strictly less than 1, i.e., $\|T\| < 1$. Then $A = I - T$ is continuously invertible. Moreover,*

$$A^{-1}\big(= (I - T)^{-1}\big) = \sum_{n=0}^{\infty} T^n, \tag{1.74}$$

where the sum on the right-hand side is defined as the uniform limit of the polynomials $S_n = I + T + T^2 + \cdots + T^n$.

Theorem 1.1   *Let $A \in L(X)$ be invertible. Then, for a given $q \in (0,1)$ and for all $B \in L(X)$ satisfying*

$$\|B\| \le q\|A^{-1}\|^{-1}, \tag{1.75}$$

the operators $A + B$ are invertible. Moreover,

$$\|(A + B)^{-1} - A^{-1}\| \leq \frac{\|B\|\|A^{-1}\|^2}{1 - \|A^{-1}\|\|B\|}.$$ (1.76)

The following theorem of Banach plays a crucial role in the study of invertible operators .

Theorem 1.2 *Let X and Y be Banach spaces, and let A be a bounded linear operator from X onto Y. If A has a unique null point in X, then*

(i) $A : X \mapsto Y$ is one-to-one, i.e., for each $y \in Y$ the equation

$$Ax = y$$ (1.77)

has a unique solution: $x =: A^{-1}y$;
(ii) the inverse mapping A^{-1} defined by (i) is a bounded linear operator from Y onto X.

Corollary 1.1 *Let X and Y be Banach spaces and let $A \in L(X,Y)$ be a bounded linear operator from X onto Y. Suppose that for some $m > 0$,*

$$\|Ax\| \geq m\|x\|, \quad x \in X.$$ (1.78)

Then there exists an inverse bounded linear operator $A^{-1} \in L(Y,X)$.

Finally, we formulate an important result of Gleason, Kahane and Zelazko (see, for example, [Dixmier (1969)]) which characterizes the so-called complex homomorphisms on the Banach algebra $L(X)$.

Theorem 1.3 *If f is a linear functional on $L(X)$ such that $f(I) = 1$ and $f(A) \neq 0$ for every invertible operator $A \in L(X)$, then f is a complex homomorphism on $L(X)$, i.e.,*

$$f(AB) = f(A) \cdot f(B)$$ (1.79)

for all A and B in $L(X)$.

1.6.6 *Spectra and resolvents of linear operators*

Definition 1.15 The **spectrum** $\sigma(A)$ of $A \in L(X)$ is the set of all complex numbers $\lambda \in \mathbb{C}$ such that $\lambda I - A$ is not invertible in $L(X)$.

The complement Ω of $\sigma(A)$ is called the **resolvent set** of A; it consists of (regular) values $\lambda \in \mathbb{C}$ for which the operator $(\lambda I - A)^{-1} =: R(\lambda, A)$ is well defined and belongs to $L(X)$.

The operator $R(\lambda, A)$ is called the **resolvent** of A.

Theorem 1.4   *The spectrum $\sigma(A)$ of $A \in L(X)$ is a nonempty, closed compact subset of $\mathbb{C}$. Hence the resolvent set Ω of A is always open.*

Theorem 1.5 (The resolvent identity) *Let $A \in L(X)$ and let $\lambda, \mu \notin \sigma(A)$. Then*

$$(\lambda I - A)^{-1} - (\mu I - A)^{-1} = (\mu - \lambda)(\lambda I - A)^{-1}(\mu I - A)^{-1}. \quad (1.80)$$

Definition 1.16   The spectral radius of A is the number

$$\rho(A) = \sup\{|\lambda| : \lambda \in \sigma(A)\}. \quad (1.81)$$

In other words, $\rho(A)$ is the radius of the smallest closed disk in $\mathbb{C}$ centered at the origin which contains $\sigma(A)$.

Theorem 1.6 (Spectral radius formula) *If $A \in L(X)$, then*

$$\rho(A) = \lim_{n \to \infty} \|A^n\|^{\frac{1}{n}} = \inf_{n \geq 1} \|A^n\|^{\frac{1}{n}}. \quad (1.82)$$

Note that the latter formula implies that

$$\rho(A) \leq \|A\|. \quad (1.83)$$

Remark 1.1   *Recall that a norm $\|\cdot\|_*$ on X is called equivalent to the original norm $\|\cdot\|$ on X if there are positive numbers m and M such that*

$$m\|x\| \leq \|x\|_* \leq M\|x\|, \quad x \in X. \quad (1.84)$$

Although the norm $\|A\|$ of an operator $A \in L(X)$ may be different for an equivalent norm on X, the spectral radius $\rho(A)$ does not depend on the choice of such a norm. Whether an element of $L(X)$ is or is not invertible in $L(X)$ is a purely algebraic property. The spectrum and the spectral radius of $A \in L(X)$ are thus defined in terms of the algebraic structure of $L(X)$, regardless of any metric (or topological) considerations.

On the other hand, the limit

$$\lim_{n \to \infty} \|A\|^{\frac{1}{n}} \quad (1.85)$$

does seem to depend on metric properties of A. This is a remarkable feature of the spectral radius formula. It asserts the equality of certain quantities which arise in entirely different ways.

Theorem 1.7 (Rutickii (see [Krasnoselskii *et al.* (1969)])) *Let X be a Banach space and let $A \in L(X)$. Then for any $\varepsilon > 0$, there is a norm $\| \cdot \|_*$ equivalent to the original norm of X, such that*

$$\rho(A) \leq \|A\|_* \leq \rho(A) + \varepsilon, \tag{1.86}$$

where $\|A\|_* = \sup_{\|x\|_*=1} \|Ax\|_*$.

The spectrum $\sigma(A)$ of a bounded linear operator A can be divided into three disjoint pieces:

$$\sigma(A) = P_\sigma(A) \cup C_\sigma(A) \cup R_\sigma(A), \tag{1.87}$$

where

- $P_\sigma(A)$ is the set of all complex numbers $\lambda \in \mathbb{C}$ such that $\lambda I - A$ has no inverse (neither bounded nor unbounded) on X; $P_\sigma(A)$ is called the **point spectrum** of A;
- $C_\sigma(A)$ is the set of $\lambda \in \mathbb{C}$ such that $\lambda I - A$ has an inverse operator which is defined on a dense subset of X, but the operator $(\lambda I - A)^{-1}$ is not bounded; $C_\sigma(A)$ is called the **continuous spectrum** of A;
- $R_\sigma(A)$ is the set of $\lambda \in \mathbb{C}$ such that the operator $(\lambda I - A)^{-1}$ is defined on a domain which is not dense in X; $R_\sigma(A)$ is called the **residual spectrum** of A.

Proposition 1.15 *A complex number λ belongs to $P_\sigma(A)$ if and only if the equation*

$$Ax = \lambda x \tag{1.88}$$

has a nonzero solution $x \neq 0$ in X.

In this case the number $\lambda \in \mathbb{C}$ is called an **eigenvalue** of A and the solution of (1.88) is called an **eigenvector** of A corresponding to λ.

The subspace

$$\mathcal{N}(\lambda I - A) = \left\{ x \in X : (\lambda I - A)x = 0 \right\} \tag{1.89}$$

is called an **eigensubspace** of A corresponding to the eigenvalue λ.

The dimension of $\mathcal{N}(\lambda I - A)$ is called the multiplicity of $\lambda \in P_\sigma(A)$.

Proposition 1.16 *Let X be a Banach space and let $A \in L(X)$. Then $\lambda \in \sigma(A)$ if and only if there is a sequence $\{x_n\} \subset X$, $\|x_n\| = 1$, such that*

$$\lim_{n \to \infty} \|\lambda x_n - A x_n\| = 0. \tag{1.90}$$

1.6.7 *Examples*

Example 1.25   Each bounded linear operator on a finite dimensional Banach space X can be represented by a matrix $A = (a_{ij})$. in this case the eigenvalues of A are solutions of the equation

$$\det(\lambda \delta_{ij} - a_{ij}) = 0, \qquad (1.91)$$

where

$$\delta_{ij} = \begin{cases} 1, & i = j \\ 0, & i \neq j. \end{cases} \qquad (1.92)$$

In fact, this is all the spectrum of A, *i.e.*, $\sigma(A) = P_\sigma(A)$.

Example 1.26   Let $X = L^2[0,1]$ be the Hilbert space of equivalence classes of complex-valued square–integrable functions on $[0,1]$, and let $A : X \mapsto X$ be defined by

$$Ax(t) = tx(t), \quad x(= x(t)) \in X. \qquad (1.93)$$

Then A has no eigenvalues, since the equation

$$(\lambda I - A)x = (\lambda - t)x(t) = 0 \qquad (1.94)$$

is satisfied for all $t \in [0,1]$ if and only if $x(t) = 0$ almost everywhere.

Thus $P_\sigma(A) = \emptyset$. Nevertheless, the spectrum $\sigma(A)$ is not empty and consists of the elements of the continuous spectrum, *i.e.*,

$$\sigma(A) = C_\sigma(A) = [0,1]. \qquad (1.95)$$

Indeed, if $\lambda \notin [0,1]$, then $(\lambda I - A)^{-1} \in L(X)$. On the other hand, if $\lambda \in [0,1]$, then one can show that there is a sequence $\{x_n\} \subset X$, $\|x_n\| = 1$, such that $(\lambda I - A)x_n \to 0$ as $n \to \infty$. Thus $(\lambda I - A)^{-1}$ is not bounded for such λ.

Example 1.27 (Volterra integral operator) Let $X = C[0,1]$ and let $V \in L(X)$ be defined by

$$Vx(s) = \int_0^s x(t)dt. \qquad (1.96)$$

It is clear that V is not invertible, so $\lambda = 0$ is a spectrum point of V. As a matter of fact, this point is the only point of $\sigma(V)$.

Indeed, it can be shown by induction that $\|V^n\| \leq \frac{1}{n!}$. Hence, $\rho(A) = \lim_{n \to \infty} \|V^n\|^{\frac{1}{n}} = 0$.

Thus $\sigma(A) = \{0\}$.

Example 1.28 (Shift mappings) Take X to be $\ell^2 = \Big\{ x = (a_0, a_1, \dots) :$ $\sum_{j=0}^{\infty} |a_j|^2 < \infty \Big\}$. The right shift R and the left shift L are defined by

$$Rx = (0, a_0, a_1, \dots) \tag{1.97}$$

and

$$Lx = (a_1, a_2, \dots,). \tag{1.98}$$

Clearly, $LR = I$, but $RL \neq I$, so neither L nor R are invertible.

Furthermore, the spectrums of both operators consist of the close unit disk $\{\lambda \in \mathbb{C} : |\lambda| \leq 1\}$.

1.7 Duality in Normed Spaces

1.7.1 *Dual spaces*

Let X be a normed space. *The space of all continuous linear functionals on X is called the **dual** (or **conjugate**) space of X and is denoted by X^*.* Thus, $X^* = L(X, \mathbb{K})$, where $\mathbb{K}$ is either $\mathbb{R}$ or $\mathbb{C}$ depending on X being either a real or a complex space. Since $\mathbb{R}$ and $\mathbb{C}$ are complete, X^* is always a Banach space.

1.7.2 *Examples*

Example 1.29 For each linear functional f on $\mathbb{C}^n$ there exist n complex numbers $a_1, \dots, a_n$ such that $f(x) = \sum_{k=1}^{n} a_k x_k$ for all $x = (x_1, \dots, x_n) \in \mathbb{C}^n$. Thus, f can be identified with $(a_1, \dots, a_n) \in \mathbb{C}^n$.

Example 1.30 For the space $L_p[a, b]$, the dual space is $L_q[a, b]$, where $q = \frac{p}{p-1}$.

1.7.3 *Weak topology and reflexivity*

A normed space X carries, along with the norm topology, topologies induced by linear functionals. If X is a normed space and X^* is its dual space, then the family of seminorms of the form

$$p_f(x) = |f(x)| \qquad (f \in X^*, \ x \in X) \tag{1.99}$$

defines a locally convex topology on X. This topology is called the **weak topology** on X, and is denoted by $\sigma(X, X^*)$.

The weak* topology on X^* is induced by the family of seminorms of the form

$$p_f(x) = |f(x)| \quad (f \in X, \ x \in X^*). \tag{1.100}$$

It is clear that the norm topology on X is finer (stronger) than the weak topology and the norm topology on X^* is stronger than the weak* topology.

In other words, *the **weak topology** on X is defined as follows: a net $\{x_d : d \in \mathcal{D}\}$ is weakly convergent if the numerical nets $\{\langle x_d, f \rangle : d \in \mathcal{D}\}$ are convergent for every $f \in X^*$.*

As we mentioned, in general, the weak topology is weaker than the norm topology. However, in the case of finite dimensional spaces these topologies coincide.

Proposition 1.17 *In a finite dimensional space the weak topology coincides with the norm topology.*

We return to the general case of a normed space.

Proposition 1.18 *If a sequence $\{x_n : n \in N\}$ is weakly convergent to x_0, then the sequence is norm bounded. Moreover, $\|x_0\| \leq \liminf \|x_n\|$.*

For $x^* \in X^*$, we will often use the following notation which pairs elements of X with elements of X^*:

$$x^*(x) = \langle x, x^* \rangle, \quad x \in X, \ x^* \in X^*. \tag{1.101}$$

(This of course is consistent with the pairing in a Hilbert space H where, by the Riesz Representation Theorem, $H = H^*$ and the value of $y(x)$ for $x \in H$, $y \in H^*$ is given by the usual inner product $\langle x, y \rangle$.)

The space $X^{**} = L(X^*, \mathbb{R})$ is called the second dual (or conjugate) space of X. If $x \in X$ is fixed, then the relation $\langle x, x^* \rangle$ defines a continuous linear functional on X^*; thus x is associated in a natural way with an element x^{**} of X^{**}. The mapping $x \mapsto x^{**}$ is called the canonical (or natural) embedding of X in X^{**}. This embedding is always a linear isometry.

*If it is also surjective, i.e., if it maps X onto X^*, then X is said to be **reflexive** and we write $X = X^{**}$.*

1.7.4 *The weak and weak* topologies*

According to our notations, the weak topology on X is the topology generated by the family of seminorms $\{p_{x^*}\}$, $x^* \in X^*$, where

$$p_{x^*}(x) = |\langle x, x^* \rangle|, \quad x \in X. \tag{1.102}$$

Similarly, the weak* topology on X^* is generated by the seminorms $\{p_x\}$, $x \in X$, where

$$p_x(x^*) = |\langle x, x^* \rangle|, \quad x^* \in X^*. \tag{1.103}$$

Both X and X^* are locally convex, linear topological spaces when endowed with their respective weak and weak* topologies. Note that X^* also has a weak topology which in general is distinct from its weak* topology. The two, of course, coincide when X is reflexive.

The weak topology on X is the weakest (coarsest) topology for which all the functionals $x^* \in X^*$ are continuous. In particular, a net $\{x_\alpha : \alpha \in A\}$ converges to an element $x \in X$ in the weak topology if and only if $\lim_\alpha \langle x_\alpha, x^* \rangle = \langle x, x^* \rangle$ for each $x^* \in X^*$. When this occurs we say that $\{x_\alpha\}$ is weakly convergent or converges weakly to x, and we write

$$w - \lim_\alpha x_\alpha = x. \tag{1.104}$$

Similarly, a net $\{x_\alpha^* : \alpha \in A\}$ in X^* converges to $x* \in X^*$ in the weak* topology if and only if for each $x \in X$, $\lim_\alpha \langle x, x_\alpha^* \rangle = \langle x, x^* \rangle$, in which case we write

$$w^* - \lim_\alpha x_\alpha^* = x^*. \tag{1.105}$$

We now collect some basic and well-known properties of the weak and weak * topologies.

Proposition 1.19 *A convex subset K of X is closed if and only if it is weakly closed.*

Proposition 1.20 *If K is a weakly compact subset of X, then $\overline{\text{conv}}\, K$ is also weakly compact.*

The above facts do not carry over to the weak* topology. However, the following fact about the weak* topology is very important.

Proposition 1.21 (Alaoglu's Theorem) *The closed unit ball B in a dual space X^* is always compact in the weak* topology.*

Note that this theorem implies that any ball or any intersection of balls in a dual space is weak* compact.

If X is reflexive, then $X = X^{**}$. Thus, in view of Alaoglu's Theorem, we have

Proposition 1.22   *If X is reflexive, then each closed ball in X is compact in the weak topology.*

To summarize, we formulate the following assertion:

Proposition 1.23 (Reflexive Banach space) *For a Banach space X, the following statements are equivalent:*

(1) The Banach space X is reflexive.

(2) The closed unit ball of X is weakly compact.

(3) The dual Banach space X^* is reflexive.

1.8   The Hahn–Banach Theorem

An important part of the study of a given normed space X is the investigation of its dual space X^* of all continuous linear functionals on X. The Hahn–Banach theorem is, together with the uniform boundedness and open mapping theorems, one of the most important theorems of Functional Analysis. This theorem assures us that there are always plenty of continuous linear functionals on any normed space.

1.8.1   *The extension theorem*

Theorem 1.8 (Hahn–Banach) *Let X be a vector space over the reals, and let p be a real-valued function defined on X, which has the following two properties:*

(i) Positive homogeneity,

$$p(ax) = ap(x) \quad \text{for all} \ \ a > 0, \tag{1.106}$$

for every x in X.

(ii) Subadditivity,

$$p(x + y) \leq p(x) + p(y), \tag{1.107}$$

for all x, y in X.

If Y is a linear subspace of X on which a linear functional ℓ is defined, and ℓ is dominated by p:

$$\ell(y) \le p(y) \quad \text{for all} \quad y \text{ in } Y, \tag{1.108}$$

then ℓ can be extended to all of X as a linear functional dominated by p:

$$\ell(x) \le p(x) \quad \text{for all} \quad x \text{ in } X. \tag{1.109}$$

The following lemma is important in the derivation of different forms of the Hahn–Banach Theorem (including the complex case).

Lemma 1.2 *Let X be a a complex vector space, and let f be a linear functional on X. Denote $q = Re(f)$. Then*

(1) q is a real linear functional on $X_{\mathbb{R}}$, the space X considered as a vector space over $\mathbb{R}$ instead of $\mathbb{C}$.
(2) $f(x) = q(x) - iq(ix)$.
(3) If p is a seminorm on X, then

$$\sup\{|f(x)| : p(x) \le 1\} = \sup\{|q(x)| : p(x) \le 1\}. \tag{1.110}$$

Theorem 1.9 (Sukhomlinoff) *Let X be a vector space over $\mathbb{K}$, p a seminorm on X, and Y a vector subspace of X. If f is a linear functional on Y such that $\|f(x)\| \le p(x)$ for all $x \in Y$, then there exists a linear functional h on X such that*

(1) $h = f$ on Y;
(2) $|h(x)| \le p(x)$ for all $x \in X$.

Theorem 1.10 (Hahn) *Let Y be a linear subspace of a normed space X and let $f \in Y^*$. Then f can be extended to $h \in X^*$ so that $\|h\| = \|f\|$.*

Corollary 1.2 *For any nonzero x in a normed space X, there exists $f \in X^*$ such that $\|f\| = 1$ and $f(x) = \|x\|$.*

Definition 1.17 The mapping $j : X \to 2^{X^*}$ defined by $j(x) = \{f \in X^* : f(x) = \|x\|, \|f\| = 1\}$ is called the duality mapping on X. Sometimes it is more convenient to define the normalized duality mapping $J : X \to 2^{X^*}$ by the formula $J(x) = \{f \in X^* : f(x) = \|x\|^2 = \|f\|^2\}$.

Corollary 1.3 *For every x in a normed space X,*

$$\|x\| = \sup\{\|f(x)\| : f \in X^*, \ \|f\| \le 1\}. \tag{1.111}$$

1.8.2 *The completion of a normed space*

Note that the bidual X^{**} of a normed space X is a Banach space. For each fixed $x \in X$ define

$$x'(f) = f(x) \tag{1.112}$$

for all $f \in X^*$. Then x' is clearly a linear functional on X^*,

$$|x'(f)| = |f(x)| \le \|f\| \cdot \|x\|, \tag{1.113}$$

and hence $x' \in X^{**}$. Thus we can define a map $\Phi : X \mapsto X^{**}$ by letting $\Phi(x) = x'$ for each x in X. This map Φ is a linear isometry, *i.e.*,

$$\|x'\| = \|x\| \tag{1.114}$$

for all $x \in X$. We shall call Φ the natural embedding of X in X^{**}. In this way X becomes a dense subspace of $\overline{\Phi(X)}$ in X^{**}. Since X^{**} is a Banach space, $\overline{\Phi(X)}$ is also a Banach space. Therefore, we have the following theorem.

Proposition 1.24 *Every normed space is isometrically isomorphic to a dense subspace of a Banach space.*

1.8.3 *Geometric Hahn–Banach separation theorems*

In spite (or perhaps because) of its nonconstructive proof, the Hahn–Banach theorem has plenty of very concrete applications. One of the most important concerns separation theorems regarding convex sets; these results are sometimes called geometric Hahn–Banach theorems.

Let X be a vector space over the reals, and let S be subset of X. A point x_0 is an internal point of S if for any y in X there is an ε depending on y, such that

$$x_0 + ty \in S \quad \text{for all real} \ \ t, \ \ |t| > \varepsilon. \tag{1.115}$$

Let K be a convex set that has an internal point, which we take to be the origin. Recall that we define the gauge p_K of K with respect to the origin as follows:

$$p_K(x) = \inf \left\{ a : a > 0, \ \frac{x}{a} \in K \right\}. \tag{1.116}$$

Since the origin is assumed to be an internal point of K,

$$p_K(x) < \infty \tag{1.117}$$

for every x.

Observe that the gauge p_K of a convex set K in a vector space over the reals is positively homogeneous and subadditive.

Theorem 1.11 *For any convex set K,*

$$p_K(x) \leq 1 \quad \text{if} \quad x \in K, \tag{1.118}$$

$$p_K(x) < 1 \quad \text{iff } x \text{ is an internal point of } K. \tag{1.119}$$

The converse of this theorem is also true:

Theorem 1.12 *Let p denote a positively homogeneous, subadditive function defined on a vector space X over the reals. Then*

(i) the set of points x satisfying

$$p(x) < 1 \tag{1.120}$$

is a convex subset of X, and 0 is an internal point of it.
(ii) The set of points x satisfying

$$p(x) \leq 1 \tag{1.121}$$

is a convex subset of X.

We turn now to the notion of a hyperplane. Suppose that ℓ is a linear functional which is not zero. For any real c, all points of X satisfy one, and only one, of the following three conditions:

$$\ell(x) < c, \quad \ell(x) = c, \quad \ell(x) > c. \tag{1.122}$$

The set of those points x which satisfy $\ell(x) = c$ is called a hyperplane; the sets where $\ell(x) < c$, $\ell(x) > c$, respectively, are called open halfspaces. The sets where

$$\ell(x) \geq c, \quad \text{or} \quad \ell(x) \leq c, \tag{1.123}$$

are called closed halfspaces.

Theorem 1.13 (Hyperplane separation theorem) *Let K be a nonempty convex subset of a vector space X over the reals; suppose that all points of K are internal. Then any point y not in K can be separated from K by a hyperplane $\{x \in X : \ell(x) = c\}$, that is, there is a linear functional ℓ, depending on y, such that*

$$\ell(x) < c \quad \text{for all} \quad x \text{ in } K; \quad \ell(y) = c. \tag{1.124}$$

Corollary 1.4 *Let K denote a convex set with at least one internal point. For any y not in K there is a nonzero linear functional ℓ that satisfies*

$$\ell(x) \le \ell(y) \quad \text{for all} \quad x \text{ in } K. \tag{1.125}$$

Theorem 1.14 (Extended hyperplane separation theorem) *Let X be a vector space over the reals, and let K and M be disjoint convex subsets of X, of which at least one has an internal point. Then K and M can be separated by a hyperplane $\{x \in X : \ell(x) = c\}$; that is, there is a nonzero linear functional ℓ and a number c such that*

$$\ell(u) \le c \le \ell(v) \tag{1.126}$$

for all u in K and all v in M.

Theorem 1.15 (Hahn–Banach separation theorem) *Let X be a Banach space (real or complex).*

(a) If C is a closed convex subset of X and $x_0 \notin C$, then there exists $f \in X^*$ such that

$$Re f(x_0) > \sup_{x \in C} Re f(x). \tag{1.127}$$

(b) If $x \in X$, then there exists $f \in X^*$, $\|f\| = 1$, such that $f(x) = \|x\|$.
(c) If $x \in X$, then $\|x\| = \displaystyle\sup_{f \in X^*,\ \|f\|=1} |f(x)|.$

1.9 Elements of Ergodic Theory

1.9.1 *Mean ergodic theorem*

Let X be a locally convex vector space and let T be a continuous linear operator mapping X into itself.

Classical ergodic theory deals with the asymptotic behavior of the arithmetic means (Cesàro averages)

$$M_n(T) = \frac{1}{n}\left(I + T + \cdots + T^{n-1}\right). \tag{1.128}$$

The mean ergodic theorem of Yosida (see [Yosida (1974)]) enables us to construct, for a class of linear operators T on X, a projection onto the null space $N(I - T)$ $(= \mathrm{Ker}(I - T))$ of the operator $I - T$.

Definition 1.18 Let X be a **locally convex vector space**. A family of operators $\{T_n\}_{n=1}^{\infty} \subset L(X)$ is said to be **equicontinuous** if for every seminorm p on X there is a continuous seminorm q on X such that

$$\sup_{n \geq 1} p(T_n x) \leq q(x) \tag{1.129}$$

for all $x \in X$.

Lemma 1.3 *Let X be a locally convex space and let $T \in L(X)$ be such that the family of iterates $\{T^n\}_{n=1}^{\infty}$ of T is equicontinuous on X. Then*

(i) *The set $\left\{ x \in X : \lim_{n \to \infty} M_n(T)x = 0 \right\} = \overline{\mathrm{Im}(I - T)}$ is the closure of the range $\mathrm{Im}(I - T) \left(= (I - T)X \right)$ of the operator $I - T$, where $M_n(T)$ is defined by* (1.129)

(ii) $\mathrm{Im}(I - T) \cap \mathrm{Ker}(I - T) = \{0\}$, *where $\mathrm{Ker}(I - T)$ is the null point set of $I - T$.*

Theorem 1.16 (Mean ergodic theorem) *Let X and $T \in L(X)$ be as above. Assume that for some $x \in X$ and for some subsequence $\left\{ M_{n_k}(T) \right\}_{n_k = 1}^{\infty}$ the weak limit*

$$w - \lim_{k \to \infty} M_{n_k}(T)x = x_0 \in X \tag{1.130}$$

exists. Then $Tx_0 = x_0$ and the strong limit $\lim_{n \to \infty} M_n(T)x = x_0$.

Corollary 1.5 *Suppose that X is a reflexive Banach space and assume that for some $T \in L(X)$ the sequence of iterates $\{T^n\}_{n=1}^{\infty}$ is uniformly bounded, i.e.,*

$$\|T^n\| \leq M < \infty, \quad n = 1, 2, \dots. \tag{1.131}$$

Then the strong limit

$$\lim_{n \to \infty} M_n(T)x = x_0 \tag{1.132}$$

exists for each $x \in X$. Moreover, the operator $P : X \to X$ defined by the equality $Px = x_0$ is a continuous linear projection of X onto $\mathrm{Ker}(I - T)$. In addition, $\mathrm{Ker}(P) = \overline{\mathrm{Im}(I - T)}$; hence

$$X = \overline{\mathrm{Im}(I - T)} \oplus \mathrm{Ker}(I - T). \tag{1.133}$$

1.9.2 *Uniform ergodic theorems in Banach spaces*

In this section we give some results concerning the uniform convergence of
the Cesàro averages as well as the structure of the fixed point set of an
operator $T \in L(X)$ satisfying the following condition:

$$\lim_{n \to \infty} \frac{1}{n} \|T^n\| = 0. \tag{1.134}$$

The following remarkable result is due to N. Dunford (see [Dunford (1943)]).

Theorem 1.17 *Let X be a Banach space, and let $T \in L(X)$. Then
the sequence $M_n(T) = \frac{1}{n} \sum_{k=0}^{n-1} T^k$ converges uniformly if and only if T satisfies* (1.134) *and the point 1 is at most a simple pole of the resolvent
$R(\lambda, T) = (\lambda I - T)^{-1}$.*

The latter condition of this theorem means that either 1 does not belong
to the spectrum $\sigma(T)$ of T or it is a simple pole of the resolvent $R(\lambda, T)$.
As a matter of fact, in our setting, this spectral condition can be replaced
by the closedness of $\mathrm{Im}(I - T)$, so one can arrive to the similar conclusion.
Namely, if we just assume that

$$\sup \|M_n(T)\| < \infty, \tag{1.135}$$

then we have

$$\mathrm{Im}(I - T) \cap \mathrm{Ker}(I - T) = \{0\}. \tag{1.136}$$

Theorem 1.18 (see [Lyubich and Zemanek (1994)]) *Let $T \in L(X)$
satisfy condition* (1.134) *and* (1.136) *(or the stronger condition* (1.135)*).
Then the following assertions are equivalent:*

(i) the number 1 is at most a simple pole of the resolvent $R(\lambda, T) = (\lambda I - T)^{-1}$;

(ii) $\mathrm{Im}(I - T)$ is closed;

(iii) $\mathrm{Im}(I - T) \oplus \mathrm{Ker}(I - T) = X$.

Thus, under the conditions of Dunford's theorem, we have that the operator
$P : X \to X$ defined by

$$P = \lim_{n \to \infty} M_n(T) \tag{1.137}$$

is a linear projection onto the null point set $\mathrm{Ker}(I - T)$ with $\mathrm{Ker}P = \mathrm{Im}(I - T)$, and hence

$$\mathrm{Ker}(I - T) \oplus \mathrm{Im}(I - T) = X. \tag{1.138}$$

Since we will be mostly interested in the condition

$$\sup \|T^n\| < \infty, \tag{1.139}$$

which is stronger than both conditions (1.134) and (1.135), we summarize the above information as follows.

Corollary 1.6 *Let $T \in L(X)$ be such that the sequence $\{T^n\}_{n=1}^{\infty}$ is uniformly bounded (i.e., (1.139) holds), and let the range $\mathrm{Im}(I - T)$ be closed. Then the Cesàro averages $M_n(T)$ converge uniformly to a projection P onto $\mathrm{Ker}(I - T)$, and $\mathrm{Ker}(I - T) \oplus \mathrm{Im}(I - T) = X$.*

Observe again that if the Cesàro averages $M_n(T)$ are uniformly convergent, then 1 is either a regular point of T (*i.e.*, $1 \notin \sigma(T)$ and $(I - T)^{-1}$ exists) or 1 is an isolated point of the spectrum such that it is a simple pole of the resolvent $R(\lambda, T) = (\lambda I - T)^{-1}$. Moreover, in this case 1 must be an eigenvalue of T (see [Lyubich and Zemanek (1994)]).

A stronger asymptotic property is the convergence of the powers T^n. A spectral criterion for uniform convergence was established by Koliha [Koliha (1974)]. His result can be reformulated as follows.

Theorem 1.19 *The sequence $\{T^n\}_{n=1}^{\infty} \subset L(X)$ converges uniformly (to a projection P onto $\mathrm{Ker}(I - T)$) if and only if it is uniformly bounded, $\mathrm{Im}(I - T)$ is closed and there is no spectral point of T (except, perhaps, 1) on the unit circle of the complex plane, i.e.,*

$$\sigma(T) \cap \{\lambda \in \mathbb{C} : |\lambda| = 1, \ \lambda \neq 1\} = \emptyset. \tag{1.140}$$

Finally, note that condition (1.139) implies $\sigma(T) \subset \bar{\Delta} = \{\lambda \in \mathbb{C} : |\lambda| \leq 1\}$. If $\sigma(T)$ lies inside the unit disk $\Delta = \{\lambda \in \mathbb{C} : |\lambda| < 1\}$, then $I - T$ is invertible and $\lim_{n \to \infty} T^n x = 0$ for all $x \in X$.

This can be easily established by using an equivalent norm $\| \cdot \|_*$ on X for which $\|T\|_* < 1$.

1.10   Lipschitzian and Nonexpansive Mappings in Metric Spaces

1.10.1   *Lipschitzian and contraction mappings*

Let M be a metric space with distance function (metric) d, and let $\mathcal{D}$ be a subset of M. *A mapping $F : \mathcal{D} \mapsto M$ is said to be **Lipschitzian** on $\mathcal{D}$ if there exists $k \geq 0$ such that for all $x, y \in \mathcal{D}$,*

$$d\big(F(x), F(y)\big) \leq kd(x, y). \tag{1.141}$$

*The smallest k, for which (1.141) holds, is said to be the **Lipschitz constant** of F.*

We will sometimes denote the respective Lipschitz constants of different mappings F and G by $k(F)$ and $k(G)$ and, where relevant, $k_d(F)$ will be used to denote the Lipschitz constant of F with respect to the metric d.

*A mapping $F : \mathcal{D} \mapsto \mathcal{D}$ which takes values in $\mathcal{D}$ is said to be a **self-mapping** of $\mathcal{D}$.*

For two self-mappings G and F of $\mathcal{D}$ one can define the **composition operation** $F \circ G$ by

$$(F \circ G)(x) = F(G(x)), \quad x \in \mathcal{D}. \tag{1.142}$$

If two self-mappings G and F of $\mathcal{D}$ are Lipschitzian, then we have

$$k(F \circ G) \leq k(F)k(G) \tag{1.143}$$

and, in particular,

$$k(F^n) \leq k^n(F), \quad n = 1, 2, \ldots, \tag{1.144}$$

where F^n denotes the n-fold **iterate** of F: $F^1 = f$, $F^n = F \circ F^{n-1}$, $n = 2, 3, \ldots.$

If $\mathcal{D} = M$ is a linear normed space whose metric is generated by a norm, then we have

$$k(F + G) \leq k(F) + k(G) \tag{1.145}$$

and, for $\alpha \geq 0$,

$$k(\alpha F) = \alpha k(F) \quad \text{for} \ \alpha \geq 0. \tag{1.146}$$

*A mapping $F : M \mapsto M$ is said to be **locally Lipschitzian** on M if there exists a family of neighborhoods $\{\mathcal{D}_\alpha\}$, covering $M, \cup \mathcal{D}_\alpha = M$, such* that F is Lipschitzian on each $\mathcal{D}_\alpha$, *i.e., there exists $k_\alpha \geq 0$ such that*

$$d(F(x), F(y)) \leq k_\alpha d(x, y) \tag{1.147}$$

for all $x, y \in \mathcal{D}_\alpha$.

*A mapping $F : M \mapsto M$ is said to be a **strict contraction** if $k(F) < 1$; more precisely, F is a k-contraction with respect to d if $k_d(F) \leq k < 1$.*

*A mapping $F : M \mapsto M$ is said to be a **contraction** if*

$$d(F(x), F(y)) < d(x, y). \tag{1.148}$$

It is clear that a strict contraction is a contraction. Both these kinds of mappings are contained in the class of nonexpansive mappings.

1.10.2 *Nonexpansive mappings*

*A mapping $F : \mathcal{D} \mapsto M$ is called **nonexpansive** if its Lipschitz constant $k(F)$ does not exceed 1.*

Thus this class of mappings contains the contractive and strictly contractive mappings; moreover, it contains all isometries (including the identity).

Explicitly, $F : \mathcal{D} \mapsto M$ is nonexpansive if

$$d(F(x), F(y)) \leq d(x, y), \quad x, y \in \mathcal{D}. \tag{1.149}$$

Let X be a Banach space with norm $\|\cdot\|$, and let M denote a nonempty, closed, convex and bounded subset of X. In this context a mapping $F : M \mapsto M$ is nonexpansive if

$$\|F(x) - F(y)\| \leq \|x - y\|, \quad x, y \in \mathcal{D}. \tag{1.150}$$

Note that (1.150) clearly implies that all the iterates F^n of F, $n = 0, 1, 2, \ldots$ (with $F^0 = I$) are nonexpansive as well.

1.10.3 *Uniformly Lipschitzian mappings*

*A self-mapping $F : \mathcal{D} \mapsto \mathcal{D}$ of a subset $\mathcal{D}$ of X is said to be **uniformly Lipschitzian** if the condition*

$$d(F^n x, F^n y) \leq k d(x, y), \quad n = 0, 1, 2, \ldots \tag{1.151}$$

holds for some $k > 0$.

Obviously, all nonexpansive mappings are uniformly Lipschitzian with $k = 1$, and the observation preceding the definition shows that all mappings which are nonexpansive with respect to some metric equivalent to the norm are uniformly Lipschitzian.

On the other hand, if X is a normed space and $F : \mathcal{D} \mapsto \mathcal{D}$ satisfies

$$\|F^n x - F^n y\| \le k\|x - y\|, \quad x, y \in \mathcal{D}, \quad n = 0, 1, 2, \ldots \qquad (1.152)$$

for some $k \ge 1$, then by setting

$$d(x, y) = \sup\{\|F^n x - F^n y\| : n = 0, 1, 2, \ldots\}, \quad x, y \in \mathcal{D}, \qquad (1.153)$$

one obtains a metric d on $\mathcal{D}$ which is equivalent to the norm and with respect to which F is nonexpansive:

$$\|x - y\| \le d(x, y) \le k\|x - y\|;$$
$$d(F(x), F(y)) \le d(x, y), \quad x, y \in \mathcal{D}. \qquad (1.154)$$

Thus the class of uniformly Lipschitzian mappings on $\mathcal{D}$ with respect to the norm is completely characterized as the class of mappings on $\mathcal{D}$ which are nonexpansive with respect to some metric on $\mathcal{D}$ which is equivalent to the norm.

1.10.4   *Firmly nonexpansive mappings*

Let $\mathcal{D}$ be a subset of a Banach space X.

A self-mapping $F : \mathcal{D} \mapsto X$ is said to be firmly nonexpansive if for each x and y in $\mathcal{D}$, the convex function $\phi : [0, 1] \mapsto [0, \infty)$ defined by

$$\phi(s) = \big\|(1 - s)x + sF(x) - \big((1 - s)y + sF(y)\big)\big\| \qquad (1.155)$$

is decreasing.

It is clear that every firmly nonexpansive mapping is nonexpansive.

Proposition 1.25   *Let $\mathcal{D}$ be a subset of a real Banach space X, J the duality mapping of X, and F a mapping from $\mathcal{D}$ into X. Then the following are equivalent:*

(a) $F : \mathcal{D} \mapsto X$ is firmly nonexpansive;
(b) for each x and y in $\mathcal{D}$,

$$\|F(x) - F(y)\| \le \big\|(1 - s)(x - y) + s\big(F(x) - F(y)\big)\big\| \qquad (1.156)$$

for all $0 \le s \le 1$;
(c) for each x and y in $\mathcal{D}$, there is $j \in J\big(F(x) - F(y)\big)$ such that

$$\|F(x) - F(y)\|^2 \le (x - y, j); \qquad (1.157)$$

(d) for each x and y in $\mathcal{D}$,

$$\|F(x) - F(y)\| \le \|r(x - y) + (1 - r)(F(x) - F(y))\| \quad (1.158)$$

for all $r > 0$.

We now characterize all firmly nonexpansive mappings in a Hilbert space H with the inner product $\langle \cdot, \cdot \rangle$. Let $\mathcal{D}$ be a convex set in H and let $F : \mathcal{D} \mapsto H$. Consider the function $\varphi : [0, 1] \mapsto R^+$ defined by

$$\varphi(s) = \|(1 - s)(x - y) + s(F(x) - F(y))\|^2. \quad (1.159)$$

Since φ is convex, a necessary and sufficient condition for F to be firmly nonexpansive is that $\varphi'(1) \le 0$. This is equivalent, in turn, to the inequality

$$\langle F(x) - F(y), x - y \rangle \ge \|F(x) - F(y)\|^2. \quad (1.160)$$

Now suppose F satisfies (1.160), and consider the mapping $G = 2F - I$. Then, for $x, y \in \mathcal{D}$,

$$\begin{aligned}
\|Gx - Gy\|^2 &= \|2F(x) - x - 2F(y) + y\|^2 \\
&= \|2F(x) - 2F(y) - (x - y)\|^2 \\
&= 4\|F(x) - F(y)\|^2 - 4\langle F(x) - F(y), x - y \rangle + \|x - y\|^2 \\
&\le \|x - y\|^2.
\end{aligned} \quad (1.161)$$

This shows that G is nonexpansive; so any firmly nonexpansive mapping F is of the form

$$F = \frac{1}{2}(I + G) \quad (1.162)$$

with G nonexpansive. Reversing this argument, we see that mappings of the above form are always firmly nonexpansive.

1.10.5 *Monotone and accretive mappings*

The study of nonexpansive mappings has been substantially motivated by the study of monotone and accretive operators, two classes of operators which arise naturally in the theory of differential equations.

The definitions given below have a very simple origin: if $\mathcal{D} \subset \mathbb{R}$ and if $\varphi : \mathcal{D} \mapsto \mathbb{R}$, then φ is monotone increasing if for each $s, t \in \mathcal{D}$

$$(s - t)(\varphi(s) - \varphi(t)) \ge 0. \quad (1.163)$$

Let X be a Banach space with dual space X^*, and let $\mathcal{D} \subset X$. As before, we use the pairing $\langle x, j \rangle$ to denote $j(x)$, $x \in X$, $j \in X^*$.

Definition 1.19   A mapping $T : \mathcal{D} \mapsto X^*$ is said to be **monotone** if for each $u, v \in \mathcal{D}$,

$$Re\langle u - v, T(u) - T(v) \rangle \geq 0. \tag{1.164}$$

*It is called **strongly monotone** if for some $c > 0$,*

$$Re\langle u - v, T(u) - T(v) \rangle \geq c\|x - y\|^2. \tag{1.165}$$

A natural analogue of the above for mappings taking values in X is the following definition.

Definition 1.20   A mapping $T : \mathcal{D} \mapsto X$ is said to be **accretive** if for all $u, v \in \mathcal{D}$ and some $j \in J(u - v)$,

$$Re\langle T(u) - T(v), j \rangle \geq 0. \tag{1.166}$$

*It is called **strongly accretive** if for some $c > 0$,*

$$Re\langle T(u) - T(v), j \rangle \geq c\|x - y\|^2. \tag{1.167}$$

Here J denotes the normalized duality mapping introduced earlier: for $x \in X$,

$$J(x) = \left\{ j \in X^* : \langle x, j \rangle = \|x\|^2 = \|j\|^2 \right\}. \tag{1.168}$$

We note first that if X is a Hilbert space then $X = X^*$, the classes of monotone (respectively, strongly monotone) and accretive (respectively, strongly accretive) mappings defined in X coincide, and (1.164) denotes the usual inner product. Thus, if X is specialized further to $X = \mathbb{R}$, (1.164) becomes $(u - v)\big(T(u) - T(v)\big) \geq 0$.

One connection between accretive mappings and nonexpansive mappings is immediate. If $F : \mathcal{D} \mapsto X$ is nonexpansive, then for $T = I - F$, $x, y \in \mathcal{D}$, and $j \in J(x - y)$,

$$\begin{aligned}
\langle T(x) - T(y), j \rangle &= \langle x - y - (F(x) - F(y)), j \rangle \\
&= \|x - y\|^2 - \langle F(x) - F(y), j \rangle \\
&\geq \|x - y\|^2 - \|F(x) - F(y)\|\|x - y\| \geq 0. \tag{1.169}
\end{aligned}$$

Thus T is accretive. In addition, if $F : \mathcal{D} \mapsto X$ is a strict contraction,

$$\|F(x) - F(y)\| \leq k\|x - y\|, \quad x, y \in \mathcal{D}, \; 0 < k < 1, \tag{1.170}$$

then $T = I - F$ is strongly accretive.

On the other hand, not all accretive mappings are of the form $I - F$ with F nonexpansive. A complete characterization of accretive mappings in metric terms has been given by Kato [Kato (1967)] (see also [Deimling (1974)].

Proposition 1.26 *Let X be a Banach space, $\mathcal{D} \subset X$, and $T : \mathcal{D} \mapsto X$. Then T is accretive if and only if for each $x, y \in \mathcal{D}$ and $\lambda \geq 0$,*

$$\|x - y\| \leq \|x - y + \lambda(T(x) - T(y))\|. \tag{1.171}$$

*Thus a mapping $T : \mathcal{D} \mapsto X$ is accretive if and only if the mapping $J_\lambda = (I + \lambda T)^{-1}$ (called the **resolvent** of T) is nonexpansive on its domain for each positive λ.*

Remark 1.2 *Using the above, it is possible to extend the definition of accretivity in a natural way to the multivalued case. For a given subset B of X, let $|B| = \inf\{\|x\| : x \in B\}$.*

*A mapping $T : \mathcal{D} \mapsto 2^X (\mathcal{D} \subset X)$ is said to be **accretive** (respectively, **strongly accretive**) if there exists $\varepsilon \geq 0$ (respectively, $\varepsilon > 0$) such that for each $x, y \in \mathcal{D}$, $z \in T(x)$, $w \in T(y)$, and $\lambda \geq 0$,*

$$(1 + \varepsilon)\|x - y\| \leq \|x - y + \lambda(z - w)\|. \tag{1.172}$$

*Again, it can be shown that the mapping $J_\lambda = (I + \lambda T)^{-1}$ is single-valued and nonexpansive (respectively, strict contraction) on its domain. If it is the case that the domain of J_λ is all of X for some (hence all) $\lambda > 0$, then T is said to be **m-accretive** (sometimes called **hyperaccretive**).*

The theory of accretive operators is extensive. The solvability of many equations involving partial differential operators (e.g., Laplacians) can be formulated as questions concerning accretive operators. We will not discuss the details here but instead refer the reader to [Browder (1976); Deimling (1992); Barbu (1976)], and [Martin (1973)]. Our motivation for including this notion here is simply to note the usefulness of fixed point theory for nonexpansive mappings in another context and to illustrate the dependence of each theory upon the other (see also [Kirk and Sims (2001)] and references therein).

It is also not difficult to see that the resolvent J_λ of an accretive mapping is firmly nonexpansive. As a matter of fact, F is firmly nonexpansive if and only if it is the resolvent $(I + A)^{-1}$ for some accretive mapping $A \subset X \times X$.

Chapter 2

Differentiable and Holomorphic Mappings in Banach Spaces

2.1 Differentiable Mappings. Fréchet Derivatives

The fundamental notions of abstract differentials, polynomials and power series were introduced by Maurice Fréchet around 1909. The crux of the theory of functions between normed spaces is the question of differentiability. In this general situation the differentials of Fréchet appear to be the most appropriate concepts. In the present chapter we develop Fréchet differentials.

Let X, Y be normed spaces and let U be an open set in X. As above, we denote by the same symbol $\| \cdot \|$ the norms of both X and Y.

Definition 2.1 A mapping $f : U \mapsto Y$ is said to be **differentiable** at $x \in U$ if there exists a linear map $A \, (= A_x) \in L(X, Y)$ such that f and the continuous affine linear map $h \in X \to f(x) + Ah \in Y$ are tangent at x, *i.e.*,

$$\lim_{h \to 0} \frac{\|f(x+h) - f(x) - Ah\|}{\|h\|} = 0 \,. \tag{2.1}$$

If there is such a linear map A_x satisfying (2.1), then it is unique.

We call $A \, (= A_x)$ the **Fréchet derivative** of f at x, and A will be denoted by $\mathcal{D}f(x)$ or $F'(x)$. The element $f'(x)h \in Y$ is called the **Fréchet differential** or the differential of f at x in the direction of $h \in X$. We note that differentiability depends only on the topologies of X and Y, and not on the particular norms used to define these topologies.

If $f : U \mapsto Y$ is differentiable at each point of U, then f is said to be differentiable on $\mathcal{D}$. In this case the mapping

$$\mathcal{D}f : x \in U \mapsto f'(x) \in L(X, Y) \tag{2.2}$$

is called the derivative (or differential) of f on U. If $\mathcal{D}$ is continuous on U, the mapping f is said to be of class C^1 on U. Note that the differential $\mathcal{D}f$ is always vector-valued and not scalar-valued even when we may be interested in the differentiation of scalar-valued functions $f : U \mapsto \mathbb{K}$ (*i.e.*, when $Y = \mathbb{K}$). In fact, in this case we have $\mathcal{D}f : U \mapsto L(X, \mathbb{K}) = X^*$. Let X, Y be normed spaces and $\mathcal{D}$ a non-empty open subset of X. Then the notion of differentiability is a local property in the following sense.

Proposition 2.1 *If $f : M \mapsto Y$ is differentiable on M, then f is differentiable on any open subset U of M, and*

$$\mathcal{D}(f|_U) = (\mathcal{D}f)|_U. \tag{2.3}$$

Proposition 2.2 *If $M = \bigcup_{i \in I} U_i$ where each U_i is open in M, then f is differentiable on M if and only if it is differentiable on each U_i.*

Theorem 2.1 (Lipschitzian property) *Let X and Y be normed spaces and U a nonempty open subset of X. If $f : U \mapsto Y$ is differentiable at $x \in U$, then there exist $C > 0$ and $\delta > 0$ such that*

$$\|f(x) - f(y)\| \le C\|x - y\| \tag{2.4}$$

for $y \in U, \|x - y\| \le \delta$. In particular, it follows that f is continuous at x.

Theorem 2.2 (Chain rule) *Let X, Y, Z be normed spaces, U an open subset of X, and V and open subset of Y. Let $f : U \mapsto V$ and $g : V \mapsto Z$. If f is differentiable at $x \in U$ and g is differentiable at $f(x) \in V$, then $g \circ f : U \mapsto Z$ is differentiable at x and*

$$(g \circ f)'(x) = g'(f(x)) \circ f'(x). \tag{2.5}$$

Corollary 2.1 *Suppose $F : \mathcal{D} \mapsto Y$ is differentiable at a point $x_0 \in \mathcal{D}$ and assume that there exists a mapping $G : Y \mapsto X$ defined on a neighborhood U of the point $y_0 = F(x_0)$, and satisfying the relations*

$$G \circ F = I_{\mathcal{D}}, \quad F \circ G = I_U \tag{2.6}$$

(we write in this case $G = F^{-1}$). If F^{-1} is differentiable at $y_0 = F(x_0) \in U$, then the linear operator $F'(x_0)$ has a continuous inverse and

$$[F'(x_0)]^{-1} = (F^{-1})'(y_0). \tag{2.7}$$

2.1.1　*Examples*

Example 2.1 (Constant mappings) Let X and Y be normed spaces. Then a constant mapping is a mapping of the form $f : x \in X \mapsto b \in Y$ for a fixed point b in Y. If $f : X \mapsto Y$ is a constant mapping, then $\mathcal{D}f = 0$, i.e., $f'(x) = 0$ for all $x \in X$.

Example 2.2 (Linear mappings) If $A \in L(X, Y)$, then $\mathcal{D}A$ is the constant mapping satisfying $A'(x) = A$ for all $x \in X$. Let $f : \mathcal{D} \mapsto Y$ be differentiable at $x \in \mathcal{D}$ and let $A \in L(Y, Z)$. Then

$$(A \circ f)'(x) = A \circ f'(x). \tag{2.8}$$

Example 2.3 (Urysohn operator) Let $K(t, s, x)$ be a function defined on $a \le t, s \le b, |x| \le r$, such that $K(t, s, x)$ and $K'_x(t, s, x)$ are everywhere continuous. Let $X = C[a, b]$ be the space of all real- (or complex-)valued continuous functions on $[a, b]$ with the norm $\|x\| = \sup\limits_{a \le t \le b} |x(t)|$. Then one can define a mapping $f : \mathcal{D} \mapsto X$, $\mathcal{D} = \{x \in X : \|x\| \le r\}$, by the formula

$$f(x(t)) = \int_a^b K(t, s, x(s)) ds. \tag{2.9}$$

This mapping is usually called the nonlinear Urysohn integral operator. The operator f takes values in $C[a, b]$ and is Fréchet differentiable on the open ball $\|x\| < r$. Moreover, for any $x_0 \in C[a, b]$ with $\|x_0\| < r$, and for every $h \in C[a, b]$, we have

$$(f'(x_0)h)(t) = \int_a^b K'_x(t, s, x_0(s)) h(s) ds. \tag{2.10}$$

Note that the Urysohn operator (2.9) can be also considered on the space $C[a, b]$ of all complex-valued continuous functions; then f is Fréchet differentiable on $C[a, b]$ if the kernel $K(t, s, x)$ has continuous partial derivatives with respect to the variable x in the disk $\{x \in \mathbb{C} : |x| < r\}$ of the complex plane.

Example 2.4 (Hammerstein operator) A particular case of the operator (2.9) is the Hammerstein operator defined by the equality

$$f(x(t)) = \int_a^b K(t, s) f(s, x(s)) ds, \tag{2.11}$$

where the functions $K(t, s)$, $f(s, x)$ and $f'_x(s, x)$ are continuous with respect to all variables for $a \le t, s \le b, |x| \le r$. Example 2.3 implies that the

Fréchet derivative of f is given by

$$(f'(x)h)(t) = \int_a^b K(t,s)f'_x(s,x(s))h(s)ds. \tag{2.12}$$

Example 2.5 (Nemytski operator) Suppose that the functions $\varphi(s,x)$ and $\varphi'_x(s,x)$ are continuous with respect to both variables for $a \leq s \leq b$, $|x| \leq r$. The Nemytski operator is the superposition operator $\Phi : \mathcal{D} \mapsto X = C[a,b]$ defined by

$$\Phi(x)(s) := \varphi(s,x(s)) \tag{2.13}$$

for each x in the ball $\mathcal{D} = \{x \in X : \|x\| < r\}$ of the space $C[a,b]$. The operator (2.13) is differentiable and

$$(\Phi'(x_0)h)(s) = \varphi'_x(s,x_0(s))h(s). \tag{2.14}$$

Example 2.6 Let $f_1(x_1\ldots,x_n),\ldots,f_m(x_1,\ldots,x_n)$ be functions defined on a domain in the Euclidean space $\mathbb{K}^n = \{x : x = (x_1,\ldots,x_n)\}$, and assume that f_j, $j = 1,\ldots,n$, and all their first order partial derivatives are continuous. Then the operator

$$f = (f_1,\ldots,f_m) : \mathbb{K}^n \mapsto \mathbb{K}^m \tag{2.15}$$

is Fréchet differentiable on the considered domain, and

$$f'(x)h = \begin{pmatrix} \frac{\partial f_1(x)}{\partial x_1} & \cdots & \frac{\partial f_1(x)}{\partial x_n} \\ \vdots & & \vdots \\ \frac{\partial f_m(x)}{\partial x_1} & \cdots & \frac{\partial f_m(x)}{\partial x_n} \end{pmatrix} \begin{pmatrix} h_1 \\ \vdots \\ h_n \end{pmatrix}. \tag{2.16}$$

On the right hand side of (2.16) we have the action of the Jacobi matrix of f on the vector $h = (h_1,\ldots,h_n) \in \mathbb{K}^n$.

2.2 Holomorphic Mappings

Definition 2.2 A complex Banach space valued function h defined on a domain (open, connected subset) $\mathcal{D}$ in a complex Banach space X is said to be **holomorphic** in $\mathcal{D}$ if for each $x \in \mathcal{D}$, the Fréchet derivative of h at x (denoted by $\mathcal{D}h(x)$ or $h'(x)$) exists as a bounded complex-linear mapping of X into the Banach space Y containing the values of h.

If $\mathcal{D}$ and Ω are domains in complex Banach spaces X and Y, respectively, then the set of holomorphic mappings from $\mathcal{D}$ into Ω is denoted by

$Hol(\mathcal{D}, \Omega)$. The notation $Hol(\mathcal{D})$ is used to denote the set $Hol(\mathcal{D}, \mathcal{D})$ of holomorphic self-mappings of $\mathcal{D}$. Note that $Hol(\mathcal{D})$ is a semigroup with respect to the composition operation.

One of the most important properties of holomorphic mappings is that they have the Cauchy integral representation, which implies many special features of holomorphic mappings such as the maximum modulus principle, power series representation, the Schwarz Lemma and others. Here we will see that a substantial portion of basic complex function theory can be extended to normed spaces in the light of the Cauchy integral formula.

2.2.1 *The Cauchy integral formula*

The classical Cauchy integral formula represents a holomorphic function defined on a domain Ω in the complex plane $\mathbb{C}$ and continuous on the closure of Ω by its values on the boundary $\partial\Omega$ of Ω. Here is a simple version of this formula.

Theorem 2.3 *Let Ω be a simply connected domain in $\mathbb{C}$ with a smooth boundary $\partial\Omega$, and let $f \in Hol(\Omega, \mathbb{C})$ be continuous on the closure of Ω. Then for each $z \in \Omega$,*

$$f(z) = \frac{1}{2\pi i} \int_{\partial\Omega} \frac{f(w)dw}{w - z}. \tag{2.17}$$

Let now $\mathcal{D}$ be a domain in a complex Banach space X. Since $\mathcal{D}$ is open, for each $x \in \mathcal{D}$ and $y \in X$, there is $r > 0$ such that $x + \zeta y \in \mathcal{D}$ for $\zeta \in \mathbb{C}$, $|\zeta| \leq r$.

Theorem 2.4 *Let X and Y be complex Banach spaces, and let $\mathcal{D}$ be a domain in X. Suppose that $f \in Hol(\mathcal{D}, Y)$, $x \in \mathcal{D}$, $y \in X$ and $r > 0$ are such that $x + \zeta y \in \mathcal{D}$ for $\zeta \in \mathbb{C}$, $|\zeta| \leq r$. Then*

$$f(x) = \frac{1}{2\pi i} \int_{|z|=r} \frac{f(x + \zeta y)d\zeta}{\zeta}. \tag{2.18}$$

For the finite-dimensional case, where $\mathcal{D}$ is a domain in $\mathbb{C}^n$, there are many different integral representations of holomorphic mappings on $\mathcal{D}$ depending on geometrical properties of $\mathcal{D}$. For instance, if $\mathcal{D}$ is the open unit ball in the Euclidian metric, the well-known representation of Martinelly–Bochner is a generalization of the Cauchy formula. Other representations, *e.g.*, for the unit polydisk, the hypercone, convex and pseudoconvex domains, can be found in [Aizenberg and Yuzhakov (1983)]. In the framework of functional analysis, however, most of the arguments may be presented by using a reduction to the one-dimensional case. This can be achieved by applying the following assertion which is based on the famous Dunford theorem (see, for example, [Khatskevich and Shoikhet (1994a)].

Theorem 2.5 *Let X and Y be complex Banach spaces, let $\mathcal{D}$ be a domain in X, and let f be a Y-valued mapping on $\mathcal{D}$. Then the following are equivalent:*

(i) The mapping $f \in \mathrm{Hol}(\mathcal{D}, Y)$;

(ii) for each $x \in \mathcal{D}$, $y \in X$ and $r > 0$ such that $x + \zeta y \in \mathcal{D}$ for $\zeta \in \Delta_r = \{\zeta \in \mathbb{C} : |\zeta| \leq r\}$, the mapping $g : \Delta_r \mapsto Y$ defined by $g(\zeta) = f(x + \zeta y)$ is holomorphic in Δ_r;

(iii) for each bounded linear functional $\ell \in Y^*$, the mapping $h = \ell \circ f : \mathcal{D} \mapsto \mathbb{C}$, is holomorphic in $\mathcal{D}$.

2.2.2   *Power series representation*

As in classical complex analysis, the power series representation, Taylor's formula, plays a crucial role in the development of the theory. As a matter of fact, the theory of power series in normed spaces is a very rich subject developed in the middle of the 20th century by E. Hille and R. Phillips, L. Nachbin, and others. Further contributions were also made by J. A. Barroso, J. Bochnak, L. A. Harris and S. Dineen (see details in [Franzoni and Vesentini (1980)] and [Chae (1985)]).

We do not intend to elaborate here on this nice theory; we just present some definitions and notations which are necessary for our further discussion.

Theorem 2.6 (Power series representation) *Let X and Y be complex Banach spaces, let $\mathcal{D}$ be a domain in X and let $f \in \mathrm{Hol}(\mathcal{D}, Y)$ be bounded. Then for each $x_0 \in \mathcal{D}$ and for each ball $B \subset\subset \mathcal{D}$ centered at x_0, the following representation holds:*

$$f(x) = \sum_{k=0}^{\infty} P^{(k)}(x_0)(x - x_0), \quad x \in B, \qquad (2.19)$$

where $P^{(k)}(x_0)$, $k = 0, 1, 2, \ldots$, are homogeneous polynomials of order k and $P^{(0)}(x_0) = f(x_0)$.

In addition, the convergence in (2.19) is uniform.

It is clear that $P^{(1)}(x_0)$ is the Fréchet derivative of f at the point x_0, i.e., $f'(x_0) = P^{(1)}(x_0)$.

Proposition 2.3 (The Cauchy inequalities) *In the setting of the above theorem, let r be the radius of the ball B, and let $\|f(x)\| \leq M$ for all $x \in B$. Then for each $k = 0, 1, 2, \ldots$ the following inequality holds:*

$$\|P^{(k)}(x_0)\| \leq Mr^{-k}. \tag{2.20}$$

Theorem 2.7 (The uniqueness theorem) *Let $f \in \mathrm{Hol}(\mathcal{D}, Y)$. Then the following assertions are equivalent:*

(i) $P^{(k)}(x_0) = 0$, $k = 0, 1, 2, \ldots$ for some $x_0 \in \mathcal{D}$;
(ii) f is identically zero in some neighborhood $U \subseteq \mathcal{D}$.
(iii) f is identically zero in $\mathcal{D}$.

2.2.3 *The maximum modulus theorem*

One consequence of the Cauchy integral formula is the so-called maximum modulus principle.

For a connected open set U in the complex plane, if $f : U \mapsto \mathbb{C}$ is holomorphic on U, then the classical maximum modulus theorem states that

- *if $|f(z)|$ has a maximum at a point of U, then $|f(z)|$ is constant on U. Furthermore, if $|f(z)|$ is a constant function, it follows that $f(z)$ is constant too.*

This theorem has a generalization for Banach space valued functions, but, unlike complex-valued functions, we cannot infer that if $\|f(z)\|$ is constant, then $f(z)$ is itself constant (see an example below).

The form of the maximum modulus theorem which claims that if $\|f(z)\|$ has a maximum at a point in U, then $f(z)$ is a constant on U is referred to as the **strong maximum modulus principle**.

Now we generalize the **weak maximum modulus theorem** for mappings defined on Banach spaces.

Theorem 2.8 *Let U be a connected open subset of $\mathbb{C}$ and $f \in H(U, Y)$. If $\|f(z)\|$ has a maximum at a point in U, then $\|f(z)\|$ is a constant on U.*

Proof. We use linear functionals in this proof. Suppose that the theorem were false, and for some $x \in U$, $\|f(z)\| \leq \|f(x)\|$ for all $z \in U$. We may assume that $\|f(x)\| \neq 0$, since otherwise the theorem holds trivially. There exists $y^* \in Y^*$ such that $\|y^*\| = 1$ and $y^*(f(x)) = \|f(x)\|$. Since f is holomorphic, $y^*(f) : U \mapsto \mathbb{C}$ is holomorphic and for all $z \in U$ we have

$$\left| y^*(f(z)) \right| \leq \|f(z)\| \leq \|f(x)\| = \left| y^*(f(x)) \right|. \tag{2.21}$$

Now the classical maximum modulus theorem implies that $y^*(f)$ is a constant map on U, and hence $y^*(f(z)) = \|f(x)\|$ for all $z \in U$. Since $\|y^*\| = 1$ and $\|f(z)\|$ is not a constant map, there exists $y \in U$ such that $\|f(y)\| < \|f(x)\|$. Thus

$$\|f(x)\| = y^*(f(y)) \leq \|f(y)\| < \|f(x)\| \tag{2.22}$$

a contradiction. This completes the proof of the theorem. $\qquad\square$

Corollary 2.2   *Let U be a connected open subset of X and $f \in \mathrm{Hol}(U, Y)$. If $\|f(x)\|$ has a maximum at a point in U, then $\|f(x)\|$ is a constant function on U.*

Proof.   Suppose that there exists a point ξ in U such that

$$\|f(\xi)\| = \sup\{\|f(x)\| : x \in U\}. \tag{2.23}$$

Since U is open, we can find r such that $B_r(\xi) \subset U$. For x in $B_r(\xi)$, $x \neq \xi$, consider the holomorphic mapping

$$g(z) := f[(1 - z)\xi + zx] \tag{2.24}$$

defined on an open set V which contains the unit disk of $\mathbb{C}$. The mapping g takes its maximum modulus at 0; thus $\|g\|$ must be constant on V. In particular, $\|g(0)\| = \|g(1)\|$ and hence $\|f(x)\| = \|f(\xi)\|$ holds for all x in $B_r(\xi)$. We now consider the following set to complete the proof:

$$A = \{x \in U : \|f(x)\| = \|f(\xi)\|\}. \tag{2.25}$$

Then A is closed in U. If $a \in A$, then, replacing in the above argument ξ with a, we see that a is an interior point of A. This proves that A is both closed and open in the connected set U; thus $A = U$. $\qquad\square$

We now present the strong maximum modulus theorem of Thorp and Whitley (1967). To do this, we recall the following definitions.

A point x in a convex subset U of a vector space X is said to be a (real) extreme point of U if for each pair of points x_1 and x_2 in U such that $x = \alpha x_1 + \beta x_2$, it follows that $x_1 = x_2 = x$ whenever $0 < \alpha, \beta$ and $\alpha + \beta = 1$.

A point x in a convex subset U of a **complex** vector space X is said to be a **complex extreme point** of U if there is no $y \neq 0$ in X such that the set $\{x + \lambda y : |\lambda| < 1\}$ is contained in U.

Thus, if U is a convex subset of a complex vector space X, then each real extreme point U is also a complex extreme point. In particular,

- Each point of the unit sphere of a complex strictly convex Banach space is a complex extreme point of its closed unit ball.

Theorem 2.9 (The strong maximum modulus principle) *Let U be a connected open subset of $\mathbb{C}$. Then every mapping $f \in \mathrm{Hol}(U, Y)$ satisfies the strong maximum modulus principle if and only if every point in the unit sphere of Y is a complex extreme point of the closed unit ball.*

A proof of this theorem can be based on a lemma of L. A. Harris (see Subsection 2.6.5, Corollaries 2.6 and 2.7).

Corollary 2.3 *Let U be a connected open subset of a complex Banach space X and let $f \in \mathrm{Hol}(U, Y)$, where Y is a complex strictly convex Banach space. Then f satisfies the strong maximum modulus principle.*

Example 2.7 If f is a complex-valued function on U such that $|f(z)|$ is constant for all $z \in U$, then f is a constant function on U. Therefore, for complex-valued functions, the strong maximum modulus theorem holds. However, this is not the case in general for vector-valued functions, as the following example shows.

Let U be the open unit disk in the complex plane, and let $Y = \mathbb{C}^2$ be normed by

$$\|x\| = \max\{|x_1|, |x_2|\}. \tag{2.26}$$

If $f : U \mapsto Y$ is defined by

$$f(z) = (1, z), \tag{2.27}$$

then f is obviously a non-constant holomorphic mapping and

$$\|f(z)\| = \max\{1, |z|\} = 1 \tag{2.28}$$

for all $z \in U$. This shows that the strong maximum modulus theorem fails to hold for the Banach space $\mathbb{C}^2$.

Example 2.8 Let c_0 be the Banach space of all complex sequences $x = (x_1, x_2 \ldots)$ such that $x_n \to 0$ as $n \to \infty$, with the norm $\|x\| = \max\{|x_1|, |x_2|, \ldots\}$. We define the mapping $f : U \mapsto c_0$ by $f(x) =$

$(1, x_2, x_3, \ldots)$ on the open unit ball U of c_0. Then we have $\|f(x)\| = 1$ for all $x \in U$ in spite of the fact that the mapping f is not constant.

As a matter of fact, it can be shown that the boundary of U has no complex extreme point of the closure of U.

Example 2.9   Since every complex (Lebesgue) space $Y = L^p[0,1]$, $1 < p < \infty$, is a strictly convex Banach space, we see that every holomorphic mapping $f \in \mathrm{Hol}(U, Y)$ on a domain U of a Banach space X with values in Y satisfies the strong maximum modulus principle.

Example 2.10   Although the complex space $L^1[0,1]$ is not strictly convex, it can be shown (see, for example, [Franzoni and Vesentini (1980)]) that each boundary point of its unit ball is a complex extreme point. Thus every holomorphic mapping $f \in \mathrm{Hol}(U, Y)$ on a domain U of a Banach space X with values in Y satisfies the strong maximum modulus principle whenever $Y = L^p[0,1]$, $1 \leq p < \infty$.

2.3   Topologies in $\mathrm{Hol}(\mathcal{D}, Y)$

2.3.1   *T-topology and compact open topology on* $\mathrm{Hol}(\mathcal{D}, Y)$

For $f \in Hol(\mathcal{D}, Y)$ and a subset K strictly inside $\mathcal{D}$ (we write $K \Subset \mathcal{D}$), we set

$$\|f(x)\|_K = \sup_{x \in K} \|f(x)\|. \qquad (2.29)$$

We will mostly consider the subspace $\widetilde{Hol}(\mathcal{D}, Y)$ of $Hol(\mathcal{D}, Y)$ consisting of all those holomorphic mappings which are bounded on each ball strictly inside $\mathcal{D}$. The system of semi-norms (2.29) induces on $\widetilde{Hol}(\mathcal{D}, Y)$ the **topology of locally uniform convergence over** $\mathcal{D}$ (or briefly, T-convergence). Thus if a net $\{f_j\}_{j \in \mathcal{A}} \subset \widetilde{Hol}(\mathcal{D}, Y)$ T-converges to $f \in \widetilde{Hol}(\mathcal{D}, Y)$, we write

$$f = T - \lim_{j \in \mathcal{A}} f_j. \qquad (2.30)$$

(For more details see [Franzoni and Vesentini (1980)] and [Isidro and Stacho (1984)].)

More generally, let $\mathcal{M} = \{K_i\}_{i \in I}$ be a family of subsets $K_i \subset \mathcal{D}$ which is closed with respect to finite unions, and for $f_0 \in \mathrm{Hol}(\mathcal{D}, Y)$, $K \in \mathcal{M}$ and $\varepsilon > 0$, define

$$U(f_0, K, \varepsilon) = \{f \in \mathrm{Hol}(\mathcal{D}, Y) : \|f - f_0\|_K < \varepsilon\}. \qquad (2.31)$$

Varying $K \in \mathcal{M}$ and $\varepsilon > 0$, we get a family $\{U(f_0, K, \varepsilon)\}$ which defines a fundamental system of neighborhoods for a locally convex topology τ_μ of $\mathrm{Hol}(\mathcal{D}, Y)$. This topology is not necessarily Hausdorff.

Thus the T-topology on $\widetilde{\mathrm{Hol}}(\mathcal{D}, Y)$ is seen to be a τ_μ-topology when $\mathcal{M}$ is the family of all finite unions of closed balls strictly inside $\mathcal{D}$.

If $\mathcal{M}$ is the family of all compact subsets of $\mathcal{D}$, then τ_μ defines the **topology of uniform convergence on compact subsets** (the so-called **compact open topology**).

The topology of local uniform convergence is finer than the compact open topology. They are one and the same if and only if X is finite dimensional.

Proposition 2.4 *The space* $\mathrm{Hol}(\mathcal{D}, Y)$ *is complete with respect to the compact open topology.*

Corollary 2.4 *The space* $\widetilde{\mathrm{Hol}}(\mathcal{D}, Y)$ *is complete with respect to the T-topology (the topology of local uniform convergence over $\mathcal{D}$).*

2.3.2 *Montel's theorem*

The classical Montel theorem states that any bounded subset of $\mathrm{Hol}(\mathcal{D}, \mathbb{C})$ is relatively compact for the compact open topology on $\mathcal{D}$.

This assertion will still be true if we replace $\mathbb{C}$ by $\mathbb{C}^n$. However, the Montel theorem is not valid if Y is infinite dimensional. It is also not valid for the topology of local uniform convergence over $\mathcal{D}$ if X is not finite dimensional.

We mention here two analogs of the Montel theorem for the infinite dimensional case which are based on the classical Ascoli theorem (see, for example, [Yosida (1974)] and [Dunford and Schwarz (1958)]).

Theorem 2.10 *Let X and Y be Banach spaces and let $\mathcal{D}$ be a domain in X. The set $\mathcal{F}_M = \{f \in \mathrm{Hol}(\mathcal{D}, Y) : \|f\|_{\mathcal{D}} \le M < \infty\}$ is relatively compact with respect to the compact open topology on $\mathrm{Hol}(\mathcal{D}, Y)$ if and only if the orbit $\mathcal{F}_M(x) = \{f(x) : f \in \mathcal{F}_M\}$ is relatively compact in Y for each $x \in \mathcal{D}$.*

Now, by the classical Alaoglu–Bourbaki theorem (see [Dunford and Schwarz (1958)]) we know that any bounded subset of a reflexive Banach space is relatively weakly compact. Hence, using the above theorem and Dunford's theorem on the equivalence of holomorphy and weak holomorphy (see Theorem 2.5) one can arrive at the following useful assertion.

Theorem 2.11 *Let Y be reflexive and let $\mathcal{F}_M = \{f \in \mathrm{Hol}(\mathcal{D}, Y) :$*

$\|f\|_{\mathcal{D}} \leq M\}$. *Then for any sequence $\{f_n\} \subset \mathcal{F}_M$, there is a subsequence $\{f_{n_k}\}$ which weakly converges to a holomorphic mapping $f \in \mathrm{Hol}(\mathcal{D}, Y)$, uniformly on each compact subset of $\mathcal{D}$.*

2.3.3   Vitali's theorem

In the one-dimensional case, Vitali's convergence theorem plays an important role in the study of analytic continuation. This theorem can be formulated as follows.

Theorem 2.12   *Let $\mathcal{D}$ be a domain in $\mathbb{C}$, and let $\{f_n\} \subset \mathrm{Hol}(\mathcal{D}, Y)$ be a sequence of uniformly bounded holomorphic mappings on $\mathcal{D}$. Then $\{f_n\}$ converges to $f \in \mathrm{Hol}(\mathcal{D}, Y)$, uniformly on each compact subset of $\mathcal{D}$ if and only if the sequence $\{f_n(x)\}$ converges at each point x in a subset U of $\mathcal{D}$ having a cluster point in $\mathcal{D}$.*

This assertion is no longer true if $\mathcal{D}$ is a domain in an infinite dimensional Banach space. Nevertheless, one can modify the assumptions of the above theorem and obtain an infinite dimensional version of Vitali's theorem.

Theorem 2.13   *Let $\mathcal{D}$ be a bounded domain in a Banach space X, and let $\mathcal{F}_M = \{f \in \mathrm{Hol}(\mathcal{D}, Y) : \|f\|_{\mathcal{D}} \leq M < \infty\}$. Then the following assertions are equivalent:*

(i) The net $\{f_j\}_{j \in \mathcal{A}} \subset \mathcal{F}_M$ converges to $f \in \widetilde{\mathrm{Hol}}(\mathcal{D}, Y)$ in the topology of local uniform convergence over $\mathcal{D}$, i.e.,

$$f = T - \lim_{j \in \mathcal{A}} f_j. \tag{2.32}$$

(ii) There exists a ball $B \subset\subset \mathcal{D}$ such that the net $\{f_j\}$ converges to $f \in \mathrm{Hol}(B, Y)$, i.e.,

$$\lim_{j \in \mathcal{A}} \|f - f_j\|_B = 0. \tag{2.33}$$

Note that this theorem remains true if we replace the T-topology by the compact open topology on $\mathcal{D}$ or even by the pointwise convergence topology on $\mathcal{D}$.

2.4 Elements of Functional Analytic Calculus

2.4.1 *Symbolic calculus on Banach algebras*

For any complex Banach algebra E one can define formally different analytic function by using the algebraic and topological structures of E. For instance, the polynomials on E can be defined by setting

$$P(A) = \sum_{i=0}^{m} \alpha_i A^i, \quad \alpha_i \in \mathbb{C}, \ A \in E. \tag{2.34}$$

The exponential function of an element $A \in E$ can be defined by

$$\exp(A) = \sum_{i=0}^{\infty} \frac{1}{n!} A^n. \tag{2.35}$$

More generally, we can define

$$f(A) = \sum_{n=0}^{\infty} \alpha_n A^n \tag{2.36}$$

for any function $f(\lambda)$ whose power series converges in a circle the radius of which exceeds the spectral radius $\rho(A)$ of A.

An equivalent (but more useful) approach to the definition of analytic functions on Banach algebras is to employ the classical Cauchy formula. The idea comes from the following simple, but very crucial assertion. For simplicity, we will consider $E = L(X)$ — the Banach algebra of all the linear bounded operators on a Banach space X.

Lemma 2.1 *Suppose that $A \in L(X)$ and $\alpha \in \mathbb{C}\backslash\sigma(A)$. Let Ω be any domain in $\mathbb{C}$ such that $\alpha \notin \Omega$ and $\sigma(A) \subset \Omega \subset \mathbb{C}$, and let Γ be a positively oriented simple closed rectifiable contour around $\sigma(A)$. Then, for all $n = 0, \pm 1, \pm 2, \ldots,$*

$$\frac{1}{2\pi i} \int_{\Gamma} (\alpha - \lambda)^n (\lambda I - A)^{-1} d\lambda = (\alpha I - A)^n. \tag{2.37}$$

This remarkable relation leads to the following notion. Note that if Ω is an open set in $\mathbb{C}$ and $\mathrm{Hol}(\Omega, \mathbb{C})$ is the algebra of all holomorphic functions on Ω, then

$$L_\Omega = \big\{ A \in L(X) : \sigma(A) \subset \Omega \big\} \tag{2.38}$$

is an open subset of E ($= L(X)$).

We define $\widehat{\mathrm{Hol}}(L_\Omega, E)$ to be the set of all E-valued functions $\hat{f}$ on L_Ω that arise from $f \in \mathrm{Hol}(\Omega, \mathbb{C})$ by the formula

$$\hat{f}(A) = \frac{1}{2\pi i} \int_\Gamma f(\lambda)(\lambda I - A)^{-1} d\lambda, \qquad (2.39)$$

where Γ is any simple positively oriented contour in Ω such that the interior domain of Γ contains $\sigma(A)$.

Remark 2.1 *Formula* (2.39) *is usually called the Riesz–Dunford representation. It calls for some comments.*

(a) *Since Γ stays away from $\sigma(A)$ and the inversion operation in E is continuous, the integrand in (2.39) is continuous. So the integral (2.39) exists and defines $\hat{f}(A)$ as an element of E.*

(b) *The integrand is a holomorphic E-valued mapping in the complement of $\sigma(T)$. The Cauchy formula implies therefore that $\hat{f}(A)$ is independent of the choice of Γ, provided only that Γ surrounds $\sigma(A)$ in Ω.*

(c) *If $A = \alpha I$ and $\alpha \in \Omega$, then the above lemma and (2.39) imply*

$$\hat{f}(\alpha I) = f(\alpha) I. \qquad (2.40)$$

In most discussions of this topic, $f(A)$ is written in place of $\hat{f}(A)$. The notation $\hat{f}$ is used here because it avoids certain ambiguities that may cause misunderstanding. We will sometimes call $f \mapsto \hat{f}$ the "shift" operation, and the function f is called the producing function of $\hat{f}$.

Theorem 2.14 *The mapping $f \mapsto \hat{f}$ is a linear algebraic isomorphism of $\mathrm{Hol}(\Omega, \mathbb{C})$ onto $\widehat{\mathrm{Hol}}(L_\Omega, E)$ which is continuous in the following sense: If $f_n \in \mathrm{Hol}(\Omega, \mathbb{C})$ converge uniformly on compact subsets of Ω to $f \in \mathrm{Hol}(\Omega, \mathbb{C})$, then*

$$\hat{f}(A) = \lim_{n \to \infty} \hat{f}_n(A), \quad A \in L_\Omega, \qquad (2.41)$$

in the operator topology on E.

Proof. The integral representation (2.39) makes it obvious that $f \mapsto \hat{f}$ is a linear mapping.

If $\hat{f} = 0$, then $f(\alpha) I = \hat{f}(\alpha I)$ (see Remark (c) above) implies $f = 0$. Thus $f \mapsto \hat{f}$ is one-to-one. The asserted continuity follows directly from formula (2.39), since $\|(\lambda I - A)^{-1}\|$ is bounded on Γ. Thus, it remains to prove that $f \mapsto \hat{f}$ is multiplicative. Indeed, if f and g are elements of $\mathrm{Hol}(\Omega, \mathbb{C})$ and $h(\lambda) = f(\lambda) \cdot g(\lambda)$, $\lambda \in \Omega$, one can show, by using Runge's approximation theorem (see, for example, [Shabat (1976)]) and continuity, that

$$\hat{h}(A) = \hat{f}(A)\hat{g}(A), \quad A \in L_\Omega. \tag{2.42}$$

Since $\mathrm{Hol}(\Omega, \mathbb{C})$ is obviously a commutative algebra, it follows that $\widehat{\mathrm{Hol}}(L_\Omega, E)$ is also commutative. This seems to be surprising, because elements of E need not be commutative in general. However, $\hat{f}(A)$ and $\hat{g}(A)$ do commute in E for each $A \in L_\Omega$ and $f, g \in \mathrm{Hol}(\Omega, \mathbb{C})$, *i.e.*,

$$\hat{f}(A)\hat{g}(A) = \hat{g}(A)\hat{f}(A). \tag{2.43}$$
$\square$

2.4.2 *The spectral mapping theorem*

One of the most important achievements of the functional calculus is the so-called **spectral mapping theorem** of Dunford.

Theorem 2.15 *Let Ω be a domain in $\mathbb{C}$ and let $A \in L_\Omega$. Then, for any $f \in \mathrm{Hol}(\Omega, \mathbb{C})$,*

$$\sigma(\hat{f}(A)) = f(\sigma(A)). \tag{2.44}$$

To prove this theorem, we need the following assertion.

Lemma 2.2 *Let $A \in L_\Omega$ and let $f \in \mathrm{Hol}(\Omega, \mathbb{C})$. Then $\hat{f}(A)$ is invertible in E ($= L(X)$) if and only if $f(\lambda) \neq 0$ for all $\lambda \in \sigma(A)$.*

Proof. If f has no zero in $\sigma(A)$, then $g = \frac{1}{f}$ is holomorphic in a neighborhood $\Omega_1 \subset \Omega$ of $\sigma(A)$. Since $fg = 1$ in Ω_1, it follows that $\hat{f}(A)\hat{g}(A) = I$ and thus $\hat{f}(A)$ is invertible.

Conversely, if $f(\alpha) = 0$ for some $\alpha \in \sigma(A)$, then there is $h \in \mathrm{Hol}(\Omega, \mathbb{C})$ such that

$$(\lambda - \alpha)h(\lambda) = f(\lambda), \quad \lambda \in \Omega. \tag{2.45}$$

This implies that

$$(A - \alpha I)\hat{h}(A) = \hat{f}(A) = \hat{h}(A)(A - \alpha I). \tag{2.46}$$

Since $\alpha \in \sigma(A)$, the operator $(A - \alpha I)$ is not invertible in E. Hence, $\hat{f}(A)$ is not invertible either.
$\square$

Proof of the spectral mapping theorem. Fix $\beta \in \mathbb{C}$. According to the above lemma, $\hat{f}(A)$ is not invertible if and only if there is $\alpha \in \sigma(A)$ such that $f(\alpha) = 0$. If in this conclusion we use $\hat{f} - \beta I$ in place of $\hat{f}$, then we obtain that $\hat{f}(A) - \beta I$ is not invertible if and only if there is $\alpha \in \sigma(A)$ such that $f(\alpha) = \beta$. Thus we have $\beta \in \sigma(\hat{f}(A))$ if and only if $\beta \in f(\sigma(A))$.

The theorem is proved. $\qquad\qquad\qquad\qquad\qquad\qquad\qquad\qquad\square$

2.4.3   *Some ∗-algebras*

Let $\mathcal{A}$ be a Banach algebra over the complex field $\mathbb{C}$, *i.e.*, $\mathcal{A}$ is a complex Banach space with the multiplication operation satisfying $\|x{\cdot}y\| \le \|x\|{\cdot}\|y\|$.

Recall that $\mathcal{A}$ is called an **algebra with involution** if there is an anti-linear mapping $x \mapsto x^*$ of $\mathcal{A}$ into itself such that $(x^*)^* = x$ and $(x \cdot y)^* = y^* x^*$. The element x^* is called the element adjoint to x.

Definition 2.3   A complex Banach algebra $\mathcal{A}$ with involution is called a C^*-**algebra** if $\|x\|^2 = \|x \cdot x^*\|$ for all $x \in \mathcal{A}$.

Standard examples of C^*-algebras are $\mathcal{A} = \mathbb{C}$, the complex plane with the conjugation $z \mapsto \bar{z}$, which is obviously an involution; $\mathcal{A} = C_0(S)$, the algebra of all complex continuous functions vanishing at infinity on a locally compact Hausdorff space S; $\mathcal{A} = L(\mathcal{H})$, the algebra of all the bounded linear operators on a complex Hilbert space $\mathcal{H}$ with the inner product $(\cdot, \cdot)$. For an element $A \in L(\mathcal{H})$, the **adjoint operator** A^* is defined by the equality $(Ax, y) = (x, A^*y)$.

At the same time, by the Gelfand–Naimark Theorem (see, for example, [Rudin (1973)]), each C^*-algebra can be realized as a closed subalgebra of $L(\mathcal{H})$ with a suitable $\mathcal{H}$. Actually, for our purpose we can restrict ourselves to these subalgebras.

An element $x \in \mathcal{A}$ is said to be Hermitian (self-adjoint) if $x^* = x$. The set of all hermitian elements will be denoted by $\mathcal{A}_H$. in particular, the elements

$$\operatorname{Re} x := \frac{1}{2}\,(x + x^*) \quad \text{and} \quad \operatorname{Im} x := \frac{1}{2i}\,(x - x^*) \tag{2.47}$$

are Hermitian.

Definition 2.4   An element $x \in \mathcal{A}$ is said to be positive if $x \in \mathcal{A}_H$ and there is $y \in \mathcal{A}_H$ such that $y^2 = x$.

It is known (see, for example, [Dixmier (1969)]) that $x \in \mathcal{A}_H$ is positive if and only if there is $y \in \mathcal{A}$ such that

$$x = y^* y. \tag{2.48}$$

If $A \in L(\mathcal{H})$, then (2.48) implies, in turn, that $A \in \mathcal{A}$ is positive if and only if

$$(A\zeta, \zeta) \geq 0 \quad \text{for all} \quad \zeta \in \mathcal{H}, \tag{2.49}$$

or if and only if the spectrum of A

$$\sigma(A) \subset \mathbb{R}^+ = [0, \infty). \tag{2.50}$$

Now we turn to the class of J^*-algebras which were introduced by L. A. Harris. This class consists of Banach spaces of operators mapping one Hilbert space into another.

Definition 2.5   Let $\mathcal{H}$ and $\mathcal{K}$ be complex Hilbert spaces and let $L(\mathcal{H}, \mathcal{K})$ be the space of all bounded linear operators from $\mathcal{H}$ into $\mathcal{K}$. A closed subspace $\mathfrak{U}$ of $L(\mathcal{H}, \mathcal{K})$ is called a J^***-algebra** if $AA^*A \in \mathfrak{U}$ whenever $A \in \mathfrak{U}$

Of course, J^*-algebras are not algebras in the ordinary sense. However, from the point of view of operator theory, they may be considered a generalization of C^*-algebras (see [Harris (1981)]). Any open unit ball of a J^*-algebra is a natural generalization of the open unit disk of the complex plane. In particular, any Hilbert space $\mathcal{H}$ may be thought of as a J^*-algebra identified with $L(\mathcal{H}, \mathbb{C})$. Also, any C^*-algebra in $L(\mathcal{H})$ is a J^*-algebra. Other important examples of J^*-algebras are the so-called Cartan factors of type I, II, II, IV, which are the sets $\mathfrak{U}_1, \mathfrak{U}_2, \mathfrak{U}_3, \mathfrak{U}_4$, respectively, where $\mathfrak{U}_1 = L(\mathcal{H}, \mathcal{K})$, $\mathfrak{U}_2 = \{A \in L(\mathcal{H}) : A^t = A\}$, $\mathfrak{U}_3 = \{A \in L(\mathcal{H}) : A^t = -A\}$ (where $A^t x = \overline{A^* \bar{x}}$ for a given conjugation $x \mapsto \bar{x}$ in $\mathcal{H}$), and $\mathfrak{U}_4$ is any closed complex subspace of $L(\mathcal{H})$ such that both $A^* \in \mathfrak{U}_4$ and $A^2 = \lambda I$ for some complex number λ whenever $A \in \mathfrak{U}_4$. (Cartan factors of type IV are variants of the spin factors.)

Thus, the four basic types of the classical Cartan domains and their infinite dimensional analogues are the open unit balls of J^*-algebras, and the same holds for any finite and infinite product of these domains.

A crucial property of J^*-algebras is that they have a kind of Jordan triple product structure and contain certain symmetrically formed products of their elements. In particular, for all elements A, B, C in a J^*-algebra $\mathfrak{U}$,

$$AB^*C + CB^*A \in \mathfrak{U}. \tag{2.51}$$

Also, we observe that every J^*-algebra is isometrically J^*-isomorphic (see [Harris (1974a); Harris (1981)]) to a J^*-algebra in $L(\mathcal{H})$ for a suitable Hilbert space $\mathcal{H}$. our further considerations will be restricted to this case.

Definition 2.6   We call a J^*-algebra **unital** if the underlying Hilbert spaces are the same and if it contains the identity operator.

Note that a closed subspace of $L(\mathcal{H})$ which contains the identity operator is a unital J^*-algebra if and only if it contains the squares and adjoints of each one of its elements.

When a unital J^*-algebra contains an operator, it contains any polynomial in that operator.

2.4.4   *l-analytic functions on unital J^*-algebras*

Let $\mathfrak{U}$ be a unital J^*-algebra and let Ω be a domain in the complex plane $\mathbb{C}$. Consider the set

$$\mathcal{D}_\Omega = \{A \in \mathfrak{U} : \sigma(A) \subset \Omega\}. \tag{2.52}$$

Since $\mathfrak{U}$ is a closed subset of $L(\mathcal{H})$, $\mathcal{D}_\Omega$ is an open set in the topology of $\mathfrak{U}$ induced by the sup-norm of $L(\mathcal{H})$.

For a function $f \in \mathrm{Hol}(\Omega, C)$ we define the function $\hat{f} : \mathcal{D}_\Omega \mapsto L(\mathcal{H})$ by using the Riesz–Dunford integral:

$$\hat{f}(A) = \frac{1}{2\pi i} \int_\Gamma f(\lambda)(\lambda I - A)^{-1} d\lambda, \tag{2.53}$$

where $\Gamma \subset \mathcal{D}_\Omega$ is a positively oriented simple closed rectifiable contour such that the interior domain of Γ contains $\sigma(A)$.

Remark 2.2   *Since $\mathfrak{U}$ is closed, the equality*

$$A^n = \frac{1}{2\pi i} \int_\Gamma \lambda^n (\lambda I - A)^{-1} d\lambda \tag{2.54}$$

and (2.51) imply that $\hat{f} : \mathcal{D} \mapsto \mathfrak{U}$, i.e., the values of $\hat{f}(A)$ for $A \in \mathcal{D}_\Omega \subset \mathfrak{U}$ are also in $\mathfrak{U}$.

It is clear that $\hat{f}$ belongs to $\mathrm{Hol}(\mathcal{D}_\Omega, \mathfrak{U})$.

Definition 2.7   Let $\mathcal{D}$ be a domain in a unital J^*-algebra $\mathfrak{U} \subset L(\mathcal{H})$. A holomorphic mapping $F : \mathcal{D} \mapsto \mathfrak{U}$ is said to be an **l-analytic function** if

(i) there is a domain $\Omega \subset \mathbb{C}$ such that $\mathcal{D} \subset \mathcal{D}_\Omega$, where $\mathcal{D}_\Omega$ is defined by (2.52);

(ii) there is a holomorhic function $f \in \mathrm{Hol}(\Omega, \mathbb{C})$ such that $F(A) = \hat{f}(A)$ for all $A \in \mathcal{D}_\Omega$, where $\hat{f}$ is defined by (2.53).

This function will be called the **producing function** of F ($= \hat{f}$). The set of all l-analytic functions on $\mathcal{D}$ will be denoted by $\widehat{\mathrm{Hol}}(\mathcal{D}, \mathfrak{U})$.

We have already mentioned that the "lifting" mapping $f \mapsto \hat{f}$ defined by (2.53) is multiplicative, *i.e.*, if f and g belong to $\mathrm{Hol}(\Omega, \mathbb{C})$ and $h = f \cdot g$, then

$$\hat{f}(A)\hat{g}(A) = \hat{h}(A). \tag{2.55}$$

Remark 2.3 *Although in general $\mathfrak{U}$ is not an algebra in the ordinary sense, i.e., for each pair of elements, does not necessarily contain their product, condition (2.55) shows that the product of values of l-analytic functions of the same element is commutative and is also an element of $\mathfrak{U}$.*

Also, if $f \in \mathrm{Hol}(\Omega, C)$, Ω_1 is a domain in $\mathbb{C}$ such that $f(\Omega) \subset \Omega_1$, and $g \in \mathrm{Hol}(\Omega_1, \mathbb{C})$, then the equality $h(\lambda) = g(f(\lambda))$ implies that

$$\hat{h}(A) = \hat{g}(\hat{f}(A)) \tag{2.56}$$

for all $A \in \mathcal{D}_\Omega$. Thus the "lifting" mapping $f \mapsto \hat{f}$ preserves the composition operation.

Remark 2.4 *If $f : \Omega \mapsto \mathbb{C}$ is univalent, then it is clear that $f(\Omega)$ is open and that f has a holomorphic inverse g which maps $f(\Omega)$ onto Ω. In this situation, the composition law (2.56) shows that the l-analytic function $\hat{f} : \mathcal{D}_\Omega \mapsto \hat{f}(\mathcal{D}_\Omega)$ is biholomorphic and, in particular, that $\hat{f}(\mathcal{D}_\Omega)$ is open.*

For more information on J^*-algebras and l-analytic functions see [Elin *et al.* (2002a)].

2.5 The Schwarz Lemma

2.5.1 *The classical Schwarz Lemma and Cartan's uniqueness theorem*

Throughout what follows we will denote by Δ the unit disk of the complex plane $\mathbb{C}$, and by $\mathrm{Hol}(\Delta)$ the set of holomorphic self-mappings of Δ. As we have mentioned, this set is a semigroup with respect to the composition operation. We begin with the classical Schwarz Lemma which plays a crucial role in geometric function theory.

Theorem 2.16 (The Schwarz Lemma) *Let $F \in \mathrm{Hol}(\Delta)$ be such that $F(0) = 0$. Then*

(i) $|F(z)| \leq |z|$ for all $z \in \Delta$;
(ii) $|F'(0)| \leq 1$.

Moreover, if equality holds in (i) for at least one $z \in \Delta$, $z \neq 0$, or $|F'(0)| = 1$, then $F(z) = e^{i\varphi} z$ for some $\varphi \in [0, 2\pi]$, and thus equality holds in (i) for all $z \in \Delta$.

Proof. Since $F(0) = 0$, the function F has the following Taylor representation:

$$F(z) = \sum_{k=1}^{\infty} a_k z^k. \tag{2.57}$$

Therefore the function

$$G(z) = \frac{F(z)}{z} = \sum_{k=1}^{\infty} a_k z^{k-1} \tag{2.58}$$

is also holomorphic in Δ. Take any $r \in (0, 1)$ and denote $\Delta_r = \{z \in \Delta : |z| < r\}$. By the maximum modulus principle we get:

$$|G(z)| = \left| \frac{F(z)}{z} \right| \leq \frac{1}{r}, \quad z \in \Delta_r. \tag{2.59}$$

Letting $r \to 1^-$ we obtain the inequality:

$$|G(z)| \leq 1 \tag{2.60}$$

for all $z \in \Delta$, which is equivalent to (i). Note also that $G(0) = F'(0)$, whereby (ii) follows too. If, in addition, $|G(z)| = 1$ for some $z \in \Delta$, then, once again, by the maximum modulus principle, $G(z) = e^{i\varphi}$ for some $\varphi \in [0, 2\pi]$; hence $F(z) = e^{i\varphi} z$ for all $z \in \Delta$. $\quad\square$

An elementary extension of the first part of this theorem to a Banach space can be formulated as follows. Let $\mathcal{D}$ be the open unit ball in a complex Banach space X, and let $\mathrm{Hol}(\mathcal{D}, X)$ be the set of all holomorphic mappings from $\mathcal{D}$ into X.

Theorem 2.17 *Let $F \in \mathrm{Hol}(\mathcal{D}, X)$ be such that $F(0) = 0$ and $\|F(z)\| \leq M$ for all $x \in \mathcal{D}$. Then*

(i) $\|F(x)\| \leq M \|x\|$ for all $x \in \mathcal{D}$;
(ii) $\|F'(0)\| \leq M$.

Proof. Apply Theorems 2.5 and 2.16. $\quad\square$

A higher dimensional analogue of the second part of the Schwarz lemma is known as Cartan's uniqueness theorem.

Theorem 2.18 *Let $\mathcal{D}$ be a bounded domain in a Banach space X and let $0 \in \mathcal{D}$. Suppose that $F \in \mathrm{Hol}(\mathcal{D})$ $(F : \mathcal{D} \mapsto \mathcal{D})$ satisfies the following assumptions:*

(1) $F(0) = 0$
and
(2) $F'(0) = I$ — the identity operator on X.

Then $F \equiv I/_{\mathcal{D}}$, i.e., $F(x) = x$ for all $x \in \mathcal{D}$.

As a matter of fact, a standard proof of this theorem shows that one can require slightly weaker assumptions to arrive to the same conclusions. Namely, *let $\mathcal{D}$ be a domain (not necessarily bounded) in a Banach space $X, \mathcal{D} \ni 0$, and let $F \in \mathrm{Hol}(\mathcal{D})$ be such that for a ball $B \subset \mathcal{D}$ centered at the origin, all iterates F^n of F are uniformly bounded, i.e.,*

$$\|F^n\|_B \leq M < \infty, \tag{2.61}$$

$n = 0, 1, 2, \ldots$. *If F satisfies conditions (1) and (2) of the theorem, then $F = I/_{\mathcal{D}}$.*

Proof.   Consider the power series expansion of the mapping $F^n : \mathcal{D} \mapsto \mathcal{D}$ around zero:

$$F^n(x) = \sum_{k=0}^{\infty} P_n^k(x), \quad n = 0, 1, 2, \ldots, \tag{2.62}$$

where $P_n^0(x) = F^n(0) = 0$, $P_n^1(x) = x$, and P_n^k are homogeneous polynomials of order $k \geq 2$. Assume that for $n = 1$, P_1^m is the first polynomial of order $m \geq 2$ which does not vanish identically, *i.e.,*

$$F(x) = x + \sum_{k=m}^{\infty} P_1^k(x), \quad P_1^m \not\equiv 0. \tag{2.63}$$

Then it can be shown by induction that

$$F^n(x) = x + nP_1^m(x) + \sum_{k=m+1}^{\infty} P_n^k(x), \tag{2.64}$$

i.e., $P_n^m(x) = nP_1^m(x)$. But it follows by the Cauchy inequalities that

$$\|P_n^m\| \leq \frac{M}{r^m}, \quad m = 1, 2, \ldots, \tag{2.65}$$

where r is the radius of the ball B. Thus we have

$$\|P_1^m\| \le \frac{M}{r^m \cdot n} \tag{2.66}$$

hence $P_1^m(x) \equiv 0$. This contradicts the choice of P_1^m.

Consequently, $F(x) = x$, and we are done. $\qquad\square$

2.6   Automorphisms

Let $\mathcal{D}$ be a domain in the complex Banach space X, and let $\mathrm{Hol}(\mathcal{D}, X)$ be the set of all holomorphic mappings from $\mathcal{D}$ into X. As above, by $\mathrm{Hol}(\mathcal{D})$ we denote the family of all holomorphic self-mappings of $\mathcal{D}$. Recall that this family is a semigroup with respect to the composition operation.

Definition 2.8   A mapping $F \in \mathrm{Hol}(\mathcal{D})$ is said to be an automorphism of $\mathcal{D}$ if F is biholomorhic self-mapping of $\mathcal{D}$ such that F^{-1} is well-defined on $\mathcal{D}$ and belongs to $\mathrm{Hol}(\mathcal{D})$. In other words, $F \in \mathrm{Hol}(\mathcal{D})$ is an automorphism if and only if F is biholomorphic and $F(\mathcal{D}) = \mathcal{D}$.

The set of all automorphisms of $\mathcal{D}$ will be denoted by $\mathrm{Aut}(\mathcal{D})$. This set is a group with respect to the composition operation.

2.6.1   *The unit disk*

As above, by Δ we denote the open unit disk in the complex plane $\mathbb{C}$. Let us fix a point $a \in \Delta$ and consider the fractional linear transformation (the so-called Möbius transformation) of the unit disk defined by

$$m_a(z) = \frac{z + a}{1 + \bar{a}z}. \tag{2.67}$$

It is easy to see that the following inequality holds:

$$1 - |m_a(z)|^2 = \frac{(1 - |a|^2)(1 - |z|^2)}{|1 + \bar{a}z|^2} > 0. \tag{2.68}$$

This inequality implies that $|m_a(z)| < 1$ whenever $z \in \Delta$ and a is a fixed element of Δ. That is, m_a maps Δ into itself and $m_a \in \mathrm{Hol}(\Delta)$. Now let h be an arbitrary element of $\mathrm{Aut}(\Delta)$. We claim that h is of the form

$$h(z) = \lambda m_a(z) \tag{2.69}$$

for some unimodular number λ. Indeed, if $h(0) = b \in \Delta$, then $m_{-b}(b) = 0$, and

$$g := m_{-b} \circ h \tag{2.70}$$

satisfies the conditions of the Schwarz Lemma, *i.e.*, $g \in \mathrm{Hol}(\Delta)$ and $g(0) = 0$. Hence $|g(z)| \leq |z|$ for all $z \in \Delta$. On the other hand, $g^{-1} = h^{-1} \circ m_b$ also belongs to $\mathrm{Hol}(\Delta)$, and $g^{-1}(0) = 0$. Therefore, $|g^{-1}(w)| \leq |w|$, $w \in \Delta$. Substituting here $w = g(z)$ we get $|g(z)| \geq |z|$, hence, in fact, $|g(z)| = |z|$. Consequently, $g(z) = e^{i\varphi}z$ for some $\varphi \in [0, 2\pi]$, and

$$h(z) = m_b(g(z)) = \frac{e^{i\varphi}z + b}{1 + e^{i\varphi}z\bar{b}} = e^{i\varphi}\frac{z + e^{-i\varphi}b}{1 + ze^{-i\varphi}\bar{b}}. \tag{2.71}$$

Setting $a = e^{-i\varphi}b$ and $\lambda = e^{i\varphi}$ we prove our claim. Thus we have proved the following assertions.

Proposition 2.5 *Each Möbius transformation m_a is an automorphism of Δ, and each automorphism of Δ is a composition of a Möbius transformation m_a and a rotation $r_\varphi : r_\varphi(z) = e^{i\varphi}z$, $\varphi \in [0, 2\pi]$.*

Proposition 2.6 ([Goebel and Reich (1984), p. 63]) *The following rules of composition hold:*

(i)

$$m_a \circ r_\varphi = r_\varphi \circ m_{r_{-\varphi}}(a); \tag{2.72}$$

(ii)

$$m_a \circ m_b = r_\varphi \circ m_{m_b}(a) \tag{2.73}$$

with $\varphi = 2\arg(1 + a\bar{b})$;
(iii) $m_a^{-1} = m_{-a}$ *and* $(-m_{-a})^{-1} = -m_{-a}$.

Let I denote the identity mapping on $\mathbb{C}$, *i.e.*, $I(z) = z$, $z \in \mathbb{C}$. For a fixed $a \in \Delta$, define a Möbius transformation M_a by the formula $M_a(z) = -m_{-a}(z)$:

$$M_a(z) = \frac{a - z}{1 - \bar{a}z}. \tag{2.74}$$

It is obvious that M_a satisfies the so-called **involution property**: $M_a \circ M_a = I$, *i.e.*, $M_a^{-1} = M_a$.

2.6.2 *The polydisk in* $\mathbb{C}^n$

The open unit ball in the Chebyshev norm of $\mathbb{C}^n$ is the polydisk Δ^n defined by

$$\Delta^n = \left\{ z = (z_1, \ldots, z_n) \in \mathbb{C}^n : |z_i| < 1, \quad i = 1, \ldots, n \right\}. \qquad (2.75)$$

Let us define functions $f_k : \Delta \mapsto \Delta$ by

$$f_k(\lambda) = \frac{a_k + \lambda}{1 + \bar{a}_k \lambda} \cdot e^{i\varphi k}, \quad \lambda \in \Delta. \qquad (2.76)$$

Since $f_k(\lambda)$, $k = 1, \ldots, n$ are automorphisms of Δ, it is clear that the mapping $f : \Delta^n \mapsto \Delta^n$ defined by $f(z_1, \ldots, z_n) = \big(f_1(z_1), \ldots, f_n(z_n)\big)$ is an automorhism of Δ^n. As a matter of fact, it turns out that each automorphism of the polydisk Δ^n can be represented as a vector function consisting of automorphisms of the unit disk. More precisely:

Proposition 2.7 *If $F : \Delta^n \mapsto \Delta^n$ is an automorphism of Δ^n, then there are automorphisms $F_1, \ldots, F_n$ of Δ, and there is a permutation $(i_1, \ldots, i_n)$ of $(1, \ldots, n)$ such that*

$$F(z_1, \ldots, z_n) = \big(F_1(z_{i_1}), \ldots, F_n(z_{i_n})\big). \qquad (2.77)$$

The group $\mathrm{Aut}(\Delta^n)$ *is a Lie group of real dimension* $3n$.

2.6.3 *The Euclidean ball in* $\mathbb{C}^n$ *and the Hilbert ball*

Let $\mathbb{B} = \mathbb{B}_H$ denote the open unit ball of a complex Hilbert space H with inner product $\langle \cdot, \cdot \rangle$. In particular, by $\mathbb{B}_{\mathbb{C}^n}$ we denote the open unit ball in the Euclidean norm of $\mathbb{C}^n$, *i.e.*,

$$\mathbb{B}_{\mathbb{C}_n} = \left\{ z = (z_1, \ldots, z_n) \in \mathbb{C}^n : \left(\sum_{j=1}^{n} |z_j|^2 \right)^{\frac{1}{2}} < 1 \right\}. \qquad (2.78)$$

The group $\mathrm{Aut}(B_n)$ for $n = 2$ was described by Poincaré (see Rudin (1980)). The case of arbitrary n is completely analogous to this particular case (*cf.*, for example, Shabat (1976) and Rudin (1980)).

The group $\mathrm{Aut}(B_n)$ is a connected Lie group of dimension $n^2 + 2n$. With each point $a \in \mathbb{B}(= \mathbb{B}_H)$ we may associate a unique automorphism φ_a enjoying the following properties:

1. $\varphi_a(0) = a$, $\varphi_a(a) = 0$;
2. $\varphi_a = \varphi_a^{-1}$;

3. φ_a has a unique fixed point.

This automorphism can be described by an explicit formula:

$$\varphi_a(z) = \begin{cases} -z, & a = 0, \\ \dfrac{a - P_a z - \left(1 - \|a\|^2\right)^{\frac{1}{2}} Q_a z}{1 - \langle z, a \rangle}, & a \in \mathbb{B} \setminus \{0\}. \end{cases} \qquad (2.79)$$

Here, P_a denotes the orthogonal projection onto the one-dimensional subspace $\mathbb{C} \cdot a$, and Q_a the projection onto its orthogonal complement, *i.e.*,

$$P_a z = \frac{\langle z, a \rangle}{\langle a, a \rangle}\, a, \qquad (2.80)$$

$$Q_a z = z - \frac{\langle z, a \rangle}{\langle a, a \rangle}\, a. \qquad (2.81)$$

The automorphism $\varphi_b \circ \varphi_a$ carries the point $a \in \mathbb{B}$ to the point $b \in \mathbb{B}$. Thus the group $\mathrm{Aut}(\mathbb{B})$ is transitive on $\mathbb{B}$. Any automorphism of the ball $\mathbb{B}$ which fixes the origin is the restriction to $\mathbb{B}$ of a unitary operator. Any automorphism $\psi \in \mathrm{Aut}(\mathbb{B})$ has a unique representation of the form

$$\psi = U \circ \varphi_a \quad (\text{respectively,} \ \ \psi = \varphi_a \circ U), \qquad (2.82)$$

where U is a unitary operator. The point $a \in \mathbb{B}$ is uniquely determined by the equation $\psi(a) = 0$ (correspondingly, $\psi(0) = a$). We present two convenient formulas in which $\psi \in \mathrm{Aut}(\mathbb{B})$ and $a = \psi^{-1}(0)$:

$$1 - \langle \psi(z), \psi(w) \rangle = \frac{\left(1 - \langle a, a \rangle\right)\left(1 - \langle z, w \rangle\right)}{\left(1 - \langle z, a \rangle\right)\left(1 - \langle a, w \rangle\right)}, \ (z, w \in \bar{\mathbb{B}}). \qquad (2.83)$$

If $\mathbb{B} = \mathbb{B}_{\mathbb{C}^n}$, then also

$$\det \psi'(z) = \left(\frac{1 - \|a\|^2}{\left|1 - \langle z, a \rangle\right|^2}\right)^{n+1}. \qquad (2.84)$$

For each $a \in \mathbb{B}$ consider also the Möbius transformation defined by

$$m_a(x) = \frac{1}{1 + \langle x, a \rangle} \left(\sqrt{1 - \|a\|^2}\, Q_a + P_a\right)(x + a), \qquad (2.85)$$

where $x \in \mathbb{B}$, P_a is the orthogonal projection of H onto the subspace $\{\lambda a : \lambda \in \mathbb{C}\}$ spanned by a, and $Q_a = I - P_a$. We note that $m_a(0) = a$, $m_a^{-1} = m_{-a}$, $m_0 = I$, and that for every $a, b \in \mathbb{B}$ the mapping $m_b \circ m_{-a}$ takes a to b. For the one-dimensional case $H = \mathbb{C}$, the mapping m_a becomes

$m_a(z) = \frac{z+a}{1+\bar{a}z}$ as defined before. Again, as in the one-dimensional case, for the case of the Hilbert ball we have the following property:

$$1 - \|m_{-x}(y)\|^2 = \sigma(x, y), \tag{2.86}$$

where

$$\sigma(x, y) = \frac{(1 - \|x\|^2)(1 - \|y\|^2)}{|1 - \langle x, y \rangle|^2} . \tag{2.87}$$

2.6.4　*Unital J^*-algebras*

Let H be a complex Hilbert space and let $L(H)$ be the (Banach) space of all the linear bounded operators on H.

Recall that a linear subspace $\mathfrak{U} \subset L(H)$ is said to be a unital J^*-algebra if AA^*A belongs to $\mathfrak{U}$ whenever $A \in \mathfrak{U}$ and I, the identity operator, belongs to $\mathfrak{U}$. The space $L(H)$, in particular, is a unital J^*-algebra. If $\mathcal{D}$ is the open unit ball in $\mathfrak{U}$ and $A \in \mathcal{D}$, then the mapping $M_A : \mathcal{D} \mapsto \mathfrak{U}$ defined by

$$M_A(T) = (I - AA^*)^{-\frac{1}{2}}(T + A)(I + A^*T)^{-1}(I - A^*A)^{\frac{1}{2}} \tag{2.88}$$

is a biholomorphic self-mapping of $\mathcal{D}$ (see [Potapov (1960)], [Harris (1974a)]) and $M_A(0) = A$.

Since $M_A^{-1} = M_{-A}$, the mapping M_A is, in fact, an automorphism of $\mathcal{D}$. We call M_A a Möbius transformation of $\mathcal{D}$.

One can note that when H is one-dimensional, then M_A becomes the standard form given in the literature for Möbius transformations of the unit disk Δ in $\mathbb{C}$.

We also observe that the operators defined by (2.88) can serve as automorphisms in a more general situation when $\mathfrak{U}$ is the Banach space $L(H, K)$ of bounded linear operators from one Hilbert space H into another Hilbert space K (see, for example, [Dineen (1989)]).

Moreover, M_A is a well-defined automorphism of the unit ball $\mathcal{D}$ of each closed subspace $\mathfrak{U}$ of $L(H, K)$ whenever $\mathcal{D}$ is homogeneous (or, which is the same, a bounded symmetric domain) (see [Harris (1974b)], [Upmeier (1986)] and [Dineen (1989)]).

2.6.5　*The Schwarz–Pick lemma*

The following assertion is the invariant form of the Schwarz Lemma.

Theorem 2.19 *Let $F \in \mathrm{Hol}(\Delta)$. Then for each pair z and w in Δ the following inequality holds:*

$$\left| m_{-F(w)}(F(z)) \right| \leq \left| m_{-w}(z) \right| \tag{2.89}$$

or, explicitly,

$$\left| \frac{F(z) - F(w)}{1 - F(z)\overline{F(w)}} \right| \leq \left| \frac{z - w}{1 - z\bar{w}} \right| \quad (\text{the Schwarz–Pick inequality}). \tag{2.90}$$

Moreover, if equality holds in (2.90) for at least one pair $z \neq w$ in Δ, then $F \in \mathrm{Aut}(\Delta)$.

Proof. Fix $w \in \Delta$ and consider the mapping $G = m_{-F(w)} \circ F \circ m_w$. It is clear that $G \in \mathrm{Hol}(\Delta)$ and $G(0) = 0$. By the Schwarz Lemma, $|G(z)| \leq |z|$, $z \in \Delta$. Replacing z by $m_{-w}(z)$ we get the required inequality. If this inequality becomes an equality for some pair $z \neq w$, then $\left| G(m_{-w}(z)) \right| = |m_{-w}(z)|$. Again by the Schwarz Lemma, $G(z) = e^{i\varphi}z$ for some $\varphi \in [0, 2\pi]$ and for all $z \in \Delta$. Hence G belongs to $\mathrm{Aut}(\Delta)$ and so does $F = m_{F(w)} \circ G \circ m_{-w}$. $\qquad\square$

Corollary 2.5 *Let $F \in \mathrm{Hol}(\Delta, \mathbb{C})$. Then F maps Δ into its closure $\bar{\Delta}$ if and only if it satisfies the condition*

$$|F'(z)| \leq \frac{1 - |F(z)|^2}{1 - |z|^2}. \tag{2.91}$$

Moreover, if F is not a constant and equality holds at one $w \in \Delta$, then $F \in \mathrm{Aut}(\Delta)$.

Proof. Indeed, if this inequality holds, then $1 - |F(z)|^2 \geq 0$ and $|F(z)| \leq 1$. Conversely. Suppose that $|F(z)| \leq 1$. If $|F(z)| = 1$ for some $z \in \Delta$, then F is a constant and our assertion is trivial. Thus we may assume that $|F(z)| < 1$ for all $z \in \Delta$. Now, rewriting the Schwarz–Pick inequality in the form

$$\left| \frac{F(z) - F(w)}{z - w} \right| \leq \left| \frac{1 - F(z)\overline{F(w)}}{1 - z\bar{w}} \right| \tag{2.92}$$

and letting $w \to z$ we get the claimed inequality. To prove the last part of the corollary, consider the mapping $G = m_{-F(w)} \circ F \circ m_w$ and note that $|G'(0)| = 1$. Thus the assertion follows from the second part of the Schwarz Lemma. $\qquad\square$

The following version of the Schwarz Lemma is a consequence of the Schwarz–Pick Lemma.

Corollary 2.6   *Let $F \in \mathrm{Hol}(\Delta)$. Then $\left|F(0) + \lambda\big(F(z) - F(0)\big)\right| \leq 1$ for all $z \in \Delta$ with $|\lambda| \leq \frac{1-|z|}{2|z|}$.*

Proof.   By the Schwarz–Pick Lemma,

$$\left| \frac{F(z) - F(0)}{1 - \overline{F(0)}F(z)} \right| \leq |z| \tag{2.93}$$

for all z in Δ. Since

$$\left|1 - \overline{F(0)}F(z)\right| \leq 1 - |F(0)|^2 + |F(0)|\left|F(z) - F(0)\right|$$
$$\leq 2\big(1 - |F(0)|\big) + |F(0)|\left|F(z) - F(0)\right| \tag{2.94}$$

we see that

$$\left|F(z) - F(0)\right| \leq 2|z|\big(1 - |F(0)|\big) + |z|\,|F(0)|\left|F(z) - F(0)\right|, \tag{2.95}$$

i.e.,

$$2|z|\,|F(0)| + \big(1 - |z|\,|F(0)|\big)\left|F(z) - F(0)\right| \leq 2|z|. \tag{2.96}$$

Since $1 - |z| \leq 1 - |z|\,|F(0)|$, we have

$$2|z|\,|F(0)| + (1 - |z|)\left|F(z) - F(0)\right| \leq 2|z|. \tag{2.97}$$

Hence if $|\lambda| \leq \frac{1-|z|}{2|z|}$, then

$$\left|F(0) + \lambda\big(F(z) - F(0)\big)\right| \leq |F(0)| + \frac{1 - |z|}{2|z|}\left|F(z) - F(0)\right| \leq 1. \tag{2.98}$$
$$\square$$

Applying the Hahn–Banach theorem one can formulate a similar assertion in a more general setting.

Corollary 2.7 ([Harris (1969)]) *Let X_0 denote the open unit ball of a Banach space X and let $f \in H(\Delta, \bar{X}_0)$. Then $\left\|f(0) + \lambda\big(f(z) - f(0)\big)\right\| \leq 1$ for all $z \in \Delta$ with $|\lambda| \leq \frac{1-|z|}{2|z|}$.*

The following corollary is another consequence of the Schwarz–Pick Lemma.

Corollary 2.8   *For each $F \in \mathrm{Hol}(\Delta)$ and $z \in \Delta$, the estimate*

$$|F'(z)| \leq \frac{1}{1 - |z|^2} \tag{2.99}$$

holds.

Remark 2.5   *Observe also that for each* $r \in (0,1)$, *we get the following uniform Lipschitz condition:*

$$\left|F(z) - F(w)\right| \leq \frac{1}{1 - r^2}\, |z - w|, \tag{2.100}$$

whenever z *and* w *belong to* $\Delta_r = \{z \in \Delta : |z| < r\}$. *Now if* $f \in \mathrm{Hol}(\Delta, \mathbb{C})$ *is bounded, i.e.,* $|f(z)| < M$ *for all* $z \in \Delta$, *then* $F = \frac{1}{M} f \in \mathrm{Hol}(\Delta)$. *Thus we have that each bounded holomorphic function on* Δ *is locally uniformly Lipschitzian with respect to the Euclidean distance in* $\mathbb{C}$.

A similar assertion holds in general.

Proposition 2.8   *Let* $\mathcal{D}$ *be a bounded domain in a complex Banach space* X *and let* $F \in \mathrm{Hol}(\mathcal{D}, X)$ *be bounded, i.e.,* $\|F(x)\| \leq M$ *for all* $x \in \mathcal{D}$. *Then for each* $x \in \mathcal{D}$,

$$\|F'(x)\| \leq \frac{M}{\mathrm{dist}(x, \partial\mathcal{D})}. \tag{2.101}$$

Consequently, each bounded holomorphic mapping on $\mathcal{D}$ *is locally Lipschitzian on* $\mathcal{D}$.

Chapter 3

Hyperbolic Metrics on Domains in Complex Banach Spaces

3.1 The Poincaré Metric on the Unit Disk

In this section we review some elements of classical hyperbolic geometry on the unit disk. Additional details can be found in [Harris (1979)], [Franzoni and Vesentini (1980)], [Goebel and Reich (1984)] and [Jarnicki and Pflug (1993)].

From the point of view of geometric function theory on the unit disk, the Euclidean metric on the disk is inappropriate. Given a class of self-mappings of a domain, it is natural to look for a metric such that each mapping of this class becomes nonexpansive with respect to it. There are different ways to define such metrics. For example, one may use the invariant metric defined by the so-called pseudohyperbolic distance.

*The **pseudohyperbolic distance** between the points z and w of Δ is determined by the function $d : \Delta \times \Delta \to \mathbb{R}^+ = [0, \infty)$ defined as follows:*

$$d(z, w) = |m_{-w}(z)|. \tag{3.1}$$

It is easy to see that, actually, d is a metric on Δ that induces the usual Euclidean topology. Thus the Schwarz–Pick Lemma *asserts the contractive property of a mapping $F \in \mathrm{Hol}(\Delta)$ with respect to this metric:*

$$d\big(F(z), F(w)\big) \le d(z, w), \quad z, w \in \Delta. \tag{3.2}$$

Moreover, F is an isometry with respect to d (i.e., $d\big(F(z), F(w)\big) = d(z, w)$) if an only if $F \in \mathrm{Aut}(\Delta)$.

Also, it is easy to show that for each $r \in (0, 1)$,

$$\lim_{|w| \to 1^-} \inf_{|z| \le r} d(z, w) = 1. \tag{3.3}$$

81

However, from the classical Riemannian metric theory point of view, it is impossible to measure the "lengths" of vectors tangent to Δ to obtain d (see details below). it turns out that the only way to eliminate this deficiency is to define the so-called **Poincaré hyperbolic metric**

$$\rho = \tanh^{-1} d, \tag{3.4}$$

which, in spite of an additional level of computation, provides the needed properties in the construction of a non-Euclidean geometry in the spirit of Riemann.

Theorem 3.1 *Let $\rho : \Delta \times \Delta \mapsto \mathbb{R}$ be a function with the following properties:*

(i) for each $F \in \mathrm{Aut}(\Delta)$ and each pair of points z, w in Δ,

$$\rho(F(z),\, F(w)) = \rho(z, w); \tag{3.5}$$

(ii) for each pair of numbers $s, t \in (0, 1),\ s < t$,

$$\rho(0, t) = \rho(0, s) + \rho(s, t); \tag{3.6}$$

(iii) $\displaystyle\lim_{t \to 0^+} \frac{\rho(0,t)}{t} = 1.$

Then

$$\rho(z, w) = \tanh^{-1} d(z, w) = \frac{1}{2} \log \frac{1 + |m_{-z}(w)|}{1 - |m_{-z}(w)|}. \tag{3.7}$$

Proof.   First we note that by (i),

$$\rho(z, w) = \rho\big(m_{-z}(z),\, m_{-z}(w)\big) = \rho(0,\, m_{-z}(w)) \tag{3.8}$$

because $m_{-z} \in \mathrm{Aut}(\Delta)$. Also, using the rotation automorphism r_φ, $r_\varphi(u) = e^{i\varphi} u$, $u \in \Delta$, with $\varphi = -\arg u$, we have

$$\rho(0, u) = \rho\big(r_\varphi(0),\, r_\varphi(u)\big) = \rho(0, |u|). \tag{3.9}$$

Comparing with (3.8), we get

$$\rho(z, w) = \rho(0, |m_{-z}(w)|). \tag{3.10}$$

So, it remains to show that for each $r \in (0, 1)$,

$$\rho(0, r) = \tanh^{-1}(r). \tag{3.11}$$

Indeed, for each $\delta > 0$ such that $r + \delta < 1$, we have, by (ii) and (i),

$$\rho(0, r + \delta) - \rho(0, r) = \rho(r, r + \delta) = \rho\big(m_{-r}(r), m_{-r}(r + \delta)\big)$$

$$= \rho\left(0, \frac{\delta}{1 - r\delta - r^2}\right). \qquad (3.12)$$

Hence, by (iii),

$$\frac{d^+\rho(0, r)}{dr} = \frac{1}{1 - r^2}. \qquad (3.13)$$

Similarly,

$$\frac{d^-\rho(0, r)}{dr} = \frac{1}{1 - r^2}, \qquad (3.14)$$

and we obtain (3.11):

$$\rho(0, r) = \int_0^r \frac{ds}{1 - s^2} = \tanh^{-1}(r) \qquad (3.15)$$

because $\rho(0, 0) = 0$. It is also clear that the function $\tanh^{-1}(|m_{-z}(w)|)$ satisfies conditions (i)–(iii). $\qquad \square$

Remark 3.1 *The following properties of the real function* $\varphi(t) = \tanh^{-1}(t)$ *are simple calculus exercises:*

(a) $\varphi : [0, 1) \mapsto [0, \infty)$ *is a strictly increasing function which tends to infinity when* $t \to 1^-$;

(b) $\frac{d\varphi}{dt}\big|_{t=0} = 1$;

(c) $\varphi\big(\frac{t+s}{1+ts}\big) = \varphi(t) + \varphi(s)$.

Theorem 3.2 *The function* $\rho(\cdot, \cdot)$ *defined by (3.7) is a metric on* Δ.

Proof.   It is clear tht $\rho(z, w) \geq 0$ for all $(z, w) \in \Delta \times \Delta$, and since $|m_{-z}(w)| = |m_{-w}(z)|$, we have $\rho(z, w) = \rho(w, z)$ and $\rho(z, w) = 0$ if and only if $z = w$.

Now we need to show the triangle inequality

$$\rho(z, w) \leq \rho(z, u) + \rho(u, w) \qquad (3.16)$$

for each triple of points z, w, u in Δ. Indeed, let $\gamma : [0, 1] \mapsto \Delta$ be a curve joining the points z and w such that $\gamma'(t)$ is piecewise continuous. We will call such a curve an **admissible curve**. Consider the function

$$\delta(z, w) = \inf\{L_\gamma : \gamma \text{ is an admissible curve joining } z \text{ and } w\}, \qquad (3.17)$$

where L_γ is the "length" of γ:

$$L_\gamma = \int_0^1 \frac{|\gamma'(t)|}{1 - |\gamma(t)|^2}\, dt. \tag{3.18}$$

We claim that $\delta(z, w) = \rho(z, w)$. Indeed, it is sufficient to show that $\delta(\cdot, \cdot)$ satisfies the conditions of Theorem 3.1. In fact, if $F \in \mathrm{Hol}(\Delta)$ and γ is an admissible curve joining z and w, then $F \circ \gamma$ is an admissible curve joining $F(z)$ and $F(w)$. It follows by the Schwarz Lemma that

$$L_{F \circ \gamma} = \int_0^1 \frac{|F'(\gamma(t))|\,|\gamma'(t)|}{1 - |F(\gamma(t))|^2}\, dt \le \int_0^1 \frac{|\gamma'(t)|}{1 - |\gamma(t)|^2}\, dt = L_\gamma. \tag{3.19}$$

Hence,

$$\delta(F(z),\, F(w)) \le \delta(z, w). \tag{3.20}$$

If now $F \in \mathrm{Aut}(\Delta)$, then applying (3.20) to F^{-1} we get the equality

$$\delta(F(z),\, F(w)) = \delta(z, w), \tag{3.21}$$

which proves (i). To prove (ii) and (iii), it is sufficient to evaluate $\delta(0, s)$ for $0 < s < 1$. If γ joins 0 and s, then so does $\beta = Re\gamma$. Hence

$$L_\gamma = \int_0^1 \frac{|\gamma'(t)|\, dt}{1 - |\gamma(t)|^2} \ge \int_0^1 \frac{\beta'(t) dt}{1 - [\beta(t)]^2} =$$

$$\int_0^s \frac{dr}{1 - r^2} = \tanh^{-1}(s), \tag{3.22}$$

i.e., $\delta(0, s) \ge \tanh^{-1}(s)$. On the other hand, for $\gamma(t) = ts$, $L_\gamma = \tanh^{-1}(s)$. Consequently, we have

$$\delta(0, s) = \tanh^{-1}(s). \tag{3.23}$$

Now it follows, by properties (b) and (c) (see Remark 3.1), that the function δ satisfies conditions (ii) and (iii) of Theorem 3.1. Thus,

$$\delta(z, w) = \rho(z, w). \tag{3.24}$$

Now it can be seen directly that (3.16) is a consequence of the definition of $\delta(\cdot, \cdot)$. $\square$

As a matter of fact, we have established somewhat more in our proof.

Theorem 3.3 *Each mapping $F \in \mathrm{Hol}(\Delta)$ is nonexpansive with respect to the metric ρ, i.e.,*

$$\rho(F(z)), F(w)) \leq \rho(z, w). \tag{3.25}$$

Moreover, if equality in (3.25) holds for some pair of points $z, w \in \Delta$, then $F \in \mathrm{Aut}(\Delta)$.

Actually, the conclusion of this theorem is also a direct consequence of the Schwarz–Pick inequality and (3.4).

This metric, the existence of which we have just established, is called the **Poincaré hyperbolic metric** on Δ and will be denoted by ρ.

Theorem 3.4 *The pair (Δ, ρ) is a complete unbounded metric space, and ρ defines on Δ the same topology as the original Euclidean topology on Δ.*

Proof. We list several topological properties of the Poincaré metric which prove our assertion.

(a) $\lim\limits_{n \to \infty} \rho(0, z_n) = \infty$ *if and only if* $|z_n| \to 1^-$. Indeed, by (3.9) we have

$$\rho(0, z_n) = \rho(0, |z_n|) = \tanh^{-1} |z_n|, \tag{3.26}$$

and the assertion follows.

(b) *Each ρ-ball $B(a, R) = \{z \in \Delta : \rho(a, z) < R\}$ is the disk $\Delta_r(a)$ defined by the formula*

$$B(a, R) = \{z \in \Delta : |z - sa| < t \cdot \tanh R\}, \tag{3.27}$$

where

$$s = \frac{1 - (\tanh R)^2}{1 - (\tanh R)^2 |a|^2}, \quad t = \frac{1 - |a|^2}{1 - (\tanh R)^2 |a|^2}. \tag{3.28}$$

In particular, if $a = 0$, then $B(0, R) = \{z \in \Delta : |z| < \tanh R\}$ is a disk centered at zero.

(c) $B(a, R) = m_a(B(0, r))$, *i.e., each ρ-ball is the image of a disk centered at zero under the corresponding Möbius transformation.*

Now, it is clear that if the sequence $\{z_n\} \subset \Delta$ converges to $z \in \Delta$ in the sense of the ρ-metric, i.e., $\rho(z_n, z) \to 0$, then $|z_n - z| \leq |z_n - sz| + (1 - s)|z| \to 0$ because $s \to 1^-$ when $R \to 0$ (see (3.28)). Conversely, if $|z_n - z| \to 0$ for $z_n, z \in \Delta$, then there is $r \in (0, 1)$ such that $z_n \in \Delta_r = \{z \in \Delta : |z| < r\}$. Hence,

$$\rho(z_n, z) = \tanh^{-1} |m_{-z}(z_n)| \leq \tanh^{-1} \frac{|z_n - z|}{1 - r^2} \to 0. \tag{3.29}$$

Finally, note that each ρ-ball is bounded away from the boundary of Δ; this fact implies the completeness of the metric space (Δ, ρ).

$\square$

3.2   The Infinitesimal Poincaré Metric and Geodesics

Let z be any point of Δ and let $u \in \mathbb{C}$ be such that $z + u \in \Delta$. If ρ is the Poincaré metric on Δ, then

$$\rho(z, z + u) = \frac{|u|}{1 - |z|^2}\,(1 + \varepsilon(u)), \tag{3.30}$$

where $\varepsilon(u) \to 0$ as $u \to 0$.

Formula (3.30) shows that the linear differential element $d\rho_z$ in the Poincaré metric is defined by the formula

$$d\rho_z = \frac{|dz|}{1 - |z|^2}\,. \tag{3.31}$$

The form

$$\alpha(z, u) = \frac{|u|}{1 - |z|^2}\,, \quad z \in \Delta, \quad u \in \mathbb{C}, \tag{3.32}$$

*is called the **infinitesimal (or differential) Poincaré hyperbolic metric on** Δ.*

The following properties are an immediate consequence of the definition:

(a) $\alpha(z, u) \geq 0, \quad z \in \Delta, \quad u \in \mathbb{C}.$
(b) $\alpha(z, tu) = |t|\alpha(z, u), \quad t \in \mathbb{C}.$

A consequence of the Schwarz–Pick Lemma is that each $F \in \mathrm{Hol}(\Delta)$ is a contraction for the infinitesimal Poincaré metric.

If $F \in \mathrm{Hol}(\Delta)$, then $d\rho_{F(z)} \leq d\rho_z$, or equivalently,

$$\alpha(F(z), dF(z)) \leq \alpha(z, dz). \tag{3.33}$$

If $F \in \mathrm{Aut}(\Delta)$, then equality in (3.33) holds for all $z \in \Delta$. Moreover, if equality in (3.33) holds for at least one $z \in \Delta$, then $F \in \mathrm{Aut}(\Delta)$.

This notion allows us, using Riemann integration, to define the "length" of any admissible curve in Δ.

Let $\gamma : [0,1] \mapsto \Delta$ be an admissible curve in Δ joining two points z and w in Δ. Then the quantity

$$L_\gamma \ (= \ L_\gamma(z,w)) = \int_\gamma d\rho_{\gamma(t)} = \int_0^1 \frac{|\gamma'(t)|}{1 - |\gamma(t)|^2} \, dt \qquad (3.34)$$

is called the hyperbolic length of γ.

We already used this notion in the previous section and saw that the hyperbolic length is greater than, or equal to, the hyperbolic distance between its end points, *i.e.*,

$$\rho(z,w) \leq L_\gamma(z,w). \qquad (3.35)$$

A curve γ joining the points z, w in Δ is called a geodesic segment in Δ if its length is equal to the hyperbolic distance between its end points z and w, *i.e.*,

$$L_\gamma(z,w) = \rho(z,w). \qquad (3.36)$$

Proposition 3.1 *For each pair of points z and w in Δ, there is a unique geodesic segment joining z and w and it is either a linear segment (if z and w lie on a diameter of Δ) or a segment of the circle in $\mathbb{C}$ which passes through z and w and is orthogonal to $\partial\Delta$, the boundary of Δ.*

Proof. Indeed, for the points 0 and s, $0 < s < 1$, the curve $\gamma_1(t) = ts$ is the unique curve joining 0 and s such that

$$\rho(0,s) = \int_{\gamma_1} d\rho_{\gamma_1} = \int_0^1 \frac{|\gamma_1'(t)|dt}{1 - |\gamma_1(t)|^2} = \tanh^{-1}(s), \qquad (3.37)$$

i.e., $\gamma_1(t)$ is the geodesic segment joining 0 and s. If z and w are arbitrary points in Δ, then the automorphism $g = r_\varphi \circ m_{-z}$, where m_{-z} is a Möbius transformation and r_φ is the rotation with $\varphi = -\arg m_{-z}(w)$, takes z into 0 and w into $s = |m_{-z}(w)|$.

If we now define $\gamma_1(t)$ as before, that is,

$$\gamma_1(t) = t|m_{-z}(w)| \qquad (3.38)$$

and

$$\gamma(t) = g^{-1}(\gamma_1(t)), \qquad (3.39)$$

we obtain $\gamma(0) = g^{-1}(\gamma_1(0)) = g^{-1}(0) = z$ and $\gamma(1) = g^{-1}(\gamma_1(1)) = g^{-1}(|m_{-z}(w)|) = m_z(e^{i\varphi}|m_{-z}(w)|) = (m_z \circ m_{-z})(w) = w$. Thus $\gamma(t)$ is an admissible curve joining z and w. In addition,

$$L_\gamma = \int_0^1 \frac{|\gamma'(t)|dt}{1 - |\gamma(t)|^2} = \int_0^1 \frac{|(g^{-1})'(\gamma_1(t))| \cdot |\gamma_1'(t)|}{1 - |g^{-1}(\gamma_1(t))|^2} \, dt$$

$$= \int_0^1 \frac{|\gamma_1'(t)|}{1 - |\gamma_1(t)|^2} \, dt = L_{\gamma_1}. \qquad (3.40)$$

But $\rho(z,w) = \rho\big(0, m_{-z}(w)\big) = L_{\gamma_1}$ and hence we obtain relation (3.36). The second part of our assertion is a direct consequence of Proposition 3.1.

$$\square$$

3.3 The Poincaré Metric on the Hilbert Ball and its Powers

Let $\mathbb{B}$ denote the open unit ball of a complex Hilbert space H. Recall that the Möbius transformation m_a on $\mathbb{B}$ is defined by

$$m_a(x) = \frac{1}{1 + \langle x, a \rangle} \left(\sqrt{1 - \|a\|}\, Q_a + P_a \right)(x + a), \qquad (3.41)$$

where $x \in \mathbb{B}$, P_a is the orthogonal projection of H onto the subspace $\{\lambda a : \lambda \in \mathbb{C}\}$, and $Q_a = I - P_a$. The **Poincaré metric** *on the Hilbert ball* $\mathbb{B}$ *is the function* $\rho_B : \mathbb{B} \times \mathbb{B} \to \mathbb{R}^+$ *given by*

$$\rho_{\mathbb{B}}(x,y) = \tanh^{-1}\|m_{-x}(y)\| = \frac{1}{2} \log \frac{1 + \|m_{-x}(y)\|}{1 - \|m_{-x}(y)\|} \,. \qquad (3.42)$$

Note that $1 - \|m_{-x}(y)\|^2 = \sigma(x,y)$, where

$$\sigma(x,y) = \frac{(1 - \|x\|^2)(1 - \|y\|^2)}{|1 - \langle x, y \rangle|^2} \,. \qquad (3.43)$$

Applying this property of Möbius transformations we get the following more explicit formula for ρ, namely,

$$\rho_{\mathbb{B}}(x,y) = \tanh^{-1}(1 - \sigma(x,y))^{\frac{1}{2}}. \qquad (3.44)$$

It is clear that each m_a is a $\rho_{\mathbb{B}}$-isometry and has a norm continuous injective extension from $\overline{\mathbb{B}}$ onto itself. In general, *if* $\mathbb{B}_1$ *and* $\mathbb{B}_2$ *are the open unit balls in complex Hilbert spaces* H_1 *and* H_2, *respectively,* ρ_1 *and* ρ_2 *are the Poincaré metrics assigned to* $\mathbb{B}_1$ *and* $\mathbb{B}_2$, *respectively, and* $f \in \mathrm{Hol}(\mathbb{B}_1, \mathbb{B}_2)$, *then*

$$\rho_2(f(x), f(y)) \leq \rho_1(x,y). \qquad (3.45)$$

In particular, each holomorphic self-mapping of $\mathbb{B}$ *is nonexpansive with respect to* $\rho_{\mathbb{B}}$. It is not difficult to see that for the Cartesian product $\mathbb{B}^n$ of n Hilbert balls, $n \geq 2$, the function $\rho_{B^n} : \mathbb{B}^n \times \mathbb{B}^n \mapsto \mathbb{R}^+$ given by

$$\rho_{\mathbb{B}^n}(x, y) = \max_{1 \leq j \leq n} \{\rho_{\mathbb{B}}(x_j, y_j)\}, \tag{3.46}$$

where $x = (x_1, \ldots, x_n)$ and $y = (y_1, \ldots, y_n)$ belong to $\mathbb{B}^n$, defines a metric on $\mathbb{B}^n$ which we call the Poincaré metric on $\mathbb{B}^n$. Again, each holomorphic self-mapping of $\mathbb{B}^n$ is nonexpansive with respect to this metric. We will see below that the Poincaré metric $\rho_{\mathbb{B}^n}$ (as well as, of course, $\rho_{\mathbb{B}}$ itself) is a particular case of the so-called Carathéodory and Kobayashi pseudometrics assigned to any domain in a complex Banach space.

3.4 The Carathéodory and Kobayashi Pseudometrics

We are now in a position to define a pseudometric on an arbitrary domain $\mathcal{D}$ which generalizes the Poincaré metric on the unit disk, the Hilbert ball and on the powers of the Hilbert ball. From now on we let $\mathcal{D}$ be a domain in an arbitrary complex Banach space X.

3.4.1 *The Carathéodory pseudometric*

For $x \in \mathcal{D}$ we set $B_r(x) := \{y \in X : \|y - x\| < r\} \subset \mathcal{D}$ and for $y \in B_r(x)$, $y \neq x$, we let $\phi : \Delta \to \mathcal{D}$ be defined by

$$\phi(\lambda) = x + r\frac{(y - x)}{\|y - x\|}\lambda, \quad \lambda \in \Delta, \tag{3.47}$$

where Δ is the open unit disk in the complex plane. If $g \in \mathrm{Hol}(\mathcal{D}, \Delta)$, then $g \circ \phi \in \mathrm{Hol}(\Delta)$ and consequently

$$\rho_\Delta\left(g(\phi(0)), \, g\left(\phi\left(\frac{\|y - x\|}{r}\right)\right)\right) \leq \rho_\Delta\left(0, \frac{\|y - x\|}{r}\right), \tag{3.48}$$

where ρ_Δ is the Poincaré metric on Δ. Hence,

$$\rho_\Delta(g(x), \, g(y)) \leq \tanh^{-1}\left(\frac{\|y - x\|}{r}\right). \tag{3.49}$$

If x and y are arbitrary points in $\mathcal{D}$ then, by connectedness, we can find $\delta > 0$ and a chain $\{x_0, x_1, \ldots, x_n\}$ of points of $\mathcal{D}$ such that $x_0 = x$, $x_n = y$, and $x_{i+1} \in B_\delta(x_i) \subset \mathcal{D}$ for $i = 0, \ldots, n-1$. By (3.49), we have

$$\rho_\Delta(g(x), g(y)) \leq \sum_{i=1}^{n} \tanh^{-1}\left(\frac{\|x_i - x_{i-1}\|}{\delta}\right) \tag{3.50}$$

for all $g \in \mathrm{Hol}(\mathcal{D}, \Delta)$. Consider the nonnegative function C $(= C_\mathcal{D})$ on $\mathcal{D} \times \mathcal{D}$ defined as follows:

$$C_\mathcal{D}(x, y) = \sup\{\rho_\Delta(f(x), f(y)) : f \in \mathrm{Hol}(\mathcal{D}, \Delta)\}. \tag{3.51}$$

By (3.50), $C_\mathcal{D}(x, y) < \infty$ for all $x, y \in \mathcal{D}$. It is easily checked that $C_\mathcal{D}$ is a pseudometric on $\mathcal{D}$.

Definition 3.1   The function C $(= C_\mathcal{D})$ is called the **Carathéodory pseudometric** on an arbitrary domain $\mathcal{D}$.

Theorem 3.5   *If C_1 and C_2 are the Carathéodory pseudometrics assigned to the domains $\mathcal{D}_1$ and $\mathcal{D}_2$, respectively, and $f \in \mathrm{Hol}(\mathcal{D}_1, \mathcal{D}_2)$, then*

$$C_2(f(x), f(y)) \leq C_1(x, y). \tag{3.52}$$

In particular, each holomorphic self-mapping of a domain $\mathcal{D}$ is nonexpansive with respect to the Carathéodory pseudometric $C_\mathcal{D}$.

Proof.   Let $f \in \mathrm{Hol}(\mathcal{D}_1, \mathcal{D}_2)$ be given. If $h \in \mathrm{Hol}(\mathcal{D}_2, \Delta)$, then $h \circ f \in \mathrm{Hol}(\mathcal{D}_1, \Delta)$ and so for any $x, y \in \mathcal{D}_1$, we have

$$\rho_\Delta\big(h(f(x)), h(f(y))\big) \leq C_1(x, y) \tag{3.53}$$

and

$$\begin{aligned} C_2(f(x), f(y)) &= \sup\{\rho_\Delta(h(f(x)), h(f(y)) : h \in \mathrm{Hol}(\mathcal{D}_2, \Delta)\} \\ &\leq C_1(x, y). \end{aligned} \tag{3.54}$$
$\square$

Proposition 3.2   *If $\mathcal{D} = \Delta$, then $C_\Delta = \rho_\Delta$ is the Poincaré metric on Δ.*

Proof.   Using the mapping $f(z) = z$ in Δ, we see that $C_\Delta(z, w) \geq \rho_\Delta(z, w)$ for $z, w \in \Delta$. Since holomorphic mappings from the disk into itself are nonexpansive with respect to the Poincaré metric, we have $C_\Delta(z, w) \leq \rho_\Delta(z, w)$, and so equality holds. $\square$

We will see below that also for the Hilbert ball $\mathbb{B}$ the Carathéodory pseudometric $C_\mathbb{B}$ assigned to $\mathbb{B}$ coincides with the Poincaré metric ρ_B on $\mathbb{B}$. Even in the one-dimensional case, however, it may happen that $C_\mathcal{D}$ is not a true metric. If, for example, $\mathcal{D} = \mathbb{C}$, then Liouville's theorem implies that every f in $\mathrm{Hol}(\mathbb{C}, \Delta)$ is a constant mapping and hence $\rho_\Delta(f(z), f(w)) = 0$ for all $z, w \in \mathbb{C}$, and consequently $C_\mathcal{D} \equiv 0$. Nevertheless, if $\mathcal{D}$ is a bounded domain in an arbitrary complex Banach space X, then $C_\mathcal{D}$ is, in fact, a metric on $\mathcal{D}$.

Proposition 3.3 *Let $\mathcal{D}$ be a bounded domain in a complex Banach space X. Then $C_\mathcal{D}$ is a metric on $\mathcal{D}$ which induces the original topology on $\mathcal{D}$ defined by the norm topology of X. Furthermore, the metric space $(\mathcal{D}, C_\mathcal{D})$ is complete.*

We will prove this assertion later in a more general setting.

3.4.2 *The Kobayashi pseudometric*

As the Carathéodory pseudometric on a domain $\mathcal{D}$ is defined by means of mappings from $\mathcal{D}$ into Δ it is natural to look at mappings from Δ into $\mathcal{D}$ to see if they also define a pseudometric on $\mathcal{D}$ which generalizes the Poincaré metric for the case of the unit disk. We will present two equivalent definitions of the Kobayashi pseudometric $K_\mathcal{D}$ on $\mathcal{D}$. The first definition is, in fact, that of the Lempert function δ ([Lempert (1982)]; see also [Dineen (1989)]). In this case we can define a pseudometric on $\mathcal{D}$ after a minor adjustment to obtain the triangle inequality. For points x and y in a domain $\mathcal{D}$ we let

$$\delta_\mathcal{D}(x, y) = \inf\{\rho_\Delta(z, w) : f \in \mathrm{Hol}(\Delta, \mathcal{D}),\ f(z) = x,\ f(w) = y\}. \quad (3.55)$$

This definition presupposes the existence of a mapping $f \in \mathrm{Hol}(\Delta, \mathcal{D})$ whose range contains both x and y. This can be shown by applying the vector-valued Stone–Weierstrass theorem.

If $\mathcal{D} = \Delta$ is the open unit disk in the complex plane, then using the homogeneity of the unit disk we see that $\delta_\Delta(z, w) = \delta_\Delta(w, z)$ for all $z, w \in \mathcal{D}$. In addition, for $z, w \in \Delta$, the identity mapping shows that $\delta_\Delta(z, w) \le \rho_\Delta(z, w)$, where ρ_Δ is the Poincaré metric on Δ. If $u, v \in \Delta$ and $f \in \mathrm{Hol}(\Delta)$ is such that $f(u) = z$ and $f(v) = w$, then we have that $\rho_\Delta(z, w) \le \rho_\Delta(u, v)$ and hence $\rho_\Delta(z, w) \le \delta_\Delta(z, w)$. So we get the following assertion.

Lemma 3.1 *If $\mathcal{D} = \Delta$ is the open unit disk in the complex plane, then $\delta_\Delta = \rho_\Delta$ is the Poincaré metric on Δ.*

Remark 3.2 *Unfortunately, some examples show that $\delta_\mathcal{D}$ does not always satisfy the triangle inequality. Therefore it cannot serve as a pseudometric generalizing the Poincaré metric. Indeed, if $\mathcal{D}\ (= \mathcal{D}_r) = \{(z_1, z_2) \in \mathbb{C}^2 :$*

$|z_1| < 1$, $|z_2| < 1$, $|z_1 z_2| < r\}$, *then one can show (see, for example, [Shabat (1976)] and [Dineen (1989)]) that there are $r > 0$ sufficiently small and points x, y in $\mathcal{D}$ $(= \mathcal{D}_r)$ such that $\delta_\mathcal{D}(z, 0) + \delta_\mathcal{D}(0, w) < \delta_\mathcal{D}(z, w)$.*

To avoid this deficiency, we define the Kobayashi pseudometric which is smaller than the Lempert function $\delta_\mathcal{D}$.

Definition 3.2 The Kobayashi pseudometric $K_\mathcal{D}$ on a domain $\mathcal{D}$ is defined by

$$K_\mathcal{D}(x, y) =$$

$$\inf\left\{ \sum_{i=1}^n \delta_\mathcal{D}(x_i, x_{i+1}) : n \in \mathbb{N}, \{x = x_1, \ldots, x_{n+1} = y\} \subset \mathcal{D} \right\}. \quad (3.56)$$

Suppose that x, y and z belong to $\mathcal{D}$, and $\{x = x_1, x_2, \ldots, x_{n+1} = y\}$ and $\{y = y_1, y_2, \ldots, y_{m+1} = z\}$ are finite subsets of $\mathcal{D}$. Then

$$K_\mathcal{D}(x, z) \le \sum_{i=1}^n \delta_\mathcal{D}(x_i, x_{i+1}) + \sum_{j=1}^m \delta_\mathcal{D}(y_j, y_{j+1}) \quad (3.57)$$

and, upon taking the infimum of each sum on the right-hand side independently, we see that

$$K_\mathcal{D}(x, z) \le K_\mathcal{D}(x, z) + K_\mathcal{D}(y, z). \quad (3.58)$$

It is clear that $K_\mathcal{D} \le \delta_\mathcal{D}$. It is also clear that if $\delta_\mathcal{D}$ does satisfy the triangle inequality then, in fact, $\delta_\mathcal{D} = K_\mathcal{D}$. As a matter of fact, we have a more general assertion.

Lemma 3.2 *If ρ is any pseudometric on $\mathcal{D}$ such that $\rho \le \delta_\mathcal{D}$, then $\rho \le K_\mathcal{D}$. In particular, $C_\mathcal{D}(x, y) \le K_\mathcal{D}(x, y)$.*

Proof. If $\{x = x_1, x_2, \ldots, x_{n+1} = y\}$ is a subset of $\mathcal{D}$, then

$$\rho(x, y) \le \sum_{i=1}^n \rho(x_i, x_{i+1}) \le \sum_{i=1}^n \delta_\mathcal{D}(x_i, x_{i+1}), \quad (3.59)$$

and upon taking the infimum of the right-hand side, we find that $\rho(x, y) \le K_\mathcal{D}(x, y)$. $\square$

We will see below that for a bounded convex domain the Caratheodory and Kobayashi metrics actually coincide. Meanwhile, we just mention this fact for the one-dimensional case.

Proposition 3.4 *If $\mathcal{D} = \Delta$, then $K_\Delta = \rho_\Delta$ is the Poincaré metric on Δ.*

Suppose now that $\mathcal{D}_1$ and $\mathcal{D}_2$ are arbitrary domains and $h \in \mathrm{Hol}(\Delta, \mathcal{D}_1)$ with $h(z) = x$ and $h(w) = y$ for some z and w in Δ. Let f belong to $\mathrm{Hol}(\mathcal{D}_1, \mathcal{D}_2)$. Then $f \circ h \in H(\Delta, \mathcal{D}_2)$ and $f(h(z)) = f(x)$, $f(h(w)) = f(y)$. Hence $\delta_{\mathcal{D}_2}(f(x), f(y)) \le \rho_\Delta(z, w)$; taking the infimum of the right-hand side of this inequality over the appropriate set of holomorphic mappings, we see that

$$\delta_{\mathcal{D}_2}(f(x),\, f(y)) \le \delta_{\mathcal{D}_1}(x, y). \tag{3.60}$$

Thus we arrive at the following assertion.

Theorem 3.6 *If K_1 and K_2 are the Kobayashi pseudometrics assigned to the domains $\mathcal{D}_1$ and $\mathcal{D}_2$, respectively, and $f \in \mathrm{Hol}(\mathcal{D}_1, \mathcal{D}_2)$, then*

$$K_2(f(x),\, f(y)) \le K_1(x, y). \tag{3.61}$$

In particular, each holomorphic self-mapping of a domain $\mathcal{D}$ is nonexpansive with respect to the Kobayashi pseudometric $K_\mathcal{D}$.

In order to present the second definition of the Kobayashi pseudometric, we need the Kobayashi infinitesimal pseudometric which is a particular case of the so-called Finsler infinitesimal pseudometric considered in the next section.

3.5 Infinitesimal Finsler Pseudometrics

In this section we describe an infinitesimal approach to the definition of a pseudometric on a domain. It is based on the idea mentioned in previous sections regarding the measurement of the lengths of smooth curves.

Definition 3.3 An **infinitesimal Finsler pseudometric** on a domain $\mathcal{D}$ in a Banach space X is a nonnegative function $\alpha : \mathcal{D} \times X \to \mathbb{R}$ such that

(a) $\alpha(x, tv) = |t| \alpha(x, v)$ for all $(x, v) \in \mathcal{D} \times X$ and $t \in \mathbb{C}$;
(b) α is upper semicontinuous.

If, in addition, $\alpha(x, v) > 0$ for $v \ne 0$, we call α an **infinitesimal Finsler metric**.

For two points x and y in $\mathcal{D}$, let $\gamma : [0, 1] \to \mathcal{D}$ denote a parametrized curve, with a piecewise continuous derivative, joining x and y. We call γ an **admissible curve joining x and y**. If α is an infinitesimal Finsler pseudometric and $\gamma : [0, 1] \to \mathcal{D}$ is an admissible curve, then the function

$t \in [0,1] \to \alpha(\gamma(t), \gamma'(t))$ is a bounded measurable function and the **length** $L_\alpha(\gamma)$ of γ is defined by

$$L_\alpha(\gamma) = \int_0^1 \alpha(\gamma(t),\, \gamma'(t))\, dt. \tag{3.62}$$

The properties

$$\alpha(x, tv) = t\alpha(x, v) \quad \text{for} \quad (x, v) \in \mathcal{D} \times X \quad \text{and} \quad t \in \mathbb{R}^+ \tag{3.63}$$

and

$$\alpha(x, v) = \alpha(x, -v) \quad \text{for} \quad (x, v) \in \mathcal{D} \times X \tag{3.64}$$

are clearly implied by condition (a) and are equivalent, respectively, to the fact that curve length is independent of the parametrization and of the orientation of the curve.

The integrated form of α is the function $d_\alpha : \mathcal{D} \times \mathcal{D} \to \mathbb{R}^+$ defined by

$$d_\alpha(x, y) = \inf\{L_\alpha(\gamma) : \gamma \text{ is an admissible curve joining } x \text{ and } y\}. \tag{3.65}$$

It is easily verified that d_α is a pseudometric on $\mathcal{D}$.

If α_1 and α_2 are infinitesimal Finsler pseudometrics on $\mathcal{D}_1$ and $\mathcal{D}_2$, respectively, and $f : \mathcal{D}_1 \to \mathcal{D}_2$ is a C^1 function, then f is called nonexpansive (respectively, an isometry) with respect to α_1 and α_2 if

$$\alpha_2(f(x), f'(x)(v)) \le \alpha_1(x, v), \tag{3.66}$$

respectively,

$$\alpha_2(f(x), f'(x)(v)) = \alpha_1(x, v), \tag{3.67}$$

for all $(x, v) \in \mathcal{D} \times X$.

Lemma 3.3   *Let α_1 and α_2 be infinitesimal Finsler pseudometrics on $\mathcal{D}_1$ and $\mathcal{D}_2$, respectively, and let d_1 and d_2 be the integrated forms of α_1 and α_2, respectively. If $f : \mathcal{D}_1 \to \mathcal{D}_2$ is a smooth function such that $f : (\mathcal{D}_1, \alpha_1) \to (\mathcal{D}_2, \alpha_2)$ is nonexpansive, then $f : (\mathcal{D}_1, d_1) \mapsto (\mathcal{D}_2, d_2)$ is also nonexpansive, i.e.,*

$$d_2(f(x), f(y)) \le d_1(x, y). \tag{3.68}$$

Proof. Let γ be an admissible curve in $\mathcal{D}_1$ joining x and y. Then $f \circ \gamma$ is an admissible curve in $\mathcal{D}_2$ joining $f(x)$ and $f(y)$. Hence

$$
\begin{aligned}
& d_2(f(x), f(y)) \\
& \leq L_{\alpha_2}(f \circ \gamma) \\
& = \int_0^1 \alpha_2((f \circ \gamma)(t), (f \circ \gamma)'(t)) dt \\
& = \int_0^1 \alpha_2(f(\gamma(t)), f'(\gamma(t))(\gamma'(t))) dt \\
& \leq \int_0^1 \alpha_1(\gamma(t), \gamma'(t)) dt = L_{\alpha_1}(\gamma)
\end{aligned}
\tag{3.69}
$$

and

$$
\begin{aligned}
\alpha_2(f(x), f(y)) & \leq \inf\{L_{\alpha_1}(\gamma) : \gamma \text{ is an admissible curve joining } x \text{ and } y\} \\
& = d_1(x, y).
\end{aligned}
\tag{3.70}
$$

$\square$

3.5.1 *Examples*

Example 3.1 (The infinitesimal Poincaré metric) Let $\alpha(z, v) = \frac{|v|}{1-|z|^2}$ for $(z, v) \in \Delta \times \mathbb{C}$. The integrated form of α is the Poincaré metric ρ on Δ since $\inf\{L_\alpha(\gamma) : \gamma \text{ is an admissible curve joining } z \text{ and } w\} = \tanh^{-1} |m_{-z}(w)|$, where

$$
m_{-z}(w) = \frac{w - z}{1 - \bar{w}z}.
\tag{3.71}
$$

This function α is called the infinitesimal Poincaré metric on Δ.

Example 3.2 (The infinitesimal Carathéodory pseudometric) The infinitesimal Carathéodory pseudometric on a domain $\mathcal{D}$ at the point (x, v) in $\mathcal{D} \times X$ is defined by

$$
c_{\mathcal{D}}(x, v) = \sup\{|f'(x)(v)| : f \in \mathrm{Hol}(\mathcal{D}, \Delta)\}.
\tag{3.72}
$$

If $\mathcal{D}$ is a domain in a complex Banach space X and $f \in \mathrm{Hol}(\mathcal{D}, \Delta)$, then $f'(x) \in X^*$ and hence

$$
c_{\mathcal{D}}(x, \lambda v) = |\lambda| c_{\mathcal{D}}(x, v) \quad \text{for all } \lambda \in \mathbb{C} \text{ and } (x, v) \in \mathcal{D} \times X.
\tag{3.73}
$$

The Cauchy estimates show that $c_{\mathcal{D}}$ is finite and locally bounded. If $\mathcal{D} = \Delta$ is the open unit disk in the complex plane $\mathbb{C}$, then it is clear that c_Δ is the infinitesimal Poincaré metric; hence the integrated form of c_Δ is the Poincaré metric on Δ.

If $\mathcal{D}_1$ and $\mathcal{D}_2$ are domains in complex Banach spaces X_1 and X_2, respectively, and $f \in \mathrm{Hol}(\mathcal{D}_1, \mathcal{D}_2)$, then

$$c_{\mathcal{D}_2}\big(f(x),\, f'(x)(v)\big) \leq c_{\mathcal{D}_1}(x, v) \tag{3.74}$$

for all $(x, v) \in \mathcal{D}_1 \times X_1$. Thus holomorphic mappings are nonexpansive with respect to the integrated forms d_1 and d_2 of $c_{\mathcal{D}_1}$ and $c_{\mathcal{D}_2}$, respectively.

In finite dimensions the definition of the pseudometric

$$d_c(x, y) = \inf\big\{ L_{c_\mathcal{D}}(\gamma) : \gamma \text{ is an admissible curve joining } x \text{ and } y \big\} \tag{3.75}$$

is due to Reiffen [Reiffen (1965)] and the pseudometric is sometimes called the Carathéodory–Reiffen–Finsler (or simply CRF) pseudometric on a domain $\mathcal{D}$.

Observe also that the infinitesimal Carathéodory pseudometric $c_\mathcal{D}$ is a continuous infinitesimal Finsler pseudometric on the domain $\mathcal{D}$.

Finally, it can be shown (see, for example, the next section) that for any domain $\mathcal{D}$ in a complex Banach space,

$$C_\mathcal{D} \leq d_c, \tag{3.76}$$

where $C_\mathcal{D}$ is the Carathéodory pseudometric on $\mathcal{D}$. We remark in passing that a strict inequality holds for some domains.

Example 3.3 (The infinitesimal Kobayashi pseudometric) The infinitesimal Kobayashi pseudometric on a domain $\mathcal{D}$ at the point (x, v) in $\mathcal{D} \times X$ is defined by

$$k_\mathcal{D}(x, v) = \inf\big\{ \eta > 0 : \exists f \in \mathrm{Hol}(\Delta, \mathcal{D}),$$
$$f(0) = x,\ f'(0)(\eta) = v \big\}. \tag{3.77}$$

In contrast with the Carathéodory pseudometric, the integrated form

$$d_k(x, y) = \inf\big\{ L_{k_\mathcal{D}}(\gamma) : \gamma \text{ is an admissible curve joining } x \text{ and } y \big\} \tag{3.78}$$

of the infinitesimal Kobayashi pseudometric coincides with the Kobayashi pseudometric $K_\mathcal{D}$ defined above for any domain $\mathcal{D}$, *i.e.*,

$$K_\mathcal{D}(x, y) = d_k(x, y), \quad x, y \in \mathcal{D}. \tag{3.79}$$

Therefore the last formula can serve as the second definition of the Kobayashi pseudometric.

3.6 Schwarz–Pick Systems of Pseudometrics

We are now in a position to define the so-called Schwarz–Pick systems of pseudometrics introduced by L. A. Harris which include the Carathéodory and Kobayashi pseudometrics as the smallest and the largest pseudometrics, respectively.

Definition 3.4 A system which assigns a pseudometric to each domain in each normed linear space is called a **Schwarz–Pick system** if the following conditions hold:

 (i) the pseudometric assigned to Δ is the Poincaré metric;
(ii) (the Schwarz–Pick inequality) if ρ_1 and ρ_2 are the pseudometrics assigned to $\mathcal{D}_1$ and $\mathcal{D}_2$, respectively, and $f \in \mathrm{Hol}(\mathcal{D}_1, \mathcal{D}_2)$, then

$$\rho_2(f(x), f(y)) \leq \rho_1(x, y) \tag{3.80}$$

for all $x, y \in \mathcal{D}_1$.

Note that condition (ii) implies that all biholomorphic mappings are isometries.

A further obvious but useful consequence of (ii) is the following fact: if $\mathcal{D}_1 \subset \mathcal{D}_2$ and ρ_1 and ρ_2 are the pseudometrics assigned to $\mathcal{D}_1$ and $\mathcal{D}_2$, respectively, then $\rho_2(x, y) \leq \rho_1(x, y)$ for all $x, y \in \mathcal{D}_1$.

Proposition 3.5 *The Carathéodory pseudometrics form the smallest Schwarz–Pick system.*

Proof. Let $\rho_\mathcal{D}$ be assigned to the domain $\mathcal{D}$ by any Schwarz–Pick system. Let $x, y \in \mathcal{D}$ be arbitrary. If $f \in \mathrm{Hol}(\mathcal{D}, \Delta)$, then

$$\rho_\Delta(f(x), f(y)) \leq \rho_\mathcal{D}(x, y). \tag{3.81}$$

Hence

$$C_\mathcal{D}(x, y) = \sup\{\rho_\Delta(f(x), f(y)) : f \in \mathrm{Hol}(\mathcal{D}, \Delta)\}$$
$$\leq \rho_\mathcal{D}(x, y). \tag{3.82}$$

$\square$

We have already seen that the Kobayashi pseudometrics $\{K_{\mathcal{D}}\}$ form a Schwarz–Pick system. Another immediate consequence of Lemma 3.2 is the following one.

Proposition 3.6 *The Kobayashi pseudometrics form the largest Schwarz–Pick system.*

Proof. Suppose that $\rho_{\mathcal{D}}$ is assigned to the domain $\mathcal{D}$ by any Schwarz–Pick system of pseudometrics and suppose that $x, y \in \mathcal{D}$. If $f \in \mathrm{Hol}(\Delta, \mathcal{D})$ and $z, w \in \Delta$ are points such that $x = f(z)$ and $y = f(w)$, then

$$\rho_{\mathcal{D}}(x, y) \leq \rho_{\Delta}(z, w). \tag{3.83}$$

Taking the infimum of the right-hand side over all such f we get $\rho_{\mathcal{D}}(x, y) \leq \delta_{\mathcal{D}}(x, y)$. By Lemma 3.2, this proves our assertion. $\qquad\square$

In proving the finiteness of the Carathéodory pseudometric we used the mapping

$$\phi : \Delta \longrightarrow B_r(x) \subset \mathcal{D}$$
$$\lambda \longrightarrow x + r\,\frac{(y - x)}{\|y - x\|}\,\lambda. \tag{3.84}$$

The same mapping shows that

$$K_{\mathcal{D}}(x, y) \leq \delta_{\mathcal{D}}(x, y) \leq \rho_{\Delta}\left(0, \frac{\|y - x\|}{r}\right) = \tanh^{-1}\left(\frac{\|y - x\|}{r}\right) \tag{3.85}$$

for any domain $\mathcal{D}$ with $B_r(x) \subset \mathcal{D}$.

We summarize the previous results by listing the relations we have established between the various pseudometricss on a domain $\mathcal{D}$:

$$C_{\mathcal{D}} \leq d_c \leq d_k = K_{\mathcal{D}}. \tag{3.86}$$

Since the Kobayashi pseudometrics form the largest Schwarz–Pick system, (3.85) implies the following proposition.

Proposition 3.7 *If $\{\rho_{\mathcal{D}}\}$ is a Schwarz–Pick system of pseudometrics, then*

$$\rho_{\mathcal{D}}(x, y) \leq \tanh^{-1}\left(\frac{\|y - x\|}{r}\right), \quad x, y \in \mathcal{D}. \tag{3.87}$$

Hence $\rho_{\mathcal{D}}$ is a continuous pseudometric on $\mathcal{D}$.

The question that arises at this point is: does ρ induce the initial topology of $\mathcal{D}$? The following theorem shows that for any bounded domain $\mathcal{D}$, a pseudometric $\rho_{\mathcal{D}}$ assigned to $\mathcal{D}$ by a Schwarz–Pick system is locally equivalent to the norm $\|\cdot\|$ of X. We denote by

$$r(x) = \mathrm{dist}_{\|\cdot\|}(x, \partial\mathcal{D}) = \inf\{\|x - z\| : z \in \partial\mathcal{D}\} \qquad (3.88)$$

the distance in X between the point x and the boundary $\partial\mathcal{D}$ of the domain $\mathcal{D}$. We also set

$$R(x) = \sup\{\|x - z\|,\ z \in \mathcal{D}\}. \qquad (3.89)$$

Theorem 3.7 *If $\mathcal{D}$ is a bounded domain in a complex Banach space X, then*

$$\tanh^{-1}\left(\frac{\|x - y\|}{R(x)}\right) \le \rho_{\mathcal{D}}(x, y) \qquad (3.90)$$

for all $x, y \in \mathcal{D}$, and

$$\rho_{\mathcal{D}}(x, y) \le \tanh^{-1}\left(\frac{\|x - y\|}{r(x)}\right) \qquad (3.91)$$

whenever $\|x - y\| < \mathrm{dist}_{\|\cdot\|}(x, \partial\mathcal{D})$.

Proof. The second inequality is a direct consequence of the previous proposition. Regarding the first inequality, it is enough to prove it for the Carathéodory pseudometrics since they form the smallest Schwarz–Pick system. So let

$$C_{\mathcal{D}}(x, y) = \sup\{\rho_{\Delta}(f(x),\, f(y)) : f \in \mathrm{Hol}(\mathcal{D}, \Delta)\} \qquad (3.92)$$

be the Carathéodory pseudometric on $\mathcal{D}$. For x and y in $\mathcal{D}$, let $\ell \in X^*$ be a bounded linear functional on X of norm 1 such that $\langle x - y, \ell \rangle = \|x - y\|$. Consider the holomorphic mapping $f : \mathcal{D} \to \Delta$ defined by

$$f(z) := \frac{1}{R(x)}\,\langle x - z, \ell \rangle. \qquad (3.93)$$

Clearly, $f(x) = 0$. Now it follows from the definition of the Carathéodory pseudometric $C_{\mathcal{D}}(x, y)$ that

$$C_{\mathcal{D}}(x, y) \ge \rho_{\Delta}(f(x), f(y)) = \rho_{\Delta}(0, f(y)) = \tanh^{-1}(|f(y)|)$$
$$= \tanh^{-1}\left(\frac{\|x - y\|}{R(x)}\right). \qquad (3.94)$$

$\square$

Definition 3.5   A domain $\mathcal{D}$ in a complex Banach space X is said to be ρ-**hyperbolic** (or just **hyperbolic**) if there exists a metric ρ assigned to $\mathcal{D}$ by a Schwarz–Pick system of pseudometrics which is equivalent to the norm of X.

Corollary 3.1   *Each bounded domain is hyperbolic.*

Definition 3.6   A domain $\mathcal{D}$ in a complex Banach space X is said to be **homogeneous** if for each pair x and y in $\mathcal{D}$ there exists an automorphism F of $\mathcal{D}$ such that $F(x) = y$.

A simple consequence of the above theorem is the following assertion which we will use in the sequel.

Corollary 3.2   *Let X be a complex Banach space such that its open unit ball B is a homogeneous domain. Then all Schwarz–Pick systems of pseudometrics (actually metrics) on B coincide. Moreover, if for $x \in B$ we denote by F_x an automorphism of B such that $F_x(x) = 0$, then*

$$\rho_B(x, y) = \tanh^{-1}(\|F_x(y)\|). \tag{3.95}$$

Proof.   Let $\rho \ (= \rho_B)$ be any metric assigned to B by a Schwarz–Pick system. Then for each $z \in B$ we have, by the above theorem, that

$$\rho(0, z) = \tanh^{-1}(\|z\|). \tag{3.96}$$

Consequently, if F_x is an automorphism of B such that $F_x(x) = 0$, then

$$\rho(x, y) = \rho(0, F_x(y)) = \tanh^{-1}(\|F_x(y)\|). \tag{3.97}$$

$$\square$$

In particular, it is well known that if X is a J^*-algebra, *i.e.*, X is a closed subspace of $L(H)$, the space of bounded linear operators on a complex Hilbert space H, such that for each $x \in X$,

$$xx^*x \in X, \tag{3.98}$$

and $\mathcal{D}$ is the open unit ball of X, then, given $x \in \mathcal{D}$, one can define the generalized Möbius transformation M_x by

$$M_x(y) = (I - xx^*)^{-\frac{1}{2}}(x - y)(I - x^*y)^{-1}(I - x^*x)^{\frac{1}{2}}, \tag{3.99}$$

where I is the identity operator in $L(H)$. Thus each metric of a Schwarz–Pick system on $\mathcal{D}$ can be represented by

$$\rho(x, y) = \tanh^{-1}(\|M_x(y)\|). \tag{3.100}$$

Remark 3.3 *Note that if X is a complex Banach space such that its open unit ball B is a homogeneous domain, then B is a bounded symmetric domain in X. The converse is also true: each bounded symmetric domain can be realized as the open unit ball of a complex Banach space. Obviously, this ball is a homogeneous domain. Since any Schwarz–Pick system is preserved under a biholomorphic mapping it follows that all Schwarz–Pick systems of pseudometrics (actually metrics) on bounded symmetric domains coincide.*

Actually, it was shown by S. Dineen, R. M. Timoney and J.-P. Vigué [Dineen et al. (1985)] (see also [Lempert (1982)] for the finite dimensional case) that this fact holds in a more general setting.

Theorem 3.8 *If $\mathcal{D}$ is a convex domain in a complex Banach space X, then $C_\mathcal{D} = K_\mathcal{D}$, i.e., all Schwarz–Pick systems of pseudometrics on $\mathcal{D}$ coincide.*

If, in addition, $\mathcal{D}$ is bounded, then, as we already know, this pseudometric is, in fact, a metric on $\mathcal{D}$. This unique metric is sometimes called the **hyperbolic metric** on $\mathcal{D}$. Finally, we observe that *a convex domain $\mathcal{D}$ in $\mathbb{C}^n$ (which is not necessarily bounded) is ρ-hyperbolic if and only if it does not contain a complex affine line* (see, for example, [Dineen (1989)]).

3.7 Bounded Convex Domains and Metric Domains in Banach Spaces

In this section we consider some special properties of bounded convex domains with respect to the hyperbolic metric. We show that this metric is compatible with the convex structure of a domain.

Lemma 3.4 *Let $\mathcal{D}$ be a bounded convex domain in a complex Banach space X endowed with the hyperbolic metric ρ. Then each ρ-ball in the metric space $(\mathcal{D}, \rho)$ is bounded away from the boundary of $\mathcal{D}$.*

Proof. We may suppose without loss of generality that $0 \in \mathcal{D}$. It suffices to show that $B_r := \{x \in \mathcal{D} : \rho(0, x) < r\}$ is bounded away from the boundary. Let $\ell \in X^*$ be such that

$$\sup_{x \in \mathcal{D}} \operatorname{Re} \ell(x) \leq 1. \qquad (3.101)$$

Let $f(x) = \frac{\ell(x)}{2 - \ell(x)}$ for $x \in \mathcal{D}$. Then $f(0) = 0$ and $|f(x)| < 1$ for all $x \in \mathcal{D}$. Hence $f \in \operatorname{Hol}(\mathcal{D}, \Delta)$. Now if $x \in B_r$, then

$$\tanh^{-1}(|f(x)|) = \rho_\Delta(f(0), f(x)) \leq \rho(0, x) < r. \qquad (3.102)$$

Hence $\|f\|_{B_r} \leq \tanh(r) < 1$ and there exists α, $0 < \alpha < 1$, such that $\operatorname{Re} \ell(x) < \alpha$ for all $x \in B_r$. Consequently, we get that $\operatorname{Re} \ell(x) < 1$ for all $x \in B_r + \frac{1}{\|\ell\|}(1 - \alpha)X_0$, where X_0 is the open unit ball of X. Since $x = 0$ is an interior point in $\mathcal{D}$, it follows that there is $M > 0$ such that $\|\ell\| < M$ for all ℓ satisfying (3.101). Hence by the Hanh–Banach separation theorem, we have that $B_r + \frac{1}{\|\ell\|}(1 - \alpha)X_0 \subset \mathcal{D}$ for some $\ell \in X^*$. Hence B_r is bounded away from the boundary of $\mathcal{D}$. $\qquad \square$

Thus we see that any ρ-Cauchy sequence is bounded away from the boundary. This yields the following important assertion.

Theorem 3.9   *A bounded convex domain $\mathcal{D}$ in a complex Banach space X endowed with the hyperbolic metric ρ is complete and ρ-hyperbolic. Moreover, a subset K of $\mathcal{D}$ is bounded away from the boundary of $\mathcal{D}$ if and only if it is ρ-bounded.*

If a subset K of $\mathcal{D}$ is bounded away from the boundary of $\mathcal{D}$ we will sometimes also say that K lies **strictly inside** $\mathcal{D}$. We need the following simple lemma, which will be useful in the sequel.

Lemma 3.5   *Let $\mathcal{D}$ be a bounded convex domain in a complex Banach space X, and let z be a point in $\mathcal{D}$. Then for a fixed $s \in [0, 1)$, the set*

$$K = \left\{ sx + (1 - s)z : x \in \overline{\mathcal{D}} \right\} \qquad (3.103)$$

is strictly inside $\mathcal{D}$. Moreover, if $r = \operatorname{dist}(z, \partial\mathcal{D})$, then $\delta = \operatorname{dist}(K, \partial\mathcal{D}) \geq (1 - s)r$.

Proof.   Let $u \in X$ be such that $\|u\| < r$. Then the element $z + u$ belongs to $\mathcal{D}$. Since $\mathcal{D}$ is convex, it follows that for all $x \in \overline{\mathcal{D}}$ the elements $sx + (1 - s)(z + u) = sx + (1 - s)z + (1 - s)u$ also belong to $\mathcal{D}$. This completes the proof. $\qquad \square$

Proposition 3.8   *Let $\mathcal{D}$ be a bounded convex domain in a Banach space X, and let ρ be the hyperbolic metric on $\mathcal{D}$.*

(i) If $x, y \in \mathcal{D}$ and $s, t \in [0, 1]$, then

$$\rho\big(sx + (1 - s)y, \ tx + (1 - t)y\big) \leq \rho(x, y). \qquad (3.104)$$

(ii) If $x, y, z \in \mathcal{D}$ and $s \in [0, 1]$, then

$$\rho\big(sx + (1 - s)z, \ sy + (1 - s)z\big)$$
$$\leq \frac{\mathrm{diam}\mathcal{D}}{\mathrm{diam}\mathcal{D} + (1 - s)\mathrm{dist}(z, \partial\mathcal{D})} \, \rho(x, y). \qquad (3.105)$$

(iii) If $x, y, w, z \in \mathcal{D}$ and $s \in [0, 1]$, then

$$\rho\big(sx + (1 - s)y, \ sw + (1 - s)z\big) \leq \max\big[\rho(x, w), \rho(y, z)\big]. \quad (3.106)$$

Proof. Assertions (i) and (iii) follow directly from the definition of the Lempert function and the Kobayashi pseudometric on a domain. To prove assertion (ii), let us fix $z \in \mathcal{D}$ and $s \in [0, 1)$. Consider the set $K = \big\{sx + (1 - s)z : x \in \overline{\mathcal{D}}\big\}$ and let $r = \mathrm{dist}(z, \partial\mathcal{D})$. Since $\mathcal{D}$ is bounded, one can find $\varepsilon > 0$ such that $\varepsilon\|x - y\| \leq (1 - s)r \leq \delta = \mathrm{dist}(K, \partial\mathcal{D})$ (say, $\varepsilon = \frac{(1-s)r}{\mathrm{diam}\mathcal{D}}$) for all x, y in $\mathcal{D}$. Hence all elements of the form $sx + (1 - s)z + s\varepsilon(x - y)$ also belong to $\mathcal{D}$. Fix any $y \in \mathcal{D}$ and consider the affine (hence holomorphic) mapping f defined as follows:

$$f(x) = sx + (1 - s)z + s\varepsilon(x - y). \qquad (3.107)$$

Let ρ be the (unique) hyperbolic metric on $\mathcal{D}$. Then ρ can be represented as the integrated form of the infinitesimal Carathéodory pseudometric on the domain $\mathcal{D}$ defined at the point (x, v) in $\mathcal{D} \times X$ by

$$c_{\mathcal{D}}(x, v) = \sup\big\{|f'(x)(v)| : f \in \mathrm{Hol}(\mathcal{D}, \Delta)\big\}, \qquad (3.108)$$

i.e.,

$$\rho(x, y) = \inf\bigg\{ L_{c_{\mathcal{D}}}(\gamma) = \int_0^1 c_{\mathcal{D}}(\gamma(t), \, \gamma'(t))dt :$$
$$\gamma \text{ is an admissible curve joining } x \text{ and } y \bigg\}. \qquad (3.109)$$

Since the mapping f is a holomorphic self-mapping of $\mathcal{D}$, we have that $c_{\mathcal{D}}(f(x), f'(x)(v)) \leq c_{\mathcal{D}}(x, v)$ for all $x \in \mathcal{D}$ and $v \in X$.

At the same time, noticing that $f'(x) = (1 + \varepsilon)sI$ (I is the identity mapping on X) does not depend on $x \in X$ and $f(y) = sy + (1 - s)z$, we see that $c_D\big(sy + (1-s)z,\ (1+\varepsilon)sIv\big) \le c_D(y,v)$, or

$$c_D\big(sy + (1-s)z,\ sIv\big) \le \frac{1}{1+\varepsilon}\, c_D(y,v). \tag{3.110}$$

Since y is arbitrary, we can replace it with any element in D. Let now $\gamma :$ $[0,1] \to D$ be any admissible curve joining x and y. Then $\gamma_1 = s\gamma + (1-s)z$ is an admissible curve joining $sx + (1-s)z$ and $sy + (1-s)z$. We have

$$c_D(\gamma_1(t), \gamma_1'(t)) = c_D\big(s\gamma(t) + (1-s)z,\ s\gamma'(t)\big)$$

$$\le \frac{1}{1+\varepsilon}\, c_D(\gamma(t), \gamma'(t)). \tag{3.111}$$

Consequently,

$$\rho\big(sx + (1-s)z,\ sy + (1-s)z\big) \le \frac{1}{1+\varepsilon}\, \rho(x,y). \tag{3.112}$$

Now substituting $\varepsilon = \frac{(1-s)\mathrm{dist}(z,\partial D)}{\mathrm{diam}D}$ we get assertion (ii). $\qquad\square$

Definition 3.7 We say that D is a metric domain in X if there exists a metric ρ on D such that

(i) for each $x \in D$ and for each $0 < \delta < \mathrm{dist}(x, \partial D)$, there are positive numbers $L\ (= L(\delta))$, $r(= r(\delta))$ and $m(= m(\delta))$ such that

$$\rho(x,y) \le L\|x - y\| \ \text{ whenever } \ \|x - y\| \le \delta, \tag{3.113}$$

and

$$\rho(x,y) \ge m\|x - y\| \ \text{ whenever } \ \rho(x,y) \le r; \tag{3.114}$$

(ii) each ρ-ball is strictly inside D.

A bounded convex domain in a complex Banach space with a metric assigned to it by the Schwarz–Pick system is a metric domain.

It is clear that if D is a metric domain, then the metric space (D, ρ) is complete.

For a bounded convex domain in a real Banach space such a metric can be induced by the complexification of X and by using a Schwarz–Pick metric ρ on the direct product of D by itself in the complex sense.

Other constructions of such domains can be given by using Hilbert's projective metric or Thompson's metric on a cone associated with a convex bounded domain D in X (see [Nussbaum (1994)]).

Definition 3.8   Let $\mathcal{D}$ be a metric convex domain in a Banach space X with a corresponding metric ρ. We say that the metric ρ is compatible with the convex structure of $\mathcal{D}$ if the following conditions hold:

(i) If $x, y, w, z \in \mathcal{D}$ and $s \in [0, 1]$, then

$$\rho\big(sx + (1 - s)y,\ sw + (1 - s)z\big) \leq \max\big[\rho(x, w),\ \rho(y, z)\big]; \quad (3.115)$$

(ii) for each $z \in \mathcal{D}$, there is a real function $\varphi : [0, 1) \to [0, 1)$ such that

$$\limsup_{s \to 1^-} \frac{1 - s}{1 - \varphi(s)} < \infty, \quad (3.116)$$

and for each pair of points $x, y \in \mathcal{D}$ and $s \in [0, 1]$, the following inequality holds:

$$\rho\big(sx + (1 - s)z,\ sy + (1 - s)z\big) \leq \varphi(s)\rho(x, y). \quad (3.117)$$

The following assertion is a direct consequence of the above propositions.

Proposition 3.9   *For each bounded convex domain $\mathcal{D}$ in a Banach space X, there is a metric ρ on $\mathcal{D}$ such that $(\mathcal{D}, \rho)$ is a complete metric space, $\mathcal{D}$ is a metric domain, and the metric ρ is compatible with the convex structure of $\mathcal{D}$.*

Proof.   Choose ρ as the hyperbolic metric on $\mathcal{D}$ and use inequality (3.112) with $\varepsilon = \frac{(1-s)\mathrm{dist}(z, \partial \mathcal{D})}{\mathrm{diam}\mathcal{D}}$. Since $\varepsilon \to 0^+$ as $s \to 1^-$, the last inequality proves our assertion. $\qquad\square$

Chapter 4

Some Fixed Point Principles

4.1 The Banach Principle

Let $\mathcal{D}$ be a topological space and assume that F is a self-mapping of $\mathcal{D}$, *i.e.*,

$$F(\mathcal{D}) \subseteq \mathcal{D}. \tag{4.1}$$

Definition 4.1 A point $z \in \mathcal{D}$ is called a **fixed point** of the mapping F if $F(z) = z$.

Condition (4.1), which is referred to as the invariance of the set $\mathcal{D}$ under the mapping F, enables us to define the iterations F^n of F by the recursive relations $F^1 = F$, $F^{n+1} = F^n \circ F$, where "$\circ$" denotes the composition operation of mappings. The subset of $\mathcal{D}$ consisting of all the fixed points of a mapping $F : \mathcal{D} \to \mathcal{D}$ will be denoted by $\text{Fix}_{\mathcal{D}}(F)$ (or simply by $\text{Fix}(F)$).

Fixed point theory is mostly concerned with the following basic issues:

(a) Existence (*i.e.*, is $\text{Fix}(F) \neq \emptyset$?).
(b) Uniqueness (or whether the fixed points are isolated).
(c) Approximation (*i.e.*, devising algorithms for locating fixed points).
(d) Properties (*e.g.*, structure, connectedness, *etc.*) of $\text{Fix}(F)$.

The following properties of F and its iterates F^n are obvious:

(i) $\text{Fix}(F) \subset \text{Fix}(F^n)$ for all $n \geq 1$;
(ii) if F is continuous and $F^n(x) \to y \in \mathcal{D}$ as $n \to \infty$, then $y \in \text{Fix}(F)$;
(iii) $\{F^n\}_{n \geq 1}$ has a natural one-parameter semigroup structure: $F^{n+m} = F^n \circ F^m$.

Recall that if (M, ρ) is a metric space and $F : M \to M$ is a mapping, then we say that F is *nonexpansive* if $\rho\big(F(x), F(y)\big) \leq \rho(x, y)$ for all

107

$x, y \in M$. If, moreover, there exists $k, 0 < k < 1$, such that $\rho\big(F(x), F(y)\big) \leq k\rho(x, y)$ for all $x, y \in M$, then we call F a *strict contraction* or a *k-contraction*. It is clear that nonexpansive mappings are continuous.

The following principle is among the more widely used fixed point theorems.

Theorem 4.1 (Banach's Contraction Principle) *Let (M, ρ) be a complete metric space and let $F : M \to M$ be a strict contraction. Then F has a unique fixed point in M, and for each $x_0 \in M$, the sequence of iterates $\{F^n(x_0)\}$ converges to this fixed point.*

Proof. Select $x_0 \in M$ and define the iterative sequence $\{x_n\}$ by $x_{n+1} = F(x_n)$ (equivalently, $x_n = F^n(x_0)$), $n = 0, 1, 2, \ldots$. Observe that for any indices $n, p \in \mathbb{N}$,

$$
\begin{aligned}
\rho(x_n, x_{n+p}) &= \rho\big(F^n(x_0), F^{n+p}(x_0)\big) = \rho\big(F^n(x_0), F^n \circ F^p(x_0)\big) \\
&\leq k(F^n)\rho(x_0, F^p(x_0)) \\
&\leq k^n \big[\rho\big(x_0, F(x_0)\big) + \rho\big(Fx_0, F^2(x_0)\big) + \cdots + \rho\big(F^{p-1}(x_0), \\
&\qquad\qquad\qquad\qquad\qquad\qquad\qquad\qquad\qquad\qquad F^p(x_0)\big)\big] \\
&\leq k^n\big(1 + k + \cdots + k^{p-1}\big)\rho\big(x_0, F(x_0)\big) \\
&= k^n \left(\frac{1 - k^p}{1 - k}\right) \rho\big(x_0, F(x_0)\big) \leq \left(\frac{k^n}{1 - k}\right) \rho\big(x_0, F(x_0)\big). \quad (4.2)
\end{aligned}
$$

This shows that $\{x_n\}$ is a Cauchy sequence, and since M is complete, there exists $x \in M$ such that $\lim_{n \to \infty} x_n = x$. To see that x is the unique fixed point of F, observe that

$$
x = \lim_{n \to \infty} x_n = \lim_{n \to \infty} x_{n+1} = \lim_{n \to \infty} F(x_n) = F(x) \qquad (4.3)
$$

and, moreover, $x = F(x)$ and $y = F(y)$ imply

$$
\rho(x, y) = \rho(F(x), F(y)) \leq k\rho(x, y), \qquad (4.4)
$$

yielding $\rho(x, y) = 0$.

Finally, letting $p \to \infty$ in (4.2), we obtain a rate of convergence:

$$
\rho(x_n, x) = \rho(F^n(x_0), x) \leq \frac{k^n}{1 - k} \rho(x_0, F(x_0)). \qquad (4.5)
$$

$\square$

The Banach fixed point theorem fails for nonexpansive mappings. A simple standard example is the shift of the real axis $\mathbb{R}$ defined by $F(x) = x + 1$.

There is a natural class of mappings which falls properly between the class of strictly contractive mappings and nonexpansive mappings.

A mapping $F : M \to M$ is called *contractive* if

$$\rho\big(F(x), F(y)\big) < \rho(x, y), \quad x, y \in M, \ x \neq y. \tag{4.6}$$

Obviously, a mapping of this type can have at most one fixed point. The mapping $F : \mathbb{R} \to \mathbb{R}$ defined by $F(x) = 1 + ln(1 + e^x)$ provides a simple example of a fixed point free contractive mapping. (In fact, $|x - F(x)| > 1$ for all $x \in \mathbb{R}$.) However, in compact spaces such mappings always have fixed points.

Theorem 4.2 *Let (M, ρ) be a compact metric space and let $F : M \to M$ be contractive. Then F has a unique fixed point in M, and for any $x_0 \in M$, the sequence $\{F^n(x_0)\}$ of iterates converges to this fixed point.*

Proof. The function $\varphi : M \to \mathbb{R}^+$ defined by $\varphi(y) = \rho(y, F(y))$ is continuous on M and hence by compactness attains its minimum, say at $x \in M$. If $x \neq F(x)$ then $\varphi(F(x)) = \rho(F(x), F^2(x)) < \rho(x, F(x))$ – a contradiction. So $x = F(x)$. Now let $x_0 \in M$ and set $a_n = \rho(F^n(x_0), x)$. Since

$$a_{n+1} = \rho(F^{n+1}(x_0), x) = \rho(F^{n+1}(x_0), Fx) \leq \rho(F^n(x_0), x) = a_n, \tag{4.7}$$

$\{a_n\}$ is a decreasing sequence of nonnegative real numbers and so has a limit, say a. Again by compactness, $\{F^n(x_0)\}$ has a convergent subsequence $\{F^{n_k}(x_0)\}$, say $\lim_{k \to \infty} F^{n_k}(x_0) = z$. Obviously, $\rho(z, x) = a$. If $a > 0$, then we obtain the contradiction:

$$a = \lim_{k \to \infty} \rho(F^{n_k+1}(x_0), x) = \rho(F(z), x) = \rho(F(z), F(x))$$
$$< \rho(z, x) = a. \tag{4.8}$$

Thus $a = 0$. Therefore any convergent subsequence of $\{F^n(x_0)\}$ must converge to x, so by compactness, $\lim_{n \to \infty} F^n(x_0) = x$. $\qquad\square$

The theorem is generally attributed to Edelstein [Edelstein (1962)] (see also [Edelstein (1964)]), who actually proved the following slightly more general result (*cf.* [Goebel and Reich (1984)]).

Theorem 4.3   *Let M be a metric space and $F : M \to M$ a contractive mapping. If there is a point z in M such that a subsequence of $\{F^n(z)\}$ converges to y, then y is the unique fixed point of F, and $\lim_{n \to \infty} F^n(z) = y$.*

Finally, we give an example which shows that additional conditions for the existence of fixed points are necessary even if M is bounded.

Example 4.1   Let $X = c_0$ and let M be the closed unit ball of X. The mapping F defined by

$$F(x) = (a, x_1, \ldots, x_n \ldots), \quad 0 < a \leq 1, \tag{4.9}$$

is an affine isometry of M, but it is fixed point free.

4.2   The Theorems of Brouwer and Schauder

Brouwer's theorem is very useful in applications in Analysis and its discovery has had a tremendous influence in the development of several branches of mathematics, most notably algebraic topology.

The simple formulation of Brouwer's theorem belies the fact that it seems to require a 'nonelementary' (nonmetric) proof. All known proofs require facts not commonly used in metric fixed point theory. On the other hand, attempts to find other versions of Brouwer's theorem have given rise to several interesting questions of metric type which will be discussed later.

Theorem 4.4 (Brouwer's Fixed Point Theorem) *If $F : \mathcal{D} \to \mathcal{D}$ is a continuous self-mapping of a compact convex subset $\mathcal{D}$ of $\mathbb{C}^n$ into itself, then F has a fixed point.*

One of the most important fixed point principles is the well-known principle due to J. Schauder, which is a generalization of the finite–dimensional fixed point principle of Brouwer.

Theorem 4.5 (J. Schauder) *Let F be a mapping which maps a closed convex subset $\mathcal{D}$ of a Banach space X into itself. If $F(\mathcal{D})$ is contained in a compact subset of $\mathcal{D}$, then F has at least one fixed point in $\mathcal{D}$.*

All the proofs of this theorem we know about rely on topological and geometric considerations. As a rule, these proofs use Brouwer's theorem and finite–dimensional approximations (see, for instance, [Trenogin (1980)] and [Krasnoselskii and Zabreiko (1984)]).

Note also that the Schauder principle has no constructive features; it fails to indicate the number of fixed points, as well as methods for their approximation.

In the following section we will consider another class of mappings, namely, those mappings which are either contractive or nonexpansive with respect to the hyperbolic metric.

4.3 Holomorphic Fixed Point Theorems

Let $\mathcal{D}$ be a domain in a complex Banach space X, and let $\mathrm{Hol}(\mathcal{D})$, as above, denote the family of all holomorphic self-mappings of $\mathcal{D}$.

Among the fixed point principles for a mapping $F \in \mathrm{Hol}(\mathcal{D})$ in general Banach space there are two useful classical criteria which ensure the existence and uniqueness of a fixed point of F. Recall that a subset S is said to lie strictly inside $\mathcal{D}$ if there exists $\varepsilon > 0$ such that the open ball $B_\varepsilon(x)$ centered at x of radius ε is a subset of $\mathcal{D}$ whenever $x \in \mathcal{D}$. In other words,

$$\inf\big\{\|x - y\| : y \in \partial\mathcal{D},\ x \in S\big\} \geq \varepsilon. \tag{4.10}$$

The following theorem, originally due to Earle and Hamilton [Earle and Hamilton (1970)], may be viewed as a holomorphic version of the Banach contraction principle. We give a slightly more general formulation of it [Harris (2003)].

Theorem 4.6 (Earle–Hamilton) *Let $\mathcal{D}$ be a nonempty domain in a complex Banach space X and let $F : \mathcal{D} \to \mathcal{D}$ be a bounded holomorphic mapping. If $F(\mathcal{D})$ lies strictly inside $\mathcal{D}$, then F has a unique fixed point in $\mathcal{D}$. Moreover, the sequence of iterates $\{F^n(x)\}_{n=1}^\infty$ converges to this point uniformly on each bounded subset of $\mathcal{D}$.*

Proof. We use a pseudometric ρ, called the Carathéodory–Reiffen–Finsler pseudometric (CRF–pseudometric), with respect to which F is a strict contraction. Let Δ be the open unit disk of the complex plane. Define

$$\alpha(x, v) = \sup\big\{|g'(x)v| : g : \mathcal{D} \to \Delta \text{ holomorphic}\big\} \tag{4.11}$$

for $x \in \mathcal{D}$ and $v \in X$, and set

$$L(\gamma) = \int_0^1 \alpha\big(\gamma(t), \gamma'(t)\big)\,dt \tag{4.12}$$

for γ in the set Γ of all curves in $\mathcal{D}$ with piecewise continuous derivative. Clearly α specifies a seminorm at each point of $\mathcal{D}$. Define a pseudometric $\rho : \mathcal{D} \times \mathcal{D} \to \mathbb{R}^+$ by

$$\rho(x, y) = \inf\{L(\gamma) : \gamma \in \Gamma, \ \gamma(0) = x, \gamma(1) = y\} \tag{4.13}$$

for $x, y \in \mathcal{D}$.

Let $x \in \mathcal{D}$ and $v \in X$. By the chain rule,

$$(g \circ F)'(x)v = g'(F(x))F'(x)v \tag{4.14}$$

for any holomorphic function $g : \mathcal{D} \to \Delta$. Hence,

$$\alpha(F(x), F'(x)v) \le \alpha(x, v). \tag{4.15}$$

By integrating this and applying the chain rule, we obtain $L(F \circ \gamma) \le L(\gamma)$ for all $\gamma \in \Gamma$ and thus the Schwarz–Pick inequality

$$\rho(F(x), F(y)) \le \rho(x, y) \tag{4.16}$$

holds for all $x, y \in \mathcal{D}$.

Now by hypothesis, there exists an $\varepsilon > 0$ such that $B_\varepsilon(F(x)) \subseteq \mathcal{D}$ whenever $x \in \mathcal{D}$. We may assume that $\mathcal{D}$ is bounded by replacing $\mathcal{D}$ with the subset

$$\cup\{B_\varepsilon(F(x)) : x \in \mathcal{D}\}. \tag{4.17}$$

Fix t with $0 < t < \varepsilon/\delta$, where δ denotes the diameter of $F(\mathcal{D})$. Given $x \in \mathcal{D}$, define

$$\hat{F}(y) = F(y) + t[F(y) - F(x)] \tag{4.18}$$

and note that $\hat{F} : \mathcal{D} \to \mathcal{D}$ is holomorphic. Given $x \in \mathcal{D}$ and $v \in X$, it follows from

$$\hat{F}'(x)v = (1 + t)F'(x)v \tag{4.19}$$

and (4.15) with F replaced by $\hat{F}$ that

$$\alpha(F(x), F'(x)v) \le \frac{1}{1 + t}\, \alpha(x, v). \tag{4.20}$$

Integrating this as before, we obtain

$$\rho(F(x), F(y)) \le \frac{1}{1 + t}\, \rho(x, y) \tag{4.21}$$

for all $x, y \in \mathcal{D}$. Since $\mathcal{D}$ is bounded, $(\mathcal{D}, \rho)$ is a complete metric space. Now our assertion is seen to follow from by the Banach fixed point theorem. $\square$

The Earle–Hamilton theorem still applies in some cases where the holomorphic mapping does not necessarily map its domain strictly inside itself. In fact, Theorem 4.8 below is a generalization of the Earle–Hamilton theorem for convex domains.

The following well known principle follows directly from a result of Krasnoselskii and Zabreiko [Krasnoselskii and Zabreiko (1984), Theorem 21.5] and a special property of holomorphic mappings [Krasnoselskii and Zabreiko (1984), Theorem 23.5].

Theorem 4.7 *Let $\mathcal{D}$ be a convex domain in X, and let $F \in \mathrm{Hol}(\mathcal{D})$ be continuous on $\overline{\mathcal{D}}$. If $F(\overline{\mathcal{D}})$ is contained in a compact subset of $\mathcal{D}$ and*

$$x \neq F(x) \tag{4.22}$$

for all $x \in \partial\mathcal{D}$, then F has a unique fixed point in $\mathcal{D}$.

Theorem 4.8 (Khatskevich–Reich–Shoikhet [Khatskevich *et al.* (1995a)]) *Let $\mathcal{D}$ be a nonempty bounded convex domain in a complex Banach space and let $F : \mathcal{D} \to \mathcal{D}$ be a holomorphic mapping having a uniformly continuous extension to $\overline{\mathcal{D}}$. If there exists an $\varepsilon > 0$ such that $\|F(x) - x\| \geq \varepsilon$ whenever $x \in \partial\mathcal{D}$, then F has a unique fixed point in $\mathcal{D}$.*

Of course, the hypotheses of this theorem are satisfied on some homothety of $\mathcal{D}$ under the assumptions of the Earle–Hamilton Theorem.

The hypothesis $\|F(x) - x\| \geq \varepsilon$ for all $x \in \partial\mathcal{D}$ is also satisfied when $F(\overline{\mathcal{D}})$ is contained in a compact subset of $\mathcal{D}$ and condition (4.22) holds. If F is not necessarily compact, this condition is also satisfied when $\mathcal{D}$ contains the origin and

$$\sup_{x \in \partial\mathcal{D}} \frac{\|F(x)\|}{\|x\|} < 1. \tag{4.23}$$

The proof we present here is due to L. A. Harris [Harris (2003)].

Proof.   Given $0 < t < 1$ and $x \in \mathcal{D}$, define a holomorphic map $f_t : \mathcal{D} \to \mathcal{D}$ by

$$f_t(y) = (1 - t)x + tF(y) \tag{4.24}$$

and let $\delta > 0$ be such that $B_\delta(x) \subseteq \mathcal{D}$. To show that $f_t(\mathcal{D})$ lies strictly inside $\mathcal{D}$, take $\varepsilon = (1 - t)\delta$. Let $y \in \mathcal{D}$ and let $w \in B_\varepsilon(f_t(y))$. Then

$$z = \frac{w - tF(y)}{1 - t} \tag{4.25}$$

is in $\mathcal{D}$ since $z \in B_\delta(x)$, so

$$w = (1 - t)x + tF(y) \in \mathcal{D}. \tag{4.26}$$

Hence $B_\varepsilon(f_t(y)) \subseteq \mathcal{D}$ for all $y \in \mathcal{D}$.

By the Earle–Hamilton theorem, f_t has a unique fixed point $g_t(x)$ in $\mathcal{D}$. Since the hyperbolic CRF-metric is continuous, the proof of the Banach contraction principle shows that the iterates of f_t at a chosen point $y_0 \in \mathcal{D}$ are holomorphic and locally uniformly Cauchy in x. Hence the limit mapping $g_t : \mathcal{D} \to \mathcal{D}$ is holomorphic. Now a point $x \in \mathcal{D}$ is a fixed point of g_t if and only if x is a fixed point of F. Thus, by the Earle–Hamilton theorem, it suffices to show that $g_t(\mathcal{D})$ lies strictly inside $\mathcal{D}$ for some $t > 0$.

Since F has a uniformly continuous extension to $\overline{\mathcal{D}}$, by hypothesis there exist $\varepsilon > 0$ and $\delta > 0$ such that $\|F(x) - x\| \geq \varepsilon$ whenever $x \in \mathcal{D}$ and

$$\operatorname{dist}(x, \partial\mathcal{D}) = \inf\{\|x - y\| : y \in \partial\mathcal{D}\} < \delta. \tag{4.27}$$

Since $\mathcal{D}$ is bounded, there is an M with $\|x\| \leq M$ for all $x \in \mathcal{D}$. If $x \in \mathcal{D}$,

$$F(g_t(x)) - g_t(x) = (1 - t)[F(g_t(x)) - x], \tag{4.28}$$

so

$$\big\|F(g_t(x)) - g_t(x)\big\| \leq 2(1 - t)M. \tag{4.29}$$

Choose t close enough to 1 so that $2(1 - t)M < \varepsilon$. If $d(g_t(x), \partial\mathcal{D}) < \delta$ for some $x \in \mathcal{D}$, then

$$\varepsilon \leq \big\|F(g_t(x)) - g_t(x)\big\|, \tag{4.30}$$

a contradiction. Thus, $B_\delta(g_t(x)) \subseteq \mathcal{D}$ for all $x \in \mathcal{D}$, as required. $\qquad\square$

Example 4.2 The hypotheses on the behavior of F on $\partial\mathcal{D}$ cannot be omitted in Theorem 4.8. This follows by considering a translate of the shift operator. Specifically, let $X = c_0$ and define

$$F(x) = \left(\frac{1}{2}, x_1, x_2 \ldots\right) \tag{4.31}$$

for $x \in X$. Clearly, F is an affine isometry on X and F maps the ball $B_r(0)$ into itself for each $r > \frac{1}{2}$. However, if $F(x) = x$, then

$$\frac{1}{2} = x_1 = x_2 = \ldots, \tag{4.32}$$

contradicting the requirement that x belongs to c_0. Thus F has no fixed point in X.

4.4 Fixed Points in the Hilbert Ball

Despite Example 4.2, it is still an open problem whether if B is the open unit ball of a separable reflexive complex Banach space and $F : B \to B$ is a holomorphic mapping with a continuous extension to $\overline{B}$, then F must have a fixed point in $\overline{B}$. However, Hayden and Suffridge [Hayden and Suffridge (1976)] have proved that $e^{i\theta}F$ has a fixed point in $\overline{B}$ for almost every θ.

Also, Goebel, Sekowski and Stachura [Goebel *et al.* (1980)] have solved the problem in the affirmative for the case where X is a Hilbert space and this has been extended by Kuczumow [Kuczumow (1984); Kuczumow (1985)] to the case where X is a finite product of Hilbert spaces (with the max norm). An example of Kakutani [Kakutani (1941)] shows that holomorphy is essential in the hypotheses since he exhibited a fixed point free homeomorphism of the closed unit ball of any infinite dimensional Hilbert space.

A related problem is to weaken the hypotheses of the Earle–Hamilton theorem by showing that if $F : B \to B$ is a holomorphic mapping such that the sequence of iterates $\{F^n(x)\}$ lies strictly inside B for some $x \in B$, then F has a fixed point is B. This has been established when X is a Hilbert space (see also [Goebel and Reich (1984)]) and when X is a finite product of Hilbert spaces in [Kuczumow (1985)] (see also [Kryczka and Kuczumow (1997)]). These results have been extended to bounded convex domains in a more general class of reflexive Banach spaces by Budzyńska [Budzyńska (2004)]. The example presented above gives a counterexample for the general case.

Let $\mathbb{B}$ be the open unit ball of a Hilbert space H. Applying the asymptotic center method due to M. Edelstein [Edelstein (1972)] and the properties of the Poincaré metric ρ on the Hilbert ball we get the following fixed point theorem.

Definition 4.2 Let $F : \mathbb{B} \to \mathbb{B}$ be a ρ-nonexpansive mapping. We shall call a sequence $\{y_n\} \subset \mathbb{B}$ an approximating sequence for F if $\lim_{n \to \infty} \rho(y_n, F y_n) = 0$.

Theorem 4.9   *(see [Goebel and Reich (1984)]) Let $F : \mathbb{B} \to \mathbb{B}$ be a ρ-nonexpansive mapping. Then the following are equivalent:*

(a) F has a fixed point;
(b) There exists a point x in $\mathbb{B}$ such that the sequence of iterates $\{F^n x\}$ is ρ-bounded;
(c) The sequence of iterates $\{F^n x\}$ is ρ-bounded for all x in $\mathbb{B}$;
(d) There exists a ρ-bounded approximating sequence for F.

The following proposition can also be found in [Goebel and Reich (1984)].

Proposition 4.1   *Let $F : \mathbb{B} \to \mathbb{B}$ be ρ-nonexpansive. If there exists x in $\mathbb{B}$ such that $\{F^n x\}$ converges weakly to a point in $\mathbb{B}$, then F has a fixed point.*

Theorem 4.10 *([Goebel (1982b)], see also [Goebel and Reich (1984)] and [Shafrir (1992b)]) If a norm continuous mapping $F : \overline{\mathbb{B}} \to \overline{\mathbb{B}}$ is ρ-nonexpansive on $\mathbb{B}$, then it has a fixed point in $\overline{\mathbb{B}}$.*

4.5   Fixed Points in Finite Powers of the Hilbert Ball

It is not difficult to observe that the hyperbolic metric in the Cartesian product $\mathbb{B}^n$ of n open unit balls $\mathbb{B}$ is given by

$$\rho_{\mathbb{B}^n}(x, y) = \max_{1 \le j \le n} \rho_{\mathbb{B}}(x_j, y_j). \tag{4.33}$$

In this section $N(\mathbb{B}^n)$ will denote the class of all $\rho_{\mathbb{B}^n}$-nonexpansive self-mapping on $\mathbb{B}^n$. The class of those mappings in $N(\mathbb{B}^n)$ which have a continuous (in norm) extension to $\overline{\mathbb{B}^n}$ will be denoted by $CN(\overline{\mathbb{B}^n})$. It will also be convenient to consider the slightly more general class of mappings $N(\overline{\mathbb{B}^n})$ which consists of all norm continuous mappings $f : \overline{\mathbb{B}^n} \to \overline{\mathbb{B}^n}$ such that $tf\big|_{\mathbb{B}^n} \in N(\mathbb{B}^n)$ for all $0 < t < 1$ [Kuczumow (1985)] and [Shafrir (1992b)].

Note that when $f \in N(\overline{\mathbb{B}^n})$ it may happen that $f(x) \in \partial(\mathbb{B}^n)$ for $x \in \mathbb{B}_H^n$. But in this case, if $f(x) = v$ with

$$\|v_{j_1}\| = \cdots = \|v_{j_r}\| = 1, \tag{4.34}$$

then

$$f(y)_{j_1} = v_{j_1}, \ldots, f(y)_{j_k} = v_{j_k} \tag{4.35}$$

for all $y \in \overline{\mathbb{B}^n}$.

The following simple generalization of Theorem 4.9 uses induction with respect to n and is based also on the asymptotic center method.

Theorem 4.11 (*cf.* [Kuczumow *et al.* (2001a)]) *Let* $F : \mathbb{B}^n \to \mathbb{B}^n$ *be a holomorphic mapping or more generally, a* $\rho_{\mathbb{B}^n}$*-nonexpansive mapping. Then the following statements are equivalent:*

(i) F *has a fixed point;*
(ii) *there exists* $x \in \mathbb{B}^n$ *such that* $\{F^k(x)\}$ *lies strictly inside* $\mathbb{B}^n$ *(this means that* $\{F^k(x)\}$ *is* $\rho_{\mathbb{B}^n}$*-bounded);*
(iii) *there exists a ball* $B(x,r)$ *in* $(\mathbb{B}^n, \rho_{\mathbb{B}^n})$ *which is* F*-invariant;*
(iv) *there exists a nonempty,* $\rho_{\mathbb{B}^n}$*-bounded,* $\rho_{\mathbb{B}^n}$*-closed and convex subset of* $\mathbb{B}^n$ *which is* F*-invariant.*

Theorem 4.12 ([Kuczumow (1985)] and [Shafrir (1992b)]) *If* $F \in N(\overline{\mathbb{B}^n})$, *then* $\mathrm{Fix}(F) \neq \emptyset$.

Results on common fixed points of commuting families of mappings in $N(\overline{\mathbb{B}^n})$ can also be found in [Kuczumow (1984)] and [Shafrir (1992b)].

Remark 4.1 *Note that Theorems 4.11 and 4.12 do not explain what happens when one tries to approximate fixed points by simply iterating* F. *In other words, do the iterations* $\{F^n(x)\}_{n=1}^{\infty}$ *converge to a fixed point of* F *(locally or globally) under the conditions of these theorems?*

This question arises, in particular, in the context of Theorem 4.12 when $\mathrm{Fix}_{\mathbb{B}^n}(F) = \emptyset$.

For the one-dimensional case, when $n = 1$ and $\mathbb{B} = \Delta$ is the open unit disk in the complex plane $\mathbb{C}$, a complete answer to this question is given by the classical theorem of Denjoy and Wolff which asserts that if $F \in \mathrm{Hol}(\Delta)$ has no interior fixed point in Δ, then the sequence $\{F^n(z)\}_{n=1}^{\infty}$ converges to a unique boundary point $\tau \in \partial\Delta$ for all $z \in \Delta$.

This result has given a powerful thrust to the study of the asymptotic behavior of iterates in different situations.

In the next section we will describe some generalizations of the Denjoy–Wolff Theorem as well as point out its crucial connections with the classical Julia Lemma, the Schwarz–Wolff boundary theorem, and geometric function theory in complex spaces.

Chapter 5

The Denjoy–Wolff Fixed Point Theory

Theorem 5.1 (Denjoy–Wolff) *Let Δ be the open unit disk in the complex plane $\mathbb{C}$. If $F \in \mathrm{Hol}(\Delta)$ is not the identity and is not an automorphism of Δ with exactly one fixed point in Δ, then there is a unique point a in the closed unit disk $\bar{\Delta}$ such that the iterates $\{F^n\}_{n=1}^{\infty}$ of F converge to a, uniformly on compact subsets of Δ.*

Let $\mathcal{D}$ be a domain in a complex Banach space X, and $F \in \mathrm{Hol}(\mathcal{D})$. The main goal of this chapter is to discuss the behavior of the iterates of F in the spirit of Theorem 5.1.

Definition 5.1 We say that a mapping $F \in \mathrm{Hol}(\mathcal{D})$ is **power convergent** to a mapping $h \in \mathrm{Hol}(\mathcal{D}, X)$ if the sequence of iterates $\{F^n\}_{n=1}^{\infty}$ converges to h uniformly on each ball strictly inside $\mathcal{D}$. If $h \equiv a$ is a constant mapping, then the point $a \in \overline{\mathcal{D}}$ will be called a **locally uniformly attractive** fixed point of F.

Recall that for $\mathcal{D} = \Delta$ an automorphism $F \in \mathrm{Aut}(\Delta)$ which has exactly one fixed point is called elliptic.

So, the **Denjoy–Wolff theorem** asserts that $F \in \mathrm{Hol}(\Delta)$ *is power convergent if and only if F is not an elliptic automorphism.*

5.1 The One-Dimensional Case

5.1.1 *Iterates of holomorphic self-mappings of Δ with an interior fixed point*

In this section we prove the Denjoy–Wolff theorem for the case when $F \in \mathrm{Hol}(\Delta)$ has an interior fixed point.

119

The Schwarz–Pick inequality provides us with information regarding the invariant behavior of a holomorphic self-mapping of a domain $\mathcal{D}$ when $F \in \text{Hol}(\mathcal{D})$ has an interior fixed point in $\mathcal{D}$. Namely, if ρ is a pseudometric on $\mathcal{D}$ assigned to $\mathcal{D}$ by a Schwarz–Pick system, then

$$F(\Omega(a)) \subseteq \Omega(F(a)), \tag{5.1}$$

where $\Omega(a)$ is any ρ-ball centered at a. Consequently,

$$F^n(\Omega(a)) \subseteq \Omega(F^n(a)) \tag{5.2}$$

for all $n = 0, 1, 2, \ldots$.

In particular, for the one-dimensional case where $\mathcal{D} = \Delta$ is the open unit disk of the complex plane $\mathbb{C}$, using a Möbius transformation, one can see that the ρ-ball Ω of radius r $(= \Omega_r(a))$ can be written in the form

$$\Omega_r(a) = \big\{ w \in \Delta : |m_{-a}(w)| < \tanh r \big\}, \quad 0 < r < \infty. \tag{5.3}$$

Solving this inequality we get that, in fact, the ρ-ball $\Omega_r(a)$ is the disk

$$\Omega_r(a) = \big\{ w \in \Delta : |w - sa| < dt \big\}, \tag{5.4}$$

where $d = \tanh r$,

$$s = \frac{1 - d^2}{1 - d^2|a|^2} \quad \text{and} \quad t = \frac{1 - |a|^2}{1 - |a|^2 d^2} \, . \tag{5.5}$$

Thus we have that for each $F \in \text{Hol}(\Delta)$,

$$F(\Omega_r(a)) \subseteq \Omega_r(F(a)), \quad r \in (0, \infty), \quad a \in \Delta, \tag{5.6}$$

i.e., F maps the disk $\Omega_r(a)$ centered at sa into the disk $\Omega_r(F(a))$ centered at $sF(a)$ with the same radius td. However, for a fixed $d \in (0, 1)$ this radius tends to zero when a tends to the boundary. We consider this case later.

The next question which naturally follows is: does the sequence $\{F^n(z)\}$ converge to a fixed point a of F for each $z \in \Delta$? The answer is "no" if F is an elliptic automorphism. But for the other cases the answer is affirmative.

Proposition 5.1 *Suppose that $F \in \text{Hol}(\Delta)$ is not an elliptic automorphism and has a fixed point $a \in \Delta$. Then the iterates F^n of the mapping F converge uniformly on compact subsets of Δ to a holomorphic mapping $\varphi \in \text{Hol}(\Delta)$. Moreover, if F is not the identity, then φ is a constant, i.e., $\varphi(z) = a$ for all $z \in \Delta$.*

Proof. If F is not an elliptic automorphism of Δ and it has a fixed point a in Δ, then the Schwarz–Pick Lemma (Proposition 1.1.3) implies that

$$|F'(a)| < 1. \tag{5.7}$$

Since $F'(z)$ is continuous in Δ there is a disk $\Delta(a,\varepsilon) \subset \Delta$ centered at a with radius $\varepsilon > 0$ such that

$$|F'(z)| < 1 \tag{5.8}$$

for all $z \in \overline{\Delta(a,\varepsilon)}$, the closure of $\Delta(a,\varepsilon)$. In turn, (5.8) implies that F satisfies the Lipcshitz condition

$$\big|F(z) - F(w)\big| \leq q|z - w|, \tag{5.9}$$

where $q = \max\{|F'(z)|,\ z \in \overline{\Delta(a,\varepsilon)}\}$. In addition, from (5.9) we have that F maps $\overline{\Delta(a,\varepsilon)}$ into itself. So, F is a q-contractive self-mapping of $\overline{\Delta(a,\varepsilon)}$, and it follows from the Banach Fixed Point Theorem that $\{F^n(z)\}$ converges to a for all $z \in \overline{\Delta(a,\varepsilon)}$. Using the Vitali property, we get our assertion. $\qquad\square$

5.1.2 *Iterates of holomorphic self-mappings of Δ with no interior fixed point*

By several technical transformations one can show that

$$\Omega_r(a) = \left\{ w \in \Delta : \frac{|1 - w\bar{a}|^2}{1 - |w|^2} < K \right\}, \tag{5.10}$$

where $d = \tanh r$ and $K = \frac{t}{s} = \frac{1-|a|^2}{1-d^2}$ (see formulas (5.4) and (5.5)).

Now suppose that $|a| = 1$. For an arbitrary $K > 0$ define the domain $\mathcal{D}(a, K)$ by the same formula as (5.10):

$$\mathcal{D}(a, K) = \left\{ z \in \Delta : \frac{|1 - z\bar{a}|^2}{1 - |z|^2} < K \right\}. \tag{5.11}$$

It is not difficult to see that $\mathcal{D}(a, K)$ is also a disk in Δ centered at the point $\frac{1}{1+K} \cdot a \in \Delta$ with radius $\frac{K}{1+K} < 1$, *i.e.*,

$$\mathcal{D}(a, K) = \left\{ z \in \Delta : \left| z - \frac{1}{K+1} a \right| < \frac{K}{K+1} \right\}. \tag{5.12}$$

This disk is internally tangent to the boundary of Δ at the point a and it is called a **horocycle**.

The following assertion establishes a property of the horocycles $\mathcal{D}(a, K)$, with respect to the family $\mathrm{Hol}(\Delta)$, which is analogous to the property of the domains $\Omega_r(a)$.

Theorem 5.2 (Julia's lemma) *Let $F \in \mathrm{Hol}(\Delta)$ and let $a \in \partial\Delta$ be a unimodular point. Suppose that there exists a sequence $\{a_n\} \subset \Delta$ converging to a such that the limits*

$$\alpha = \lim_{a_n \to a} \frac{1 - |F(a_n)|}{1 - |a_n|} \tag{5.13}$$

and

$$b = \lim_{n \to \infty} F(a_n) \tag{5.14}$$

exist. Then the following inequality holds for each $z \in \Delta$:

$$\frac{|1 - \bar{b}F(z)|^2}{1 - |F(z)|^2} \leq \alpha \, \frac{|1 - \bar{a}z|^2}{1 - |z|^2} \tag{5.15}$$

Proof.   It is clear that $|b| = 1$. For $z, w \in \Delta$ we define the function

$$\sigma(z, w) = \frac{(1 - |z|^2)(1 - |w|^2)}{|1 - z\bar{w}|^2} \, . \tag{5.16}$$

It is a simple exercise to show that the Schwarz–Pick Inequality is equivalent to

$$\sigma(z, w) \leq \sigma\big(F(z), F(w)\big) \tag{5.17}$$

for all $F \in \mathrm{Hol}(\Delta)$ and $z, w \in \Delta$. In particular, we have

$$\sigma(z, a_n) \leq \sigma\big(F(z), F(a_n)\big) \tag{5.18}$$

or

$$\frac{\left|1 - F(z)\overline{F(a_n)}\right|^2}{1 - |F(z)|^2} \leq \frac{\big(1 - |F(a_n)|^2\big)|1 - \bar{a}_n z|^2}{\big(1 - |a_n|^2\big)\big(1 - |z|^2\big)} \, . \tag{5.19}$$

Letting n tend to infinity we get (5.15). $\square$

The following assertion obtained by J. Wolff [Wolff (1926c)] (see also [Wolff (1926a)] and [Wolff (1926b)]) is often called the Wolff–Schwarz Boundary Lemma.

Theorem 5.3 (Wolff) *Let $F \in \mathrm{Hol}(\Delta)$ have no fixed point in Δ. Then there is a unique unimodular point $a \in \partial\Delta$ such that for each $K > 0$, the horocycle*

$$\mathcal{D}(a, K) = \left\{ z \in \Delta : \frac{|1 - \bar{a}z|^2}{1 - |z|^2} < K \right\}, \tag{5.20}$$

internally tangent to $\partial\Delta$ at a, is F-invariant. Moreover, there is a number $\alpha \in (0, 1]$ such that

$$F^n(\mathcal{D}(a, K)) \subset \mathcal{D}(a, \alpha^n K) \tag{5.21}$$

for all $n = 0, 1, 2, \ldots$.

Proof. It is sufficient to show that there is a sequence $\{a_n\}_{n=1}^{\infty} \subset \Delta$ converging to a, which satisfies the conditions of Julia's lemma, *i.e.*, the limits

$$\alpha = \lim_{a_n \to a} \frac{1 - |F(a_n)|}{1 - |a_n|} \tag{5.22}$$

and $\lim_{n \to \infty} F(a_n) = a$ exist. Moreover, we will show that $\alpha \leq 1$. Then our assertion will be seen to be a consequence of Julia's lemma. Take an arbitrary positive sequence $\{r_n\}$ increasing to 1 and consider the mappings $F_n = r_n F$. It is clear that F_n is continuous on $\bar{\Delta}_{r_n} = \{z \in \Delta : |z| \leq r_n\}$ and that it maps this set into itself. Thus F_n has a fixed point $a_n \in \bar{\Delta}_{r_n} \subset \Delta$. Passing to a subsequence, we may assume that $\{a_n\}$ converges to a point $a \in \bar{\Delta}$. If $a \in \Delta$, then, by continuity, we have

$$a = \lim_{n \to \infty} a_n = \lim_{n \to \infty} r_n F_n(a) = F(a), \tag{5.23}$$

contradicting the assumption that F has no fixed point in Δ. Consequently, $|a| = 1$.

In addition,

$$\frac{1 - |F(a_n)|}{1 - |a_n|} = \frac{1 - \frac{1}{r_n}|a_n|}{1 - |a_n|} \leq 1 \tag{5.24}$$

for all $n \in \mathbb{N}$.

Once again, perhaps by passing to a subsequence, we conclude that α in (5.22) exists and is less than or equal to 1. It is clear that $F(a_n) = \frac{1}{r_n} a_n$ converges to a and by inequality (5.15) of Julia's Lemma, we obtain

$$\frac{|1 - \bar{a}F(z)|^2}{1 - |F(z)|^2} \leq \alpha \frac{|1 - \bar{a}z|^2}{1 - |z|^2}, \tag{5.25}$$

which implies (5.21). $\qquad\qquad\qquad\qquad\qquad\qquad\qquad\qquad$ $\square$

Definition 5.2   Given a point $a \in \partial\Delta$ and a real number $k > 1$, a **nontangential approach region** at a is the set

$$\Gamma(a, k) = \{z \in \Delta : |z - a| < k(1 - |z|)\}. \qquad (5.26)$$

Definition 5.3   Given a function $h \in \mathrm{Hol}(\Delta, \mathbb{C})$, we say that h has the **angular limit** L at a boundary point $a \in \partial\Delta$ if $h(z) \to L$ as $z \to a$, $z \in \Gamma(a, k)$ for each $k > 1$. We write in this case

$$L = \angle \lim_{z \to a} h(z). \qquad (5.27)$$

Remark 5.1   *It is easy to see that h has the angular limit L at a point $a \in \partial\Delta$ if and only if $f(z) \to L$ as $z \to a$ for each angle region*

$$\widehat{S} = \Big\{z \in \Delta : |\arg(1 - \bar{a}z)| < \beta, \ |z - a| < r,$$

$$\beta \in \left(0, \frac{\pi}{2}\right), \ r \in (0, 2\cos\beta)\Big\}, \qquad (5.28)$$

which is a sector in Δ bounded between two straight lines in Δ that meet at a and are symmetric about the radius to a. This set is usually called a **Stolz angle** *at a.*

Definition 5.4   A mapping $F \in \mathrm{Hol}(\Delta)$ is said to have an **angular derivative** at $a \in \partial\Delta$ if the angular limit

$$\angle \lim_{z \to a} \frac{F(z) - b}{z - a} := \angle F'(a) \qquad (5.29)$$

exists for some $b \in \partial\Delta$.

The following essential contribution to the Denjoy–Wolff Theory, which is a complement of the Julia Lemma, was made by C. Carathéodory [Carathéodory (1929)] and [Carathéodory (1954)].

Theorem 5.4 (Julia–Carathéodory theorem) *Let $F \in \mathrm{Hol}(\Delta)$ and $a \in \partial\Delta$. The following are equivalent:*

(i) $\delta = \delta(F) = \liminf\limits_{z \to a} \frac{1 - |F(z)|}{1 - |z|} < \infty$, *where the limit is taken unrestrictedly in Δ.*

(ii) F *has a finite angular derivative at a.*

(iii) F *has the angular limit $b \in \partial\Delta$ and* $\angle \lim\limits_{z \to a} F'(z) = \delta a \bar{b}$.

We will prove this theorem in the next section in a more general setting.

An important consequence of this theorem and Theorems 5.2 and 5.3 is the following version of the so-called Julia–Wolff–Carathéodory theorem (*cf.* [Shapiro (1993)] and [Cowen and MacCluer (1995)]).

Theorem 5.5 *Let $F \in \mathrm{Hol}(\Delta)$. The following are equivalent;*

(i) F has no fixed point in Δ.
(ii) There is a boundary point $a \in \partial\Delta$ such that $0 < \lim_{z \to a} F'(z) \leq 1$.

Such a point $a \in \partial\Delta$, which satisfies the conclusion of Theorem 5.3 of Wolff, will be called a **sink point** of F on $\partial\Delta$ (or **Wolff's point**).

A mapping $F \in \mathrm{Hol}(\Delta)$ with a sink point $a \in \partial\Delta$ is said to be of **hyperbolic type** if the number α in (5.22) (Julia's number) is strictly less than 1. Otherwise ($\alpha = 1$) the mapping F is said to be of **parabolic type**.

Of course, if $0 < \alpha < 1$, then F is power convergent.

The question is whether this point is also attractive when $\alpha = 1$. The affirmative answer to this question is given in the next assertion, following Wolff and Denjoy [Wolff (1926a); Wolff (1926b); Wolff (1926c)], [Denjoy (1926)].

Theorem 5.6 *If $F \in \mathrm{Hol}(\Delta)$ has no fixed points in Δ, then there is a unique unimodular point $a \in \partial\Delta$ which is a sink point of F and the iterates $\{F^n\}$ converge locally uniformly on Δ to a constant mapping $\varphi(z) \equiv a$.*

Proof. If $F \in \mathrm{Hol}(\Delta)$ is of parabolic type, and an automorphism of Δ, then the assertion is an exercise requiring just simple calculations. Thus we may assume that $F \notin \mathrm{Aut}(\Delta)$. Since $\{F^n\}$ is a normal family, there is a subsequence $\{F^{n_j}\}$ which converges to a mapping $\varphi \in \mathrm{Hol}(\Delta, \mathbb{C})$. First, we show that φ must be constant. In fact, assuming the contrary we deduce from the maximum modulus principle that $\varphi \in \mathrm{Hol}(\Delta)$. By passing to a subsequence, if necessary, we may assume that the sequences of integers $p_j = n_{j+1} - n_j$ and $q_j = p_j - 1$ tend to infinity, and that the corresponding sequences $\{F^{p_j}\}$ and $\{F^{q_j}\}$ converge to holomorphic mappings, say h and g respectively, which belong to $\mathrm{Hol}(\Delta, \mathbb{C})$. It follows from the continuity of the composition operation that

$$h \circ \varphi = \lim_{j \to \infty} F^{p_j} \circ F^{n_j} = \lim_{j \to \infty} F^{n_{j+1}} = \varphi. \tag{5.30}$$

Since φ is not constant, h is also not constant and it has more than one fixed point in Δ. Hence h is the identity. At the same time,

$$g \circ F = \lim_{j \to \infty} F^{q_j} \circ F = \lim_{j \to \infty} F^{p_j} = I = \lim_{j \to \infty} F \circ F^{q_j} = F \circ g. \qquad (5.31)$$

Thus, F is an automorphism. Having reached a contradiction, we conclude that $\varphi(z) \equiv a$ is a constant after all.

Now, if there is a subsequence $\{n_j\} \subset \mathbb{N}$ such that $\{F^{n_j}(z)\}$ converges for $z \in \Delta$ to a constant $b \in \bar{\Delta}$ different from a, then one can find $K > 0$ such that the closure of the horocycle $\mathcal{D}(a, K)$ does not contain b.

Since $F^{n_j}(\mathcal{D}(a, K)) \subset \mathcal{D}(a, K)$, it is impossible that for $z \in \mathcal{D}(a, K)$, the sequence $\{F^{n_j}(z)\}$ converges to b. Hence $\{F^n(z)\}$ converges to a for any $z \in \Delta$. $\qquad\square$

Thus the classical Denjoy–Wolff theorem is, in fact, a summary of the following three assertions due to Denjoy, Wolff and Julia. Each one of them has been extended to different situations.

- If $F \in \mathrm{Hol}(\Delta)$ is not an automorphism of Δ and has a fixed point c in Δ, then this point is unique in Δ, and the sequence $\{F^n\}_{n=1}^{\infty}$ converges to c, uniformly on compact subsets of Δ.

- **(The Wolff–Schwarz Lemma)** If $F \in \mathrm{Hol}(\Delta)$ has no fixed point in Δ, then there is a unique unimodular point $a \in \partial\Delta$ such that every disk $\mathcal{D}_a$ in Δ, internally tangent to $\partial\Delta$ at a, is F-invariant, *i.e.*,

$$F(\mathcal{D}_a) \subset \mathcal{D}_a. \qquad (5.32)$$

- If $F \in \mathrm{Hol}(\Delta)$ has no fixed point in Δ, then there is a unique unimodular point $b \in \partial\Delta$ such that the sequence $\{F^n\}_{n=1}^{\infty}$ converges to b, uniformly on compact subsets of Δ.

The limit point of the sequence $\{F^n(z)\}_{n=1}^{\infty}$, $z \in \Delta$, will be called the **Denjoy–Wolff point** of F.

The point a and the point b in the last two assertions are, of course, one and the same. In other words, the sink point of F is also the Denjoy–Wolff point of F. However, this is not always the case in higher dimensional situations.

In fact, there are many situations in the higher dimensional case when a holomorphic fixed point free mapping has a sink point, but is not power convergent (see the next section).

Returning to the one-dimensional case, we refer the reader to the paper by R. Burckel [Burckel (1981)] and to the books [Shapiro (1993)] and

[Cowen and MacCluer (1995)] for a modern interpretation of the Denjoy–Wolff Theorem and its applications. Here we only mention the following observation concerning this case.

Remark 5.2 *By using the Schwarz Lemma, Proposition 5.1 can be rephrased in the following manner:*

- *Let $F \in \mathrm{Hol}(\Delta)$ have a fixed point $a \in \Delta$. If F is not the identity, then F is power convergent if and only if $|F'(a)| < 1$.*

It turns out that by using the notion of the derivative and its spectral properties, one can also study power convergent mappings in higher dimensional spaces. See Sections 5.4 and 5.5 below.

We are now in a position to formulate several generalizations of Theorems 5.2 and 5.3.

5.2   The Unit Hilbert Ball

Let H be a complex Hilbert space with the inner product $\langle \cdot, \cdot \rangle$, and let $\mathbb{B}$ be the open unit ball in H.

The following generalization of the Wolff–Schwarz Boundary Lemma (Theorem 5.3) is due to K. Goebel [Goebel (1982b)]. For the finite dimensional case, $H = \mathbb{C}^n$, this result was independently obtained by B. MacCluer [MacCluer (1983)] and G. Chen [Chen (1984)].

Theorem 5.7   *If $F \in \mathrm{Hol}(\mathbb{B})$ has no fixed point, then there exists a unique point $a \in \partial \mathbb{B}$ such that for each $0 < R < \infty$, the set*

$$E(a, R) = \left\{ x \in \mathbb{B} : \frac{|1 - \langle x, a \rangle|^2}{1 - \|x\|^2} < R \right\} \tag{5.33}$$

is F-invariant.

Geometrically, the set $E(a, R)$ is an ellipsoid the closure of which intersects the unit sphere $\partial \mathbb{B}$ at the point a. It is a natural analogue of the horocycle $\mathcal{D}(a, R)$. As a matter of fact, Theorem 5.7 holds in a more general setting (*cf.* Theorem 25.2 in [Goebel and Reich (1984)]).

Theorem 5.8   *Let ρ be the Poincaré hyperbolic metric on $\mathbb{B}$, and let $F : \mathbb{B} \to \mathbb{B}$ be a ρ-nonexpansive mapping. If F is fixed point free, then there is a unique boundary point $a \in \partial \mathbb{B}$ such that all the ellipsoids $E(a, k)$, $k > 0$,* *are F-invariant. In other words,*

$$\varphi_a(F(x)) \le \varphi_a(x), \qquad (5.34)$$

where

$$\varphi_a(x) = \frac{|1 - \langle x, a \rangle|^2}{1 - \|x\|^2} \,. \qquad (5.35)$$

To prove this theorem we use the following generalized version of Julia's Lemma.

Lemma 5.1   *Let $F : \mathbb{B} \to \mathbb{B}$ be a ρ-nonexpansive self-mapping of $\mathbb{B}$. Suppose that for some sequence $\{a_n\}_{n=1}^{\infty} \subset \mathbb{B}$ which converges to a boundary point $a \in \partial\mathbb{B}$, the following conditions hold:*

$$\delta(F) = \lim_{n \to \infty} \frac{1 - \|F(a_n)\|}{1 - \|a_n\|} < \infty \qquad (5.36)$$

and $b = \lim_{n \to \infty} F(a_n) \in \partial\mathbb{B}$. Then for each $x \in \mathbb{B}$,

$$\varphi_b(F(x)) \le \delta(F)\varphi_a(x). \qquad (5.37)$$

Proof.   Since F is ρ-nonexpansive, we have

$$\rho\big(F(x), F(a_n)\big) \le \rho(x, a_n) \qquad (5.38)$$

or, equivalently,

$$\sigma\big(F(x), F(a_n)\big) \ge \sigma(x, a_n), \qquad (5.39)$$

where

$$\sigma(x, y) = \frac{(1 - \|x\|^2)(1 - \|y\|^2)}{|1 - \langle x, y \rangle|^2} \,, \qquad (5.40)$$

$x, y \in \mathbb{B}$.

Writing the latter inequality explicitly we get

$$\frac{(1 - \|F(x)\|^2)\big(1 - \|F(a_n)\|^2\big)}{\big|1 - \langle F(x), F(a_n) \rangle\big|^2} \ge \frac{(1 - \|x\|^2)(1 - \|a_n\|^2)}{|1 - \langle x, a_n \rangle|^2} \qquad (5.41)$$

or

$$\frac{\big|1 - \langle F(x), F(a_n) \rangle\big|^2}{1 - \|F(x)\|^2} \le \frac{1 - \|F(a_n)\|^2}{1 - \|a_n\|^2} \cdot \frac{\big|1 - \langle x, a_n \rangle\big|^2}{1 - \|x\|^2} \,. \qquad (5.42)$$

$$\square$$

Proof of Theorem 5.8 Let $F : \mathbb{B} \to \mathbb{B}$ be a ρ-nonexpansive self-mapping of $\mathbb{B}$. Consider a mapping $F_t : \mathbb{B} \to \mathbb{B}$ defined by $F_t = tF$, $0 \le t < 1$. We have

$$\rho\big(F_t(x), F_t(y)\big) \le t\rho\big(F(x), F(y)\big) \le t\rho(x, y). \tag{5.43}$$

So, for each $t \in [0, 1)$ the mapping $F_t : \mathbb{B} \to \mathbb{B}$ is a strict contraction with respect to the metric ρ. Hence it has a unique fixed point $x_t = F_t(x_t)$.

In addition,

$$x_t - F(x_t) = F_t(x_t) - F(x_t) = (t - 1)F(x_t). \tag{5.44}$$

So $\|x_t - F(x_t)\| \to 0$ when t tends to 1. Thus, if F has no fixed point in $\mathbb{B}$ we must have that

$$\lim_{t \to 1^-} \|x_t\| = 1. \tag{5.45}$$

On the other hand, since $\overline{\mathbb{B}}$ is weakly compact, there is a sequence $t_n \to 1^-$ such that $\{x_{t_n}\}$ weakly converges to an element $a \in \overline{\mathbb{B}}$. But if we assume that $\|a\| < 1$, then we have

$$\rho\big(F(x_{t_n}), F(a)\big) \le \rho(x_{t_n}, a) \tag{5.46}$$

or, equivalently,

$$\sigma\big(F(x_{t_n}), F(a)\big) \ge \sigma(x_{t_n}, a), \tag{5.47}$$

where

$$\sigma(x, y) = \frac{(1 - \|x\|^2)(1 - \|y\|^2)}{|1 - \langle x, y \rangle|^2}. \tag{5.48}$$

Explicitly the latter inequality implies

$$1 \ge \frac{1 - \frac{1}{t_n^2}\|x_{t_n}\|^2}{1 - \|x_{t_n}\|^2} \ge \frac{(1 - \|a\|^2)\big|1 - \frac{1}{t_n}\langle x_{t_n}, F(a) \rangle\big|^2}{(1 - \|F(a)\|^2)|1 - \langle x_{t_n}, a \rangle|^2}. \tag{5.49}$$

Letting $n \to \infty$, we get $\sigma(a, F(a)) \ge 1$. Hence, $a = F(a)$. This contradiction shows that a must lie on the boundary of $\mathbb{B}$, *i.e.*, $\|a\| = 1$.

It follows that the convergence of x_{t_n} to a is, in fact, strong, that is,

$$\|x_{t_n} - a\| \to 0 \quad \text{as} \quad n \to \infty. \tag{5.50}$$

In addition, the strong limit

$$\lim_{n \to \infty} F(x_{t_n}) = \lim_{n \to \infty} \frac{1}{t_n} x_{t_n} = a \tag{5.51}$$

and (by passing to a subsequence, if necessary)

$$\delta(F) = \lim_{n \to \infty} \frac{1 - \|F(x_{t_n})\|}{1 - \|x_{t_n}\|} \leq 1. \tag{5.52}$$

Setting $a_n = x_{t_n}$ and $a = b$ in the above version of Julia's Lemma, we obtain the required inequality:

$$\varphi_a(F(x)) \leq \varphi_a(x). \tag{5.53}$$

Finally, we claim that a point $a \in \partial\mathbb{B}$ satisfying $\varphi_a(F(x)) \leq \varphi_a(x)$ must be unique. Indeed, assume that there is $b \in \partial\mathbb{B}$ such that $\varphi_b(F(x)) \leq \varphi_b(x)$. If $b \neq a$, then for each k the ellipsoids $E(a, k)$ and $E(b, k)$ are F-invariant. It is clear that one can find $k > 0$ (large enough) such that $\mathcal{D} := E(a, k) \cap E(b, k) \neq \emptyset$ and lies strictly inside $\mathbb{B}$. Then for all $n = 0, 1, 2 \ldots$, $F^n(\mathcal{D}) \subset \mathcal{D}$. Hence for a point $x \in \mathcal{D}$, the sequence $\{F^n(x)\}$ is bounded away from the boundary of $\mathbb{B}$. It follows that F must have a fixed point in $\mathbb{B}$. A contradiction. Theorem 5.8 is proved. $\qquad\square$

Now Theorem 4.10 (see Chapter 4) follows immediately:

Corollary 5.1   *Let $F : \mathbb{B} \to \mathbb{B}$ be a ρ-nonexpansive self-mapping of $\mathbb{B}$. If F has a continuous extension to $\overline{\mathbb{B}}$, then it has a fixed point $a \in \overline{\mathbb{B}}$. Moreover, if F has no interior fixed point in $\mathbb{B}$, then its sink (Wolff's) point a must be a boundary fixed point of F. In addition, the approximating curve $\{x_t\}_{0 \leq t < 1}$ defined by*

$$x_t = tF_t(x_t) \tag{5.54}$$

converges to a as t goes to 1.

Remark 5.3   *As a matter of fact, it can be shown (see [Goebel and Reich (1982)], [Goebel and Reich (1984)] and [Shafrir (1992b)]) that for each $y \in \mathbb{B}$ the approximating curve*

$$x_t = (1 - t)y + tF(x_t) \tag{5.55}$$

is well defined and converges to a fixed point of F. If F has no interior fixed point in $\mathbb{B}$, then for each $y \in \mathbb{B}$, this curve converges to the same (Wolff's) sink boundary point $a \in \partial\mathbb{B}$.

Now the question is whether the sink point a in Theorem 5.7 is also the Denjoy–Wolff point of F, *i.e.*, is it attractive?

For the finite dimensional case ($\mathbb{B}$ is then the open Euclidean ball in $H = \mathbb{C}^n$) and when $F \in \mathrm{Hol}(\mathbb{B})$ the affirmative answer was given by B. Mac-Cluer [MacCluer (1983)].

Theorem 5.9 *Let $H = \mathbb{C}^n$ be a Euclidean complex space (finite dimensional Hilbert space) and let $\mathbb{B}$ be the open unit ball in H. If $F \in \mathrm{Hol}(\mathbb{B})$ has no fixed point in $\mathbb{B}$, then there is a point a of norm 1 ($a \in \partial\mathbb{B}$) such that the iterates $F^n(z)$ converge to a for all $z \in \mathbb{B}$, uniformly on compact subsets of $\mathbb{B}$.*

For infinite dimensional Hilbert balls, A. Stachura [Stachura (1985)] has constructed a counterexample which shows that this convergence result fails even for biholomorphic self-mappings.

Nevertheless, some restrictions on a mapping from $\mathrm{Hol}(\mathbb{B})$ lead to a generalization of Theorem 5.6.

The following result was obtained by C.–H. Chu and P. Mellon [Chu and Mellon (1997)].

Theorem 5.10 *Let $\mathbb{B}$ be the open unit ball in a Hilbert space H, and let $F \in \mathrm{Hol}(\mathbb{B})$ be a compact mapping with no fixed point in $\mathbb{B}$. Then the sink point a in Theorem 5.7 is attractive, i.e., the sequence $\{F^n\}$ of iterates of F converges locally uniformly on $\mathbb{B}$ to the constant mapping taking the value a.*

Although this theorem contains Theorem 5.6 (as well as the above-mentioned finite-dimensional result of B. MacCluer), it does not apply to the automorphisms of $\mathbb{B}$ because of the compactness restriction. In this connection we have to mention a result by T. L. Hayden and T. J. Suffridge [Hayden and Suffridge (1971)] which complements our information (see also [Suffridge (1974)]).

Theorem 5.11 *Let $\mathbb{B}$ be as above, and let $F \in \mathrm{Hol}(\mathbb{B})$ be an automorphism of $\mathbb{B}$ with exactly two fixed points on $\partial\mathbb{B}$. Then one of them is an attractive sink point of F.*

Remark 5.4 *It is known that each automorphism of $\mathbb{B}$ can be extended to an automorphism of $\overline{\mathbb{B}}$ and that it has a fixed point in $\overline{\mathbb{B}}$ (see [Hayden and Suffridge (1971)]). Moreover, if it does not have a fixed point in $\mathbb{B}$, then it has one or two fixed points on the boundary.*

For holomorphic self-mappings of the Hilbert ball the following versions of the Julia–Carathéodory and the Julia–Wolff–Carathéodory theorems are very important complements to the Denjoy–Wolff theorem.

We need some additional notions which are well–known in the finite–dimensional case (see, for example, [Rudin (1980)]).

Definition 5.5   A curve $\Lambda : [0,1) \to \mathbb{B}$ is said to be **asymptotically normal** at a point $\tau \in \mathbb{B}$ if

$$\lim_{s \to 1^-} \frac{\|\Lambda(s) - \lambda(s)\|^2}{1 - \|\lambda(s)\|^2} = 0; \tag{5.56}$$

$$\frac{\|\lambda(s) - \tau\|}{1 - \|\lambda(s)\|} \le M \le \infty, \quad 0 \le s < 1, \tag{5.57}$$

where $\lambda(s)$ is the ortogonal projection of $\Lambda(s)$ onto the complex line through 0 and τ:

$$\lambda(s) = \langle \Lambda(s), \tau \rangle \tau. \tag{5.58}$$

Let h be a holomorphic function on $\mathbb{B}$ with values in the complex plane. We say that h has a **restricted limit** L at $\tau \in \partial \mathbb{B}$ if h has the limit L along every curve which is asymptotically normal at τ.

Definition 5.6   Let $f : \mathbb{B} \to H$ be a holomorphic mapping on $\mathbb{B}$, and let $\tau \in \partial \mathbb{B}$. We say that f has a finite **angular derivative** at τ if for some element $y \in \mathbb{B}$, the function $h : \mathbb{B} \to \mathbb{C}$ defined by

$$h(x) = \frac{\langle y - f(x), \tau \rangle}{1 - \langle x, \tau \rangle} \tag{5.59}$$

has a finite restricted limit at τ. We denote this limit $\angle f'(\tau)$.

Theorem 5.12   *Let $\mathbb{B}$ be the open unit ball in the finite dimensional Hilbert space $H = \mathbb{C}^n$, let F be a holomorphic self-mapping of $\mathbb{B}$, and let a be a boundary point. The following are equivalent:*

(i) $\delta(F) = \lim\limits_{x \to a} \inf \frac{1 - \|F(x)\|}{1 - \|x\|} < \infty$, where the limit is taken as x approaches a unrestrictedly in $\mathbb{B}$.

(ii) F has finite angular derivative at a;

(iii) F has a restricted limit $b \in \partial \mathbb{B}$ and $\langle F'(x)a, b \rangle$ has a restricted limit K at $a \in \partial \mathbb{B}$.

Moreover, if one of these conditions holds, then $\delta(F) = K$.

We will call the number $\delta(F)$ the **Julia number** of F at a.

Theorem 5.13   *Let $\mathbb{B}$ be the open unit ball in the finite dimensional Hilbert space $H = \mathbb{C}^n$, and let F be a holomorphic self-mapping of $\mathbb{B}$ with* *no fixed point in* $\mathbb{B}$*. Then there is a unique boundary point* $a \in \partial\mathbb{B}$ *such that*

$$0 < \delta(F) = \left(\lim_{x \to a} \inf \frac{1 - \|F(x)\|}{1 - \|x\|} \right) = \angle F'(a) \le 1. \tag{5.60}$$

Remark 5.5 *As a matter of fact, a similar assertion also holds for the infinite dimensional case if we replace the angular derivative by the so-called radial derivative of F at its boundary sink point*

$$\uparrow F'(a) = \lim_{r \to 1^-} \frac{1 - \langle F(ra), a \rangle}{1 - r}. \tag{5.61}$$

Moreover, rather surprisingly, this derivative turns out to exist even if F is not necessarily holomorphic, but only ρ-nonexpansive.

Theorem 5.14 *Let $\mathbb{B}$ be the open unit ball in a complex Hilbert space H and let $F : \mathbb{B} \to \mathbb{B}$ be a ρ-nonexpansive self-mapping of $\mathbb{B}$ with no fixed point in $\mathbb{B}$. Assume that $a \in \partial\mathbb{B}$ is the sink point of F and $\delta = \delta(F)$ is the Julia number at a. Then the radial derivative*

$$\uparrow F'(a) := \lim_{r \to 1^-} \frac{1 - \langle F(ra), a \rangle}{1 - r} \tag{5.62}$$

exists and equals δ.

Proof. First, by the generalized Julia's Lemma 5.1, we have

$$\frac{|1 - \langle F(ra), a \rangle|^2}{1 - \|F(ra)\|^2} \le \delta \frac{(1 - r)^2}{1 - r^2} = \delta \frac{1 - r}{1 + r}. \tag{5.63}$$

This implies that

$$\lim_{r \to 1^-} \langle F(ra), a \rangle = 1. \tag{5.64}$$

Now we calculate

$$\frac{1 - |\langle F(ra), a \rangle|}{1 - r} = \frac{\left(1 - |\langle F(ra), a \rangle|\right)^2}{(1 - r)\left(1 - |\langle F(ra), a \rangle|\right)}$$

$$\le \frac{|1 - \langle F(ra), a \rangle|^2}{1 - |\langle F(ra), a \rangle|^2} \cdot \frac{1 + |\langle F(ra), a \rangle|}{1 - r}. \tag{5.65}$$

Therefore it follows from (5.63) that

$$\frac{1 - |\langle F(ra), a\rangle|}{1 - r} \le \delta \, \frac{1 + |\langle F(ra), a\rangle|}{1 + r}. \tag{5.66}$$

Thus, by using (5.64), we obtain

$$\lim_{r \to 1^-} \frac{1 - |\langle F(ra), a\rangle|}{1 - r} \le \delta. \tag{5.67}$$

On the other hand,

$$\frac{1 - \|F(ra)\|}{1 - r} \le \frac{1 - |\langle F(ra), a\rangle|}{1 - r}. \tag{5.68}$$

Since

$$\delta = \liminf_{x \to a} \frac{1 - \|F(x)\|}{1 - \|x\|}, \tag{5.69}$$

where the limit is taken when x approaches a unrestrictedly, we get that

$$\lim_{r \to 1^-} \frac{1 - \|F(ra)\|}{1 - r} = \lim_{r \to 1^-} \frac{1 - |\langle F(ra), a\rangle|}{1 - r} = \delta. \tag{5.70}$$

Returning to inequality (5.63), we can rewrite it as follows:

$$\frac{|1 - \langle F(ra), a\rangle|^2}{(1 - r)^2} \le \delta \, \frac{1 - \|F(ra)\|^2}{1 - r^2}. \tag{5.71}$$

Letting r tend to 1^- and using (5.70), we get

$$\lim_{r \to 1^-} \frac{|1 - \langle F(ra), a\rangle|}{1 - r} \le \delta. \tag{5.72}$$

But again, $|1 - \langle F(ra), a\rangle| \ge 1 - |\langle F(ra), a\rangle|$ and therefore we have, by (5.70) and (5.72),

$$\lim_{r \to 1^-} \frac{|1 - \langle F(ra), a\rangle|}{1 - r} = \delta. \tag{5.73}$$

Combining now (5.70) and (5.73), we get

$$\lim_{r \to 1^-} \frac{|1 - \langle F(ra), a\rangle|}{1 - |\langle F(ra), a\rangle|} = 1. \tag{5.74}$$

It follows that $\arg\big(1 - \langle F(ra), a\rangle\big)$ tends to 0 as r tends to 1^-. Thus we obtain

$$\lim_{r-1^-} \frac{1 - \langle F(ra), a\rangle}{1 - r} = \delta. \tag{5.75}$$

The theorem is proved. $\qquad\qquad\qquad\qquad\qquad\qquad\qquad\qquad\square$

Now, using the Lindelöf principle (see [Rudin (1980)]) and Theorem 5.14, we obtain Theorem 5.13 as well as Theorem 5.12 by applying an appropriate rotation of the unit ball.

Combining now Theorem 5.14 with the proof of Theorem 5.8, we obtain a stronger version of the latter theorem; it is a generalization of the Julia–Wolff–Carathéodory theorem.

Theorem 5.15 *Let $F : \mathbb{B} \to \mathbb{B}$ be a ρ-nonexpansive self-mapping of $\mathbb{B}$.*

If F has no fixed point in $\mathbb{B}$, then there is a unique point $a \in \partial\mathbb{B}$ such that

$$\varphi_a(F^n(x)) \le \delta^n \varphi_a(x), \tag{5.76}$$

where

$$0 < \delta = \lim_{r \to 1^-} \frac{\big|1 - \langle F(ra), a\rangle\big|}{1 - r} = \liminf_{x \to a} \frac{1 - \|F(x)\|}{1 - \|x\|} \le 1 \tag{5.77}$$

is the radial derivative of F at this point a.

Definition 5.7 A mapping $F \in \mathcal{N}_\rho(\mathbb{B})$ is said to be of **hyperbolic type** if it is fixed point free and its radial derivative at its sink (boundary Wolff's) point is strictly less than 1.

Corollary 5.2 *Each ρ-nonexpansive mapping of hyperbolic type is power convergent to its boundary sink point.*

5.3 Convex Domains in $\mathbb{C}^n$

In 1941 M. H. Heins [Heins (1941)] extended the Denjoy–Wolff Theorem to a finitely connected domain bounded by Jordan curves in $\mathbb{C}$. His approach is specific to the one-dimensional case.

Another look at the Denjoy–Wolff Theorem is provided by a useful result of P. Yang [Yang (1978)] concerning a characterization of the horocycle in terms of the Poincaré hyperbolic metric in Δ. More precisely, he established the following formula:

$$\lim_{\mu \to a} \left[\rho(\lambda, \mu) - \rho(0, \mu) \right] = \frac{1}{2} \log \frac{|1 - \lambda \bar{a}|^2}{1 - |\lambda|^2} . \tag{5.78}$$

So, in these terms the horocycle $\mathcal{D}_a$ in Δ can be described by the formula

$$\mathcal{D}(a, R) = \left\{ z \in \Delta : \lim_{\omega \to a} \left(\rho(z, \omega) - \rho(0, \omega) \right) < \frac{1}{2} \log R \right\}. \tag{5.79}$$

Since a hyperbolic metric can be defined in each bounded domain in $\mathbb{C}^n$, one can try to extend this formula and use it as the definition of the horosphere in a domain in $\mathbb{C}^n$. Unfortunately, in general the limit in (5.79) does not exist.

Therefore, the new idea of M. Abate [Abate (1988a)] was to study two kinds of horospheres. More precisely, he defined the small horosphere $E_{z_0}(x, R)$ of center x, pole z_0 and radius R by the formula

$$E_{z_0}(x, R) = \left\{ z \in \mathcal{D} : \limsup_{\omega \to x} \left[K_{\mathcal{D}}(z, w) - K_{\mathcal{D}}(z_0, w) \right] < \frac{1}{2} \log R \right\},$$
$$\tag{5.80}$$

and the big horosphere of center x, pole z_0 and radius R by the formula

$$F_{z_0}(x, R) = \left\{ z \in \mathcal{D} : \liminf_{\omega \to x} \left[K_{\mathcal{D}}(z, w) - K_{\mathcal{D}}(z_0, w) \right] < \frac{1}{2} \log R \right\}, \tag{5.81}$$

where $\mathcal{D}$ is a bounded domain in $\mathbb{C}^n$ and $K_{\mathcal{D}}$ is its Kobayashi metric. For the Euclidean ball in $\mathbb{C}^n$, $E_{z_0}(x, R) = F_{z_0}(x, R)$ (see [Abate (1988a); Mercer (1992); Mercer (1993)]).

Thus each assertion which states for a domain $\mathcal{D}$ in $\mathbb{C}^n$ the existence of a point $a \in \partial \mathcal{D}$ such that

$$f^n(E_z(a, R)) \subset F_z(a, R) \tag{5.82}$$

for all $z \in \mathcal{D}$, $R > 0$, $f \in \text{Hol}(\mathcal{D})$ and $n = 1, 2, \ldots$ is a generalization of Wolff's Theorem 5.3.

This is true, for example, for a bounded convex domain in $\mathbb{C}^n$. A more general result was established by M. Abate in another work [Abate (1989b)].

Theorem 5.16   *Let $\mathcal{D}$ be a bounded complete hyperbolic domain with a simple boundary, and let $z \in \mathcal{D}$. Suppose that for some $f \in \text{Hol}(\mathcal{D})$ the sequence of iterates $\{f^n\}$ is compactly divergent. Then there is $a \in \partial \mathcal{D}$ such that for all $z \in \mathcal{D}, R > 0$ and $n = 1, 2, \ldots$ the inclusion (5.82) holds.*

In fact, the assumptions of Theorem 5.16 are not sufficient to ensure a convergence result. Indeed, in order to generalize the notion of a sink point we give the following definition.

Let $\mathcal{D}$ be a domain in a Banach space X and let $F \in \mathrm{Hol}(\mathcal{D})$. We will say that a point $x \in \partial\mathcal{D}$ is a boundary sink point for F if there exist two sets of neighborhoods $\{U_\alpha\}$ and $\{V_\alpha\}$, $\alpha \in \mathcal{A}$ (a directed set), in $\mathcal{D}$ such that the following conditions hold:

(i) $U_\alpha \subset V_\alpha \subset \mathcal{D}$;
(ii) $x \in \overline{U}_\alpha$;
(iii) $\cap_{\alpha \in \mathcal{A}} U_\alpha = \cap_{\alpha \in \mathcal{A}} V_\alpha = \emptyset$;
(iv) $F(U_\alpha) \subset V_\alpha$;
(v) For $\alpha_1 < \alpha_2$, $\alpha_1, \alpha_2 \in \mathcal{A}$, $U_{\alpha_2} \subset U_{\alpha_1}$ and $V_{\alpha_2} \subset V_{\alpha_1}$.

The following example is due to C.-H. Chu and P. Mellon [Chu and Mellon (1997)].

Example 5.1 Let $\mathcal{D}$ be the open unit bidisk in $\mathbb{C}^2$, *i.e.*, $\mathcal{D} = \Delta \times \Delta$, and let $h \in \mathrm{Hol}(\Delta)$ be a fixed point free mapping with the Denjoy–Wolff point $\zeta \in \partial\Delta$. Consider the mapping $F \in \mathrm{Hol}(\mathcal{D})$ defined as follows:

$$F(z, w) = (e^{i\varphi} z, h(w)), \tag{5.83}$$

where $0 < \varphi < 2\pi$. It is clear that if F has a sink point $a \in \partial\mathcal{D}$, then $a = (0, \zeta) \in \partial\mathcal{D}$. At the same time, the sequence of iterates $\{F^n\}$ does not converge to any boundary point of $\mathcal{D}$.

Nevertheless, the convergence result does hold for bounded strongly convex C^2 domains, and for strongly pseudo–convex hyperbolic domains with a C^2 boundary [Abate (1988a); Abate (1988b)].

To generalize these facts, M. Abate defined the following notion [Abate (1989b)]:

A domain $\mathcal{D} \subset\subset \mathbb{C}^n$ is said to be F-convex at $x \in \partial\mathcal{D}$ if for all $z \in \mathcal{D}$ and $R > 0$,

$$\overline{F_z(x, R)} \cap \partial\mathcal{D} = \{x\}. \tag{5.84}$$

The domain $\mathcal{D}$ is said to be F-convex if (5.84) holds for each $x \in \partial\mathcal{D}$.

His result is the following one:

Theorem 5.17 *Suppose that in addition to the assumptions of Theorem 5.16 the domain $\mathcal{D}$ is F-convex. Then there is a point $a \in \partial\mathcal{D}$ which is attractive for f.*

Certain types of domains are known to be F-convex. We mention, for example, strictly pseudo-convex domains with a C^2 boundary, and the so-called m-convex domains (see [Abate (1988a); Abate (1988b); Mellon (1996)]).

5.4 Domains in Banach Space.

Unfortunately, our knowledge does not include positive results regarding extensions of Theorems 5.3 and 5.6 to general Banach spaces. Indeed, a simple example shows that there is a situation where even a sink point does not exist.

Example 5.2   Consider the Banach space c_0 of all complex sequences $x = (x_1, x_2, \ldots, x_n, \ldots)$ such that $x_n \to 0$ as n tends to infinity, with the max norm, and the affine (hence holomorphic) mapping $F : c_0 \to c_0$ defined by $F(x) = (a, x_1, x_2 \ldots)$, where $a \neq 0$, $|a| < 1$. It is clear that F maps the open unit ball $\mathcal{D}$ of c_0 into itself and that it is continuous on $\overline{\mathcal{D}}$. If F had a sink point on $\partial\mathcal{D}$, then it would necessarily be a fixed point of F in $\overline{\mathcal{D}}$. But F has no fixed point there.

However, the question is still open for reflexive Banach spaces. Moreover, if $\mathcal{D}$ has a strictly convex boundary and F is compact, then the convergence result holds [Kapeluszny *et al.* (1999a) and (1999b)].

Proposition 5.2   *Let X be a strictly convex Banach space, and let F be a holomorphic compact self-mapping of the open unit ball $\mathcal{D}$ of X without a fixed point in $\mathcal{D}$. Then F is power convergent on $\mathcal{D}$.*

The situation is more fully understood when F has a fixed point in the domain.

From now on we will assume that $\mathcal{D}$ is a bounded domain in a Banach space X and that $F \in \text{Hol}(\mathcal{D})$ has an interior fixed point $a \in \mathcal{D}$. Of course, as simple examples show, one cannot expect that a is always an attractive point even if F is not an automorphism (consider Example 5.1 with $h(w) = w^2$). Nevertheless, rephrasing the Denjoy–Wolff Theorem 5.6 in the form of Remark 5.2, one can point out several generalizations of this result.

Proposition 5.3   ([Vesentini (1983); Vesentini (1985); Khatskevich and Shoikhet (1984)]). *Let $\mathcal{D}$ be a bounded domain in X and let $F \in \text{Hol}(\mathcal{D})$. Suppose that F has an interior fixed point $a \in \mathcal{D}$ and let $F'(a)$ be its Fréchet* *derivative at this point. Then a is a locally uniformly attractive point of F if and only if the spectral radius of $F'(a)$ is strictly less than 1.*

Proof. The necessity of the assertion immediately follows from the Cauchy inequality if we note that

$$(F^n)'(a) = [F'(a)]^n \tag{5.85}$$

by the chain rule.

To establish sufficiency, denote $F'(a)$ by A. If $\rho(A) < 1$, then it follows from the Rutickii theorem that there is a norm $\|\cdot\|_1$ equivalent to the original norm of X such that $\|A\|_1 \left(= \sup\limits_{\|x\|_1=1} \|Ax\|_1 \right) < 1$. Thus there is a neighborhood (a ball) $U \subset\subset \mathcal{D}$ of the point a such that

$$\sup_{x\in\overline{U}} \|F'(x)\|_1 = q < 1. \tag{5.86}$$

The mean–value theorem implies that for all x and $y \in U$,

$$\|F(x) - F(y)\|_1 \le q\|x - y\|_1. \tag{5.87}$$

Now, by the Banach Fixed Point Principle, we get that the sequence $\{F^n\}_{n=1}^{\infty}$ converges uniformly to a on $\overline{U}$. Using the Vitali Theorem we obtain our assertion. $\qquad\square$

Remark 5.6 *As a matter of fact, we will see below that this assertion is a consequence of a much more general assertion due to E. Vesentini (see Theorem 5.18).*

Combining Proposition 5.3 with the Earle–Hamilton [Earle and Hamilton (1970)] fixed point theorem, one can formulate a geometrical characterization of the attractive fixed point.

Proposition 5.4 ([Khatskevich and Shoikhet (1984); Khatskevich and Shoikhet (1994a)]). *Let $\mathcal{D}$ be a bounded domain in X and let $F \in \mathrm{Hol}(\mathcal{D})$. Then F has an attractive fixed point in $\mathcal{D}$ if and only if there exist a domain $\widetilde{\mathcal{D}} \subseteq \mathcal{D}$ and an integer $n > 0$ such that F^n maps $\widetilde{\mathcal{D}}$ strictly inside itself.*

If a is an attractive fixed point of F, then by definition it is unique and F is power convergent.

For $\mathcal{D} \subset \mathbb{C}$, the converse is also true:
If $F \in \mathrm{Hol}(\mathcal{D})$ is not the identity and has an interior fixed point, and F is power convergent, then a is unique.

However, this is not always the case in higher dimensions. Indeed, the following example exhibits a function F, mapping the bidisk $\mathbb{B}_2$ in $\mathbb{C}^2$ into itself, which is power convergent, but its fixed point set consists of an infinite number of points (actually it is a submanifold of $\mathcal{D}$ of dimension 1).

Example 5.3   Let $\mathcal{D} = \mathbb{B}_2 = \Delta \times \Delta$ and $F = \left(z_1, \frac{1}{2}\left(z_2^2 + z_1\right)\right)$. Then F is a power convergent mapping, but all the points of the form $\left(z_1, 1 - \sqrt{1 - z_1}\right)$ are fixed points of F.

A full description of such a situation was obtained by E. Vesentini [Vesentini (1983); Vesentini (1985)].

Theorem 5.18 (Vesentini) *Let $\mathcal{D}$ be a bounded convex domain in a Banach space X, and let F belong to $\mathrm{Hol}(\mathcal{D})$. Suppose that F has a fixed point $a \in \mathcal{D}$, and denote the spectrum of the linear operator $F'(a)$ by $\sigma(F'(a))$. Then F is power convergent if and only if the following two conditions hold:*

(i) $\sigma(F'(a)) \subset \Delta \cup \{1\}$; and
(ii) 1 is a pole of the resolvent of $F'(a)$ of order at most 1.

Comments.   Condition (ii) is actually equivalent to the condition

$$\mathrm{Ker}(I - F'(a)) \oplus \mathrm{Im}(I - F'(a)) = X \tag{5.88}$$

(see, for example, [Gohberg and Markus (1960)] and [Lyubich and Zemanek (1994)]). It is also known that conditions (i) and (ii) are equivalent to $F'(a)$ being power-convergent to a projection P onto $\mathrm{Ker}(I - F'(a))$. So, if $R \in \mathrm{Hol}(\mathcal{D})$ is the limit point of $\{F^n\}$ under these conditions, then $R = a$ is constant if and only if $P = 0$.

In general, there is a neighborhood U of a such that $R^2 = R$ on U, that is, R is an idempotent of the algebraic semigroup $\mathrm{Hol}(U)$. In this case the mapping R is said to be a **local holomorphic retraction**, and its image $R(U)$ is called a **holomorphic retract** of U.

In our setting this means that $\mathrm{Fix}F \cap U = R(U)$ is a submanifold of U tangent to $\mathrm{Ker}(I - F'(a))$ (see [Cartan (1986)]). This fact describes the general structure of the fixed point set. Actually, only condition (ii) or (5.88) is sufficient for $\mathrm{Fix}F$ to be a local retract (see, for example, [Shoikhet (1986); Shoikhet (1993)], [Vigué (1986); Vigué (1991a)], [Mazet and Vigué (1991); Mazet and Vigué (1992)]). Moreover, if $\mathcal{D}$ is convex, $\mathrm{Fix}F$ is a global retract of $\mathcal{D}$, and hence a connected complex submanifold of $\mathcal{D}$.

To prove this assertion, as well as Theorem 5.18, we need the following lemma.

Lemma 5.2 *Let F be a holomorphic mapping defined on a neighborhood $\mathcal{D}$ of the origin and suppose $F(0) = 0$. Assume that all the iterates $\{F^n\}_{n=0}^{\infty}$ are well defined on $\mathcal{D}$ and bounded, i.e.,*

$$\|F^n(x)\| \leq M < \infty, \quad x \in \mathcal{D}. \tag{5.89}$$

If

$$\mathrm{Ker}(I - F'(0)) \oplus \mathrm{Im}(I - F'(0)) = X, \tag{5.90}$$

then there is a neighborhood $U \subset \mathcal{D}$ such that $\mathrm{Fix}(F) \cap U$ is a complex submanifold of U tangent to $\mathrm{Ker}(I - F'(0))$. Moreover, there exists a holomorphic mapping Ψ defined on U such that $\Psi^2 = \Psi$ and

$$\mathrm{Fix}_U(\Psi) = \mathrm{Fix}_U(F). \tag{5.91}$$

Proof. Let P be a linear projection onto $\mathrm{Ker}(I - F'(0))$ and Q its complement, *i.e.*, $P + Q = I$. Let U_1 be a neighborhood of the origin in the space PX, and consider any holomorphic mapping f defined on U_1 with values in $\mathrm{Ker}P \left(= \mathrm{Im}(I - F'(0))\right)$. We claim that if f satisfies the condition

$$\|f(u)\| = o(\|u\|), \tag{5.92}$$

then

$$PF(u + f(u)) = u \tag{5.93}$$

for all $u \in U_1$ such that $u + f(u) \in U$. Indeed, fix such $u \in U_1$ and consider the vector-functions

$$\varphi_n(\lambda) = PF^n(\lambda u + f(\lambda u)) \tag{5.94}$$

with $\lambda \in \bar{\Delta} = \{\lambda \in \mathbb{C} : |\lambda| \leq 1\}$. By our assumption the sequence $\{\varphi_n\}$ is uniformly bounded, *i.e.*,

$$\|\varphi_n(\lambda)\| \leq \|P\| \cdot M, \quad u = 0, 1, 2, \ldots, \quad \lambda \in \bar{\Delta}. \tag{5.95}$$

Since $\varphi_n : \Delta \to PX$ is holomorphic, one can represent it in the form

$$\varphi_n(\lambda) = \varphi_n'(0) \cdot \lambda + h_{m_n} \lambda^{m_n} + \ldots, \tag{5.96}$$

where $h_{m_n} \in X$ is the first nonvanishing coefficient in the power series expansion of φ_n.

Observe now that for all $n = 0, 1, 2, \ldots$ we have

$$P[F'(0)]^n = P. \tag{5.97}$$

Denoting $S = F - F'(0)$, we get by a direct computation,

$$\varphi_1(\lambda) = PF(\lambda u + f(\lambda u)) = \lambda PF'(0)u + PF'(0)f(\lambda u)$$
$$+PS(\lambda u + f(\lambda u)) = \lambda u + h_{m_1}\lambda^{m_1} + \dots . \qquad (5.98)$$

Similarly, by using an induction argument we obtain

$$\varphi_n(\lambda) = \lambda u + n h_{m_1}\lambda^{m_1} + \dots . \qquad (5.99)$$

Thus we get that $m_n = m_1$ and $h_{m_n} = n \cdot h_{m_1}, \quad n = 1, 2, \dots .$ But on the other hand, it follows from the Cauchy inequalities that

$$\|h_{m_n}\| \le \|P\| \cdot M. \qquad (5.100)$$

This inequality shows that h_{m_1} must be zero, and therefore $\varphi_1 : \bar{\Delta} \to PX$ has the form

$$\varphi_1(\lambda) = \lambda u. \qquad (5.101)$$

Setting $\lambda = 1$, we obtain

$$\varphi_1(1) = PF(u + f(u)) = u. \qquad (5.102)$$

Since u is arbitrary, our claim is proved.

Now consider the set $\mathrm{Fix}F \cap U$ defined by the equation

$$x = F(x), \quad x \in U. \qquad (5.103)$$

Using the projections P and Q, one can rewrite this equation in the form

$$\begin{cases} u = PF(u + v) \\ v = QF(u + v) \end{cases} \qquad (5.104)$$

where $u = Px$ and $v = Qx$.

Since the operator $I - QF'(0)$ is invertible on $\mathrm{Im}(I - F'(0))$, it follows by the implicit function theorem that the second equation of the latter system has a unique solution $v = f(u)$ for all u sufficiently small. It is also easy to see that $\|f(u)\| = o(\|u\|)$. By using our previous claim we have that the first equation is satisfied with the substitution $v = f(u)$, *i.e.*,

$$u \equiv PF(u + f(u)). \qquad (5.105)$$

In other words, all the solutions of the equation $x = F(x)$ can be represented in the form

$$x = u + f(u). \qquad (5.106)$$

Thus, shrinking U if necessary, we obtain that $\text{Fix}(F) \cap U$ is a complex submanifold of U.

Finally, if we define the mapping $\Psi : U \to X$ by

$$\Psi(x) = Px + f(Px), \tag{5.107}$$

we get $\Psi^2 = \Psi$ and $\text{Fix}(\Psi) \cap U = \text{Fix}(F) \cap U$. This concludes our proof.$\square$

Proof of Theorem 5.18   First we note that by Lemma 5.2 and the Vitali property of holomorphic mappings in the topology of local uniform convergence over $\mathcal{D}$, we can assume that $\mathcal{D} = U$ is a convex domain in X. Then there is a retraction $\psi : D \to \text{Fix}(F)$ which satisfies the condition

$$\psi \circ F = \psi. \tag{5.108}$$

In addition, we have shown (see also the H. Cartan Theorem [Cartan (1986)]) that in a neighborhood U of the fixed point a of F we can find a local chart $g : U \to V$ such that $g(a) = 0$ and such that

$$g \circ \psi \circ g^{-1} = P \tag{5.109}$$

is a linear projection.

Now consider the mapping

$$G = g \circ F \circ g^{-1}, \tag{5.110}$$

defined on some neighborhood W of zero, together with its iterates $G^n = g \circ F^n \circ g^{-1}$. (Indeed, by the boundedness of $\{F^n\}$ this sequence is uniformly Lipshitzian in some neighborhood of a. Hence, since $F^n(a) = a$, we can find a neighborhood W such that $F^n(g^{-1}(W)) \subseteq U$.) We now have

$$PG = g \circ \psi \circ g^{-1} g \circ F \circ g^{-1} = g \circ \psi \circ g^{-1} = P. \tag{5.111}$$

In addition, $G'(0) = g'(a) \circ F'(a) \circ [g'(a)]^{-1}$ and therefore $\sigma(G'(0)) = \sigma(F'(a))$, and P is a projection onto $\text{Ker}(I - G'(0))$. Thus, if $u = Px$ and $v = (I - P)x$, $x \in X$, we have $G(u,v) = (u, G_2(u,v))$, where $\sigma\left(\frac{\partial G_2}{\partial v}(0,0)\right) \subset\subset \Delta$, and $G_2(0,0) = 0$. Hence, for u small enough, the iterates $G^n(u,v) = \left(u, G_2\left(u, G_2^{(n-1)}(u,v)\right)\right)$ converge locally uniformly to the mapping $h \in \text{Hol}(g(W), W_1)$, where W_1 is a neighborhood of zero in X (see Proposition 5.3). But then it follows that the iterates $F^n = g^{-1} \circ G^n \circ g$ also converge locally uniformly to the mapping $\varphi = g^{-1} \circ h \circ g$ in W. Using the Vitali property once again, we arrive at our assertion. $\square$

However, once again, a retraction onto the fixed point set can be obtained by this simple iteration method only if condition (i) holds, *i.e.*, only if $F'(a)$ has no spectrum points on the unit circle except, possibly, 1. This brings us to another interesting aspect of fixed point theory: the study of the holomorphic retracts of a domain.

5.5 Holomorphic Retracts and the Structure of the Fixed Point Sets.

Let, as above, X be a complex Banach space, and let $\mathcal{D}$ be a domain (open connected subset) in X.

Lemma 5.2 in the previous section provides us with information on the local structure of the fixed point set of a holomorphic self-mapping of $\mathcal{D}$ under a certain condition of regularity. To elaborate on this result in more detail we give the following definition.

Definition 5.8 Let $F \in \mathrm{Hol}(\mathcal{D})$. A point $a \in \mathrm{Fix}(F)$ is said to be **quasi-regular** if the following condition holds:

$$(*)\quad \mathrm{Ker}(I - F'(a)) \oplus \mathrm{Im}(I - F'(a)) = X. \qquad (5.112)$$

If, in addition, $\mathrm{Ker}(I - F'(a)) = \{0\}$, *i.e.*, the linear operator $I - F'(a)$ is invertible, then we say that a is a **regular fixed point** of F.

By the implicit function theorem, it is clear that a regular fixed point is an isolated point of the set $\mathrm{Fix}(F)$, and that in the case of a finite–dimensional X each fixed point is quasi-regular (or, in particular, regular).

Thus, in these terms Lemma 5.2 can be reformulated as follows:

Theorem 5.19 *Let $\mathcal{D}$ be a domain in a complex Banach space X, and let $F \in \mathrm{Hol}(\mathcal{D})$ have a quasi-regular fixed point $a \in \mathrm{Fix}(F) \subset \mathcal{D}$. If $F(\mathcal{D})$ is bounded, then there is a neighborhood $U \subset \mathcal{D}$ with $a \in U$ such that $\mathrm{Fix}(F) \cap U$ is a complex submanifold of U tangent to $\mathrm{Ker}(I - F'(a))$.*

If, in addition, $\mathcal{D}$ is a bounded convex domain, then it can be shown (see Theorems 5.22 and 5.23 below) that $\mathrm{Fix}(F)$ is a connected subset of $\mathcal{D}$.

This result for finite dimensional Banach spaces was obtained by J.-P. Vigué [Vigué (1986)] and for the general case by P. Mazet and J.-P. Vigué [Mazet and Vigué (1991)] (see also [Shoikhet (1986)]) by using a retraction method.

Earlier, W. Rudin [Rudin (1978)] (see also [Rudin (1980)]) established a surprising result regarding the global structure of holomorphic self-mappings of the Hilbert ball. See also [Goebel and Reich (1984), p. 121].

Theorem 5.20 *Let $\mathbb{B}$ be the open unit ball in a complex Hilbert space H, and let $F \in \mathrm{Hol}(\mathbb{B})$. Then $\mathrm{Fix}(F)$ is an affine submanifold of $\mathbb{B}$.*

Proof. First we note that by direct calculation one can easily verify that each automorphism of the unit ball translates an affine subset of $\mathbb{B}$ onto another affine subset of $\mathbb{B}$. Therefore it is sufficient to prove the theorem under the assumption that $\mathrm{Fix}(F)$ contains the origin. In other words, we assume that $F(0) = 0$. We show that in this case $\mathrm{Fix}(F) = \mathrm{Fix}(A) \cap \mathbb{B}$, where A is the linear operator defined by the Fréchet derivative of F at the origin, *i.e.*, $A = F'(0)$.

Indeed, it follows by the Schwatz Lemma that

$$\|F(x)\| \le \|x\|, \quad x \in \mathbb{B}, \tag{5.113}$$

and

$$\|A\| \le 1. \tag{5.114}$$

Further, one can write each $x \in \mathbb{B}$ in the form $x = ru$, where $u \in \partial\mathbb{B}$ and $0 \le r < 1$.

Let Δ be the open unit disk in the complex plane $\mathbb{C}$ and consider the holomorphic function g on Δ defined by

$$g(\lambda) = \langle F(\lambda u), u \rangle, \tag{5.115}$$

where $\langle \cdot, \cdot \rangle$ denotes the inner product in H. Clearly, $g(0) = 0$ and $g'(0) = \langle Au, u \rangle$.

If now $x\ (= ru) \in \mathrm{Fix}(F)$, then $g(r) = r$ and, by the one-dimensional Schwarz Lemma, $g(\lambda) = \lambda$ for all $\lambda \in \Delta$. Consequently, $g'(0) = \langle Au, u \rangle = 1$ which implies $Au = u$. Hence $Ax = x$.

Conversely, assume that $x\ (= ru) \in \mathrm{Fix}(A)$. Then again $g'(0) = \langle Au, u \rangle = 1$ and, by the Schwarz Lemma , $g(\lambda) = \lambda \in \Delta$. In particular, $g(r)\ (= \langle F(ru), u \rangle) = r$. On the other hand, we have that $\|F(ru)\| \le \|r\|$. Setting $y = r^{-1}F(ru)$, we have $\|y\| \le 1$ and $\langle y, u \rangle = 1$. Therefore $y = u$.

This means that $F(ru) = ru$, or, which is the same, $F(x) = x$. The theorem is proved. $\qquad\square$

The simple example from the previous section (Example 5.3) shows that this result does not hold even in $\mathbb{C}^2$ if we replace the Euclidian ball by the polydisk (*i.e.*, the unit ball of $\mathbb{C}^2$ in the Chebyshev norm). Nevertheless, as we have already mentioned, the global structure of the fixed point set can be described by using the retraction method and H. Cartan's theorem.

Definition 5.9 A mapping $\varphi \in \mathrm{Hol}(\mathcal{D})$ is called a **holomorphic retraction** of $\mathcal{D}$ if it is an idedmpotent of the semigroup $\mathrm{Hol}(\mathcal{D})$, *i.e.*, $\varphi^2 = \varphi$. In other words, if we let $V = \varphi(\mathcal{D})$, then $\varphi|_V = I|_V$. In this case the set $V \subset \mathcal{D}$ is called a **holomorphic retract** of $\mathcal{D}$.

Note that once the existence of a holomorphic retraction onto $\mathrm{Fix}(F)$ is established, it follows that $\mathcal{F} = \mathrm{Fix}(F)$ is a complex analytic submanifold of D [Cartan (1986)] (see also a result of Rossi in [Fischer (1976)]).

Indeed, if $\varphi \in \mathrm{Hol}(\mathcal{D})$ is a holomorphic retraction of $\mathcal{D}$, *i.e.*, $\varphi^2 = \varphi$, then $V = \varphi(\mathcal{D}) = \mathrm{Fix}(\varphi)$. If now $a \in V$, then it follows by the chain rule that $P = \varphi'(a)$ is a linear projection in X ($P^2 = P$). Thus we have that $\varphi'(a)$ satisfies condition (5.112) of Definition 5.8. That is, $a \in V$ is a quasi-regular fixed point of φ. Then, by Theorem 5.19, V is a local submanifold of $\mathcal{D}$ tangent to $\mathrm{Ker}(I-P) = \mathrm{Im}P$. Furthermore, V is a connected subset of $\mathcal{D}$. Indeed, taking two different points $a \in V$ and $b \in V$ and any continuous curve $\Gamma \subset \mathcal{D}$ ending at these points, we get that $\gamma = \varphi(\Gamma) \subset V$ is also continuous with $\gamma(a) = a$ and $\gamma(b) = b$.

Thus we have proved the following assertion:

Proposition 5.5 ([Cartan (1986)]) *A holomorphic retract of a domain $\mathcal{D}$ in a complex Banach space is a connected complex analytic submanifold of $\mathcal{D}$.*

In the context of the study of $\mathrm{Fix}(F) = \{x \in \mathcal{D} : x = F(x)\}$ one can ask several questions. The first one is: *if $\mathcal{F} = \mathrm{Fix}(F) \neq \emptyset$, when is it a holomorphic retract of $\mathcal{D}$?* This question in its turn leads to the second one: *if $\mathcal{F} = \mathrm{Fix}(F)$ is a holomorphic retract of $\mathcal{D}$, how can one determine a retraction of $\mathcal{D}$ onto $\mathcal{F}$?*

The second question mentioned above is also of great interest. This is because the construction of a retraction onto $\mathcal{F} = \mathrm{Fix}(F)$ is equivalent to finding a method (explicit, implicit or approximate) for solving the equation $x = F(x)$. A holomorphic retraction may not exist in general even in the one-dimensional case.

Example 5.4 ([Mazet and Vigué (1991)]) The mapping F defined by $F(z) = \frac{1}{z}$ takes the annulus $\mathcal{D} = \{z \in \mathbb{C} : \frac{1}{2} < |z| < 2\}$ into itself and has two different fixed points $z = 1$ and $z = -1$ in $\mathcal{D}$.

Even if the retraction onto $\mathrm{Fix}(F)$ exists, it need not be the limit of the iterates of the operator, even in the case of a linear operator A such that $\{0\} \neq \mathrm{Ker}(I - A) \neq \mathrm{Ker}(I - A)^2$.

We will now mention several results concerning both these questions.

Recall that a net $\{F_j\}_{j \in \mathcal{A}} \subset \mathrm{Hol}(\mathcal{D}, X)$ is said to converge to a mapping $F \in \mathrm{Hol}(\mathcal{D}, X)$ in the topology of locally uniform convergence over $\mathcal{D}$ (or, briefly, T-converge) if for every ball $\mathbb{B} \subset\subset \mathcal{D}$,

$$\lim_{j \in \mathcal{A}} \sup_{x \in \mathbb{B}} \|F_j(x) - F(x)\| = 0. \tag{5.116}$$

In this case we also write

$$F = T - \lim_{j \in \mathcal{A}} F_j. \tag{5.117}$$

It is obvious that if $F \in \mathrm{Hol}(\mathcal{D})$ is power convergent, then $\mathcal{F} = \mathrm{Fix}(F) \neq \emptyset$ and $\varphi = T - \lim_{n \to \infty} F^n$ is a retraction onto $\mathcal{F}$. So, by using a result due to J. J. Koliha [Koliha (1974)] Vesentini's theorem (Theorem 5.18) can be reformulated as follows.

Theorem 5.21 *Let $\mathcal{D}$ be a bounded convex domain in X, and let $F \in \mathrm{Hol}(\mathcal{D})$ with $\mathcal{F} = \mathrm{Fix}(F) \neq \emptyset$. If $a \in \mathrm{Fix}(F)$ is a quasi-regular fixed point and $A = F'(a)$, then F is power convergent to a retraction $\varphi : \mathcal{D} \to \mathcal{F} = \mathrm{Fix}(F)$ if and only if A is power convergent in the norm topology of $L(X)$ to a projection $P : X \to \mathrm{Ker}(I - A)$.*

Moreover, we will see below that if the domain $\mathcal{D}$ is bounded, then the set $\mathcal{F}$ consists of only quasi-regular (or regular) points.

The converse, of course, is not true. To see this, consider, for example, a rotation $F(x) = e^{i\theta}x$, $0 < \theta < 2\pi$, of the open unit disk Δ in $\mathbb{C}$. This mapping has a unique and regular fixed point (the origin) in Δ, but F is not power convergent.

Remark 5.7 *Let $\sigma(A)$ denote the spectrum of the linear operator $A : X \to X$, and let Δ be the open unit disk in $\mathbb{C}$. As above, assume that $\mathcal{F} = \mathrm{Fix}(F) \neq \emptyset$ for some $F \in \mathrm{Hol}(\mathcal{D})$ and that $a \in \mathcal{D}$. Setting $A = F'(a)$ and using the Cauchy integral formula, the chain rule and the boundedness of $\mathcal{D}$, we see that the powers of A are uniformly bounded, i.e.,*

$$\|A^n\| \leq M < \infty. \tag{5.118}$$

This implies that $\sigma(A) \subset \bar{\Delta}$.

It is clear that if in the setting of the above theorem $\sigma(A) \subset \Delta$, then $a \in \mathcal{F}$ is a regular point. In addition, it was shown that such a point is the unique fixed point of F in $\mathcal{D}$. As a matter of fact, as we will see in the sequel, for a bounded convex domain, the regularity of a point $a \in \mathrm{Fix}(F)$ is already sufficient for its uniqueness, i.e., if a is regular, then $\mathrm{Fix}(F) = \{a\}$.

Remark 5.8   *Vesentini's theorem is also true for bounded domains which are not necessarily convex, but satisfy the following maximum modulus principle:*

For each $f \in \mathrm{Hol}(\mathcal{D}, \overline{\mathcal{D}})$ such that $f(\mathcal{D}) \cap \partial\mathcal{D} \neq \emptyset$, it follows that $f(\mathcal{D}) \subset \partial\mathcal{D}$.

However, we will mainly consider convex domains in X because this is all that is needed in order to describe $\mathrm{Fix}(F)$ locally in each bounded $\mathcal{D}$. Indeed, it has been shown by P. Mazet [Mazet (1992)] that if $\mathcal{D}$ is bounded, then for each $F \in \mathrm{Hol}(\mathcal{D})$ and each $a \in \mathrm{Fix}(F)$, there is a convex neighborhood $U \subset \mathcal{D}$ such that $a \in U$ and $F(U) \subset U$, i.e., $F \in \mathrm{Hol}(U)$.

In addition, we need the convexity to consider different types of mean ergodic procedures.

As a matter of fact, if the conditions of Theorem 5.20 hold for at least one point of $\mathcal{F}$, then they hold for all the points of $\mathcal{F}$. Moreover, if $1 \in \sigma(A)$, then $\mathcal{F}$ contains infinitely many points because it is a retract of $\mathcal{D}$, hence a connected submanifold of $\mathcal{D}$ tangent to $\mathrm{Ker}(I - A)$. We will see below that the latter fact is true whenever condition (5.112) holds, even if the spectrum $\sigma(A)$ contains other points on the boundary $\partial\Delta$ of the open unit disk Δ different from 1. But in this case, of course, by Vesentini's theorem, F is not power convergent and therefore the question of approximating its fixed points is still open.

Nevertheless, if $\mathcal{D}$ is a bounded convex domain in X, and $F \in \mathrm{Hol}(\mathcal{D})$ has at least one quasi-regular fixed point in $\mathcal{D}$, then there is another mapping $\varphi \in \mathrm{Hol}(\mathcal{D})$ with $\mathrm{Fix}(F) = \mathrm{Fix}(\varphi)$ which is power convergent. Hence, in this case $\mathrm{Fix}(F)$ is a holomorphic retract of $\mathcal{D}$. For the finite dimensional case this was established by J.-P Vigué [Vigué (1986); Vigué (1991a)], and in the general case by P. Mazet and J.-P Vigué [Mazet and Vigué (1991)]. They used nonlinear analogues of mean ergodic constructions. More precisely, they considered the Cesàro averages

$$C_n = \frac{1}{n} \sum_{k=0}^{n-1} F^k \tag{5.119}$$

and proved the following results.

Theorem 5.22 *Let $\mathcal{D}$ be a bounded convex domain in $X = \mathbb{C}^n$, and let $F \in \mathrm{Hol}(\mathcal{D})$ with $\mathcal{F} = \mathrm{Fix}(F) \neq \emptyset$. Then there is a subsequence $\{C_{n_k}\}$ of $\{C_n\}$ which T-converges to a power–convergent holomorphic mapping $\varphi : \mathcal{D} \to \mathcal{D}$ with $\mathrm{Fix}(\varphi) = \mathcal{F}$.*

Theorem 5.23 *Let $\mathcal{D}$ be a bounded convex domain in an arbitrary complex Banach space X, and let $F \in \mathrm{Hol}(\mathcal{D})$ with $\mathcal{F} = \mathrm{Fix}(F) \neq \emptyset$. If $\mathcal{F}$ contains at least one quasi-regular point, then there exists an integer p such that the mapping C_p defined by (5.119) is power convergent to a holomorphic retraction onto $\mathcal{F}$.*

Remark 5.9 *Both Theorems 5.22 and 5.23 state that $\mathcal{F} = \mathrm{Fix}(F)$ is a holomorphic retract of $\mathcal{D}$. But a deficiency of Theorem 5.22 is that we do not know how to choose a convergent subsequence of $\{C_n\}$ in order to approximate fixed points.*

Theorem 5.23 (which also covers the finite–dimensional case) would improve this situation if we could determine the minimal number p for which the mapping C_p defined by (5.119) is power convergent.

We give below, inter alia, the simplest possible answer to this question, namely, $p = 2$.

A partial generalization of Theorem 5.23 to unbounded domains was obtained by Do [Do (1992)] who has proved that if $\mathcal{D}$ is a convex hyperbolic domain in X, and $F \in \mathrm{Hol}(\mathcal{D})$ is such that $F(\mathcal{D})$ is contained in some compact convex subset of X with $\mathcal{F} = \mathrm{Fix}(F) \neq \emptyset$, then this set $\mathcal{F}$ is a holomorphic retract of $\mathcal{D}$. (We recall, in passing, that a domain $\mathcal{D}$ in a complex Banach space X is said to be hyperbolic if the Kobayashi pseudometric $K_{\mathcal{D}}$ generates the relative topology of $\mathcal{D}$ in X (see Chapter 3.))

This result is only a partial generalization of Theorem 5.23 because of its compactness hypothesis. Indeed, according to a theorem of Krasnoselskii [Krasnoselskii and Zabreiko (1984)], $F'(x)$ is compact for all $x \in \mathcal{D}$ and hence each point in $\mathrm{Fix}(F)$ is quasi-regular. However, we will show below that this compactness hypothesis is unnecessary. Moreover, we present a simple method for constructing a retraction onto $\mathrm{Fix}(F)$. This will also provide a proof of Theorem 5.23 [Reich and Shoikhet (1998c)].

Theorem 5.24 *Let $\mathcal{D}$ be a hyperbolic convex domain in X, and let $F \in \mathrm{Hol}(\mathcal{D})$ be bounded on each subset which is strictly inside $\mathcal{D}$.*

(i) If $\mathcal{F} = \mathrm{Fix}(F)$ contains at least one quasi-regular point $a \in \mathcal{D}$, then for

each $\lambda \in (0,1)$, the averaged mapping

$$F_\lambda = \lambda I + (I - \lambda)F \tag{5.120}$$

is power convergent.

(ii) If for at least one $\lambda \in (0,1)$ the mapping F_λ defined by (5.120) is power convergent, then $\mathrm{Fix}(F)$ consists of quasi-regular points.

First we observe that it is sufficient to establish this theorem for bounded domains. Indeed, if a is a fixed point of F (hence of F_λ), the hyperbolicity of $\mathcal{D}$ implies the existence of a ball $\mathcal{B}(a, R)$ (with respect to the $K_\mathcal{D}$ pseudometric) centered at a which is bounded and strictly inside $\mathcal{D}$. Since F_λ is nonexpansive with respect to $K_\mathcal{D}$, this ball is invariant under F_λ. Suppose now that (i) is established for that ball and let $\mathcal{B}(c, r)$ be any ball (with respect to the norm) which is strictly inside $\mathcal{D}$. Then the family $\{F_\lambda^n\}$ is bounded on the convex hull of $\mathcal{B}(a, R) \cup \mathcal{B}(c, r)$ (which is also strictly inside $\mathcal{D}$), and therefore F_λ is power convergent on $\mathcal{B}(c, r)$ by the Vitali property.

Proof of Theorem 5.24   (i) Let $a \in \mathrm{Fix}(F)$ be a quasi-regular point, and let $A_\lambda = \lambda I + (1 - \lambda)F'(a)$ for $\lambda \in [0, 1)$. It is clear that $a \in \mathrm{Fix}(F_\lambda)$, where F_λ is defined by (5.120) and

$$A_\lambda = (F_\lambda)'(a), \quad \lambda \in [0, 1), \tag{5.121}$$

with $A_0 = A = F'(a)$.

We intend to show that for each $\lambda \in (0, 1)$ the mapping F_λ satisfies the conditions of Vesentini's Theorem. First we note that condition $(*)$ (see formula (5.112) in Definition 5.8) is obvious because

$$I - A_\lambda = (1 - \lambda)A. \tag{5.122}$$

Now we must show that the set $\sigma(A_\lambda) \backslash \{1\}$ lies inside the open unit disk Δ for each $\lambda \in (0, 1)$. Once again, the Cauchy integral formula shows that the operator A_λ is power bounded. Suppose now that there exists $\zeta \in \partial\Delta \cap \sigma(A_\lambda)$ and $\zeta \neq 1$. Then we have for such ζ

$$\zeta I - A_\lambda = (1 - \lambda)(tI - A), \tag{5.123}$$

where

$$t = \frac{\zeta - \lambda}{1 - \lambda} \in \sigma(A). \tag{5.124}$$

It is clear that $|t| \geq 1$. But on the other hand, $|t| \leq 1$, since $\sigma(A) \subseteq \bar{\Delta}$ (see (5.118)). So, $|t| = 1$ and we have, by (5.124),

$$\zeta = \lambda + (1 - \lambda)t \in \partial\Delta. \tag{5.125}$$

But this is possible only if $\zeta = t = 1$. This contradiction proves our assertion.

(ii) Now, if for some $\lambda \in (0,1)$ the mapping F_λ is power convergent, then it follows from the Cauchy inequalities that for each $a \in \text{Fix}(F)$ the operator $A_\lambda = (F_\lambda)'(a)$ is power convergent too. This fact in turn implies condition (*) for A_λ. Hence, it follows from (5.122) that a is quasi-regular. The proof is complete. $\qquad\qquad\square$

So, setting $\lambda = \frac{1}{2}$ in Theorem 5.24, we have that the Cesàro average (see (5.119))

$$C_2 = \frac{1}{2}\,(I + F) \tag{5.126}$$

is power convergent, *i.e.*, it is sufficient to take $p = 2$ in Theorem 5.23. As a matter of fact, it turns out that the Cesàro averages defined by (5.126) are power convergent for all $p = 2, 3 \ldots$. Moreover, we are able to show that all proper convex combinations of the iterates of F are also power convergent. We will describe below a general scheme for constructing a retraction onto $\text{Fix}(F)$.

A simple but very important consequence of Theorem 5.24 is the so-called Cartan's uniqueness theorem, which is a generalization of the second part of the classical Schwarz Lemma.

Corollary 5.3 *Let $\mathcal{D}$ be a hyperbolic convex domain in X, and let $F \in \text{Hol}(\mathcal{D})$ be a bounded on each subset which is strictly inside $\mathcal{D}$. If $a \in \mathcal{D}$ is a fixed point of F such that $F'(a) = I$, the identity operator on X, then $F(x) = x$ for all $x \in \mathcal{D}$.*

Now we return to a general approach to find a retraction onto $\text{Fix}(F)$. It follows by Theorem 5.21 that to construct a retraction onto $\text{Fix}(F)$ by using a power convergent mapping, we must find a mapping $\Phi \in \text{Hol}(\mathcal{D})$ such that $\mathcal{F} = \text{Fix}(F) = \text{Fix}(\Phi)$ and for some point $a \in \mathcal{F}$ the operator $\Phi'(a)$ is power convergent.

Lemma 5.3 *Let A be a bounded linear operator in X, and let $\lambda_0 \in \sigma(A)$ be such that*

$$\text{Ker}(\lambda_0 I - A) \oplus \text{Im}(\lambda_0 I - A) = X. \tag{5.127}$$

Suppose that there exist a domain $\Omega \subset \mathbb{C}$ and a holomorphic function f defined in a neighborhood $\widetilde{\Omega}$ of Ω with the following properties:

(a) $\overline{\Omega} \supset \sigma(A)$ and $\lambda_0 \in \partial\Omega$;
(b) $f(\Omega) \subseteq \overline{\Delta}$;
(c) λ_0 is a simple root of the equation

$$f(\lambda_0) = 1; \tag{5.128}$$

(d) $|f(\lambda)| \neq 1$ for all $\lambda \in \partial\Omega, \ \lambda \neq \lambda_0$.

Then the linear operator $B = f(A) : X \to X$ defined by the formula

$$B = \frac{1}{2\pi i} \int_\Gamma f(\lambda)(\lambda I - A)^{-1} d\lambda, \tag{5.129}$$

where $\Gamma \subset \widetilde{\Omega}$ is a closed path around $\sigma(A)$, is power convergent.

Proof.   It follows from Dunford's Spectral Mapping Theorem that

$$\sigma(B) = f(\sigma(A)). \tag{5.130}$$

Hence, conditions (a), (b) and (d) imply that

$$\sigma(B)\backslash\{1\} \subset \Delta. \tag{5.131}$$

Consider now the function

$$g(\lambda) = [1 - f(\lambda)](\lambda_0 - \lambda)^{-1}. \tag{5.132}$$

Condition (c) implies that $g(\lambda)$ is holomorphic in a neighborhood of the point $\lambda_0 \in \partial\Omega$, and $f(\lambda_0) \neq 0$. In addition, $g(\lambda) \neq 0$ for all $\lambda \in \sigma(A)$, by condition (d), and therefore the operator $C = g(A)$, defined by the formula

$$C = \frac{1}{2\pi i} \int_\Gamma g(\lambda)(\lambda I - A)^{-1} d\lambda, \tag{5.133}$$

is invertible in X. Furthermore, it follows from the multiplicative property of the calculus of $L(X)$-valued functions defined by (5.129) and (5.133) (see [Rudin (1973)]) and by formula (5.132) that

$$I - B = C(\lambda_0 I - A) = (\lambda_0 I - A)C. \tag{5.134}$$

This implies $\mathrm{Ker}(I - B) = \mathrm{Ker}(\lambda_0 I - A)$ and $\mathrm{Im}(I - B) = \mathrm{Im}(\lambda_0 I - A)$. Now (5.127) and (5.128) imply that B is power convergent. The lemma is proved.

$$\square$$

Theorem 5.25 *Let $\mathcal{D}$ be a hyperbolic convex domain in X, and let $F \in \text{Hol}(\mathcal{D})$ with $\mathcal{F} = \text{Fix}(F) \neq \emptyset$. If $\mathcal{F}$ contains a quasi-regular point $a \in \mathcal{D}$, then each mapping Φ of the form $\Phi = \sum_{k=0}^{p} a_k F^k$, where $\sum_{k=0}^{p} a_k = 1$ and $0 \leq a_k \neq 1$ for all k, is power convergent.*

Proof. As in the proof of Theorem 5.24, we may assume that $\mathcal{D}$ is bounded. Let $\Phi \in \text{Hol}(\mathcal{D})$ be defined by

$$\Phi = \sum_{k=0}^{p} a_k F^k, \tag{5.135}$$

where $\sum_{k=0}^{p} a_k = 1$ and $0 \leq a_k \neq 1$ for all $k = 0, 1, \ldots p$.

Consider the holomorphic function (polynomial) $f : \bar{\Delta} \to \bar{\Delta}$ defined by the following formula:

$$f(\lambda) = \sum_{k=0}^{p} a_k \lambda^k, \quad \lambda \in \bar{\Delta}. \tag{5.136}$$

It is clear that f satisfies conditions (a)–(d) of Lemma 5.3, with $\Omega = \Delta$, and therefore the operator B defined by (5.129) is power convergent to a projection onto $\text{Ker}(I - A)$, where $A = F'(a)$. But it follows from the chain rule that $\Phi'(a) = B = f(A)$. Hence, Φ is power convergent onto $\text{Fix}(\Phi)$ (see Remark 5.10). Furthermore, (5.135) implies that $\text{Fix}(F) \subseteq \text{Fix}(\Phi)$. At the same time, these sets are connected submanifolds in $\mathcal{D}$ tangent to the same subspace $\text{Ker}(I - A)$. Therefore they coincide in $\mathcal{D}$. This completes the proof. $\square$

Remark 5.10 *If $\mathcal{D} \ni 0$, then it can be shown that each infinite convex combination*

$$\Phi = \sum_{k=0}^{\infty} a_k F^k, \tag{5.137}$$

where $\sum_{k=0}^{\infty} s_k = 1$ and $0 \leq a_k \neq 1$ for all k, belongs to $\text{Hol}(\mathcal{D})$ and is also power convergent to a retraction onto $\text{Fix}(F)$.

We now describe an implicit method for approximating fixed points of holomorphic mappings which has been used many times in the theory of nonexpansive mappings (see, for example, [Goebel and Reich (1984)]).

Let $\mathcal{D}$ be a bounded convex domain in X, and let F belong to $\text{Hol}(\mathcal{D})$. For $t \in [0, 1)$ and a fixed $y \in \mathcal{D}$, consider the mapping $\Phi(x) = tF(x) + (1-t)y$, $x \in \mathcal{D}$. Since Φ maps $\mathcal{D}$ strictly inside itself, the Earle–Hamilton theorem [Earle and Hamilton (1970)] implies that there exists a unique fixed point $z = z_t(y) \in \mathcal{D}$ of the mapping Φ. Moreover,

$$z = T - \lim_{n \to \infty} \Phi^n. \tag{5.138}$$

The X-valued function $z_t(y)$, $0 \leq t < 1$ is called an approximating curve. It was shown in [Kuczumow and Stachura (1990)] that if $X = \mathbb{C}^n$ and $\text{Fix}(F) \neq \emptyset$, then there is a sequence $t_n \in (0, 1)$, $t_n \to 1$ such that for each $y \in \mathcal{D}$, the sequence $z_{t_n}(y)$ converges to a fixed point of F in $\mathcal{D}$.

At the same time, changing our point of view, for each $t \in [0, 1)$, $z_t = z_t(y)$ holomorphically depends on $y \in \mathcal{D}$, by (5.138). We denote this mapping by $\mathcal{T}_t$. In other words, $\mathcal{T}_t$ is the unique solution of the nonlinear operator equation

$$\mathcal{T}_t = tF \circ \mathcal{T}_t + (1-t)I. \tag{5.139}$$

The mapping $\mathcal{T}_t$ belongs to $\text{Hol}(\mathcal{D})$ and it is easy to check that

$$\text{Fix}(\mathcal{T}_t) = \text{Fix}(F) \quad t \in (0, 1). \tag{5.140}$$

For the Hilbert ball $\mathbb{B}$ (as we have already mentioned) the situation is as follows:

The net $\{\mathcal{T}_t\}$ is convergent in $\mathbb{B}$ as $t \to \infty$ to a mapping $\Psi : \mathbb{B} \to \overline{\mathbb{B}}$. If $\text{Fix}(F) \neq \emptyset$, then Ψ is a holomorphic retraction onto $\text{Fix}(F)$ (otherwise, Φ is a constant $a \in \partial\mathbb{B}$ which is the sink point of F). Also, since for each $t > 0$ the mapping $\{\mathcal{T}_t\}$ is a firmly holomorphic self-mapping of $\mathbb{B}$, the iterates $\{\mathcal{T}_t^n\}_{n=1}^{\infty}$ converge weakly to a holomorphic mapping $\psi : \mathbb{B} \to \overline{\mathbb{B}}$, which is a retraction onto $\text{Fix}(F)$ if it is not empty.

A similar situation occurs in general Banach spaces.

Theorem 5.26   *Let $\mathcal{D}$ be a bounded convex domain in a complex Banach space X, and let $F \in \text{Hol}(\mathcal{D})$ with $\mathcal{F} = \text{Fix}(F) \neq \emptyset$. If $\mathcal{F}$ contains a quasi-regular point in $\mathcal{D}$, then the mapping $\mathcal{T}_t$ defined by (5.139) is power convergent to a retraction onto $\mathcal{F}$.*

Proof. Let $a \in \mathcal{F}$, and let $B = (\mathcal{T}_t)'(a)$, $t \in (0,1)$. Then B satisfies the operator equation

$$B = tA \circ B + (1-t)I, \tag{5.141}$$

where $A = F'(a)$.

Setting $r = \frac{1-t}{1}$, we have

$$B = [I + r(I - A)]^{-1}. \tag{5.142}$$

Therefore, B can be defined by formula (5.129) with

$$f(\lambda) = \frac{1}{1 + r(1 - \lambda)}, \quad \lambda \in \bar{\Delta}. \tag{5.143}$$

Since $r > 0$, this function maps $\overline{\Omega} = \bar{\Delta}$ into itself and satisfies all the condition of Lemma 5.3. Thus, by this lemma, B is power convergent to a projection onto $\mathrm{Ker}(I - A)$ and so $\mathcal{T}_t$ is power convergent to a retraction onto $\mathrm{Fix}(F)$. The theorem is proved. $\qquad\square$

Remark 5.11 *Note that it follows from the Neumann series representation of the operator B in (5.129) that the mapping Φ_t defined by (5.137),*

$$\Phi_t = \sum_{k=0}^{\infty} a_k F^k, \tag{5.144}$$

with $a_k = t(1-t)^k$, $t \in (0,1)$, has $B = [I + r(I-A)]^{-1}$ as its Fréchet derivative at the point a. However, generally speaking, Φ_t and $\mathcal{T}_t$ are different in the nonlinear case.

Finally, we observe that P. Mazet and J.-P. Vigué ([Mazet and Vigué (1991)]) have shown that under certain conditions on the Banach space X (for example, reflexivity), condition $(*)$ of Definition 5.8 is not necessary. In particular, they proved the following result:

Theorem 5.27 ([Mazet and Vigué (1991)]) *Let $\mathcal{D}$ be a bounded convex domain in a reflexive complex Banach space X, and let $F \in \mathrm{Hol}(\mathcal{D})$ with $\mathcal{F} = \mathrm{Fix}(F) \neq \emptyset$. Then $\mathcal{F}$ is a holomorphic retract of $\mathcal{D}$.*

The scheme of the proof is as follows. Since $\mathcal{D}$ is a bounded convex domain in X, the Cesàro averages

$$C_n = \frac{1}{n} \sum_{k=0}^{n-1} F^k \tag{5.145}$$

are well defined and bounded on $\mathcal{D}$. Then the Montel property implies that $\{C_n\}$ contains a subsequence $\{C_{n_k}\}$ which weakly converges to a holomorphic mapping $\Psi \in \mathrm{Hol}(\mathcal{D})$. It is clear that if $a \in \mathrm{Fix}(F) \subset \mathcal{D}$, then $C_n(a) = a$ and $\Psi(a) = a$. At the same time, since X is reflexive, it follows by the mean ergodic theorem that $(C_n)'(a) = \frac{1}{n} \sum_{k=0}^{n-1} (F^k)'(a)$ strongly converges to $P = (\Psi)'(a)$, which is a projection onto $\mathrm{Ker}(I - F'(a))$. Hence the point $a \in \mathrm{Fix}(F) \subset \mathrm{Fix}(\Psi)$ is a quasi-regular fixed point of Ψ. Moreover, by Theorem 5.22 the mapping Ψ is power convergent to a retraction $\Psi \in \mathrm{Hol}(\mathcal{D})$ onto $\mathrm{Fix}(\Psi)$. Without loss of generality, assume now that $\mathcal{D}$ is a neighborhood of the origin and $a = 0$. The crucial point of the proof is the following technical auxiliary lemma:

Lemma 5.4 ([Mazet and Vigué (1991)]) *Let $S_n(h)$ denote the Taylor polynomial in the power series expansion of a holomorphic mapping h at a neighborhood of the origin. Then for the mappings F and Ψ defined above, the following equality holds:*

$$S_n(F \circ \Psi^n) = S_n(\Psi^n) = S_n(\Psi^{n+1}). \qquad (5.146)$$

Now this lemma implies that $F \circ \Phi = \Phi$, hence $\mathrm{Fix}(\Phi) \subset \mathrm{Fix}(F)$. On the other hand, it follows by the constructions of the mappings Ψ and Φ that $\mathrm{Fix}(F) \subset \mathrm{Fix}(\Psi) = \mathrm{Fix}(\Phi)$. Thus we see that $\mathrm{Fix}(F) = \mathrm{Fix}(\Psi) = \mathrm{Fix}(\Phi)$ is a retract of $\mathcal{D}$, hence a connected submanifold of $\mathcal{D}$ tangent to $\mathrm{Ker}(I - F'(a)) = \mathrm{Ker}(I - P)$.

Chapter 6

Generation Theory for One–Parameter Semigroups

6.1 Continuous and Discrete One-Parameter Semigroups on Metric Spaces

6.1.1 *Discrete and continuous flows on a domain*

Let D be a topological space. A family $M(D)$ of self-mappings of D forms a semigroup with respect to the composition operation on D if

$$F \circ G \in M(D) \tag{6.1}$$

whenever F and G belong to $M(D)$.

If, in particular, D is a metric space with a metric ρ on D, then the set $N_\rho(D)$ of ρ-nonexpansive mappings on D forms a semigroup with respect to the composition operation.

If D is a complex domain in a Banach space X, then the set $\mathrm{Hol}(D)$ of all holomorphic self-mappings of D also forms a semigroup with respect to the composition operation. The set $\mathrm{Aut}(D)$ of all the automorphisms of D forms a group with respect to the composition operation because of the inclusion

$$F^{-1} \in \mathrm{Hol}(D) \tag{6.2}$$

which holds whenever $F \in \mathrm{Aut}(D)$. Hence $F^{-1} \in \mathrm{Aut}(D)$.

Let now A be a topological (additive) semigroup with zero, and let there exist a natural ordering of A, *i.e.*, $r \geq t$ if and only if there is $s \in A$ such that $r = t + s$. We will mostly consider two cases:

(a) $A = \mathbb{N} \cup \{0\} = \{0, 1, 2, \dots\}$ (respectively, $A = \mathbb{Z} = \{\cdots - 3, -2, -1, 0, 1, 2, 3 \dots\}$);
(b) $A = [0, T), \quad 0 < T \leq \infty$ (respectively, $A = (-T, T), \ 0 < T \leq \infty$).

Definition 6.1   Let A be as above and let $M(D)$ be the semigroup of self-mappings with respect to the composition operation of a topological space D. A family $S = \{F_t\} \subset M(D)$ is called a **one-parameter semigroup** on D with respect to A if the following conditions hold:

(i) $F_{t+s} = F_t \cdot F_s$,

    whenever s, t and $s + t$ belong to A;

(ii) $F_0 = I$, the identity operator on D.

In other words, a one-parameter semigroup $S \subset M(D)$ is defined by a mapping $A \mapsto M(D)$ which preserves the additive structure of A with respect to the composition operation on $M(D)$.

If $A = \mathbb{N} \cup \{0\} = \{0, 1, 2, \dots\}$, then $S = \{F_0, F_1, F_2, \dots, F_n, \dots\}$, $F_n \in M(D)$, is called a one-parameter discrete semigroup. Actually, such a semigroup consists of the iterates of a self-mapping $F = F_1$, because of conditions (i) and (ii), *i.e.*, $F_0 = I$, $F_n = F^n$, $n = 1, 2, \dots$. If, in particular, (D, ρ) is a metric space and $M(D) = N_\rho(D)$, then for each $F \in N_\rho(D)$ the set $S = \{F^n\} \cup \{I\}$ forms a one-parameter semigroup of ρ-nonexpansive mappings on D.

If for each $n \geq 0$, F^n is invertible and $F^{-n} \in N_\rho(D)$, then S can be extended to a one-parameter group $G_T = \{F^n\} \cup \{F^{-n}\}$, $n \geq 0$. In this case condition (i) holds for all t and s in $A = \mathbb{Z} = \{\dots -2, -1, 0, 1, 2 \dots\}$. This set G is called a one-parameter discrete group of isometries on D because of the relation

$$\rho(F(x), F(y)) = \rho(x, y), \quad x, y \in D. \tag{6.3}$$

We have already mentioned that for $F \in \text{Hol}(D)$ the family of iterates $S = \{F^n\}_{n=0}^\infty$, $F_0 = I$, can be considered a one-parameter discrete-time semigroup. In this case the vector field $f = I - F$ is referred to as the generator of S (see [Kato (1966)]).

It is clear that S can be extended to a one-parameter discrete group on D if and only if $F \in \text{Aut}(D)$.

Definition 6.2   Let D be a topological space. A family $S = \{F_t : t \in (0, T)\}$, $T > 0$, of self-mappings F_t of D is called a **(one-parameter) continuous semigroup** if

(i)

$$F_{s+t} = F_t \circ F_s, \quad 0 < s + t < T; \tag{6.4}$$

(ii) for each $x \in D$,

$$\lim_{t \to 0^+} F_t(x) = x, \tag{6.5}$$

where the limit is taken with respect to the topology of D.

6.1.2 *Examples*

Example 6.1   Let X be a Banach space and let $L(X)$ be the space of all bounded linear operators on X with the norm

$$\|A\| = \sup_{\|x\|=1} \|Ax\|, \quad A \in L(X), \quad x \in X. \tag{6.6}$$

For a linear operator $A \in L(X)$ we can define the family $\{B_t\}_{t \in \mathbb{R}}$ of bounded linear operators by using the following exponential expansion:

$$B_t = e^{-tA} := \sum_{k=0}^{\infty} \frac{(-1)^k}{k!} t^k A^k. \tag{6.7}$$

It is clear that the family $\{B_t\}$ satisfies the semigroup property (i) and (ii), *i.e.*,

$$B_t \circ B_s = e^{-tA} \circ e^{-sA} = e^{-(t+s)A} = B_{t+s}, \tag{6.8}$$

and $B_0 = I$, the identity operator on X. So if $D = X$ this family forms a one–parameter group of self-mappings of D because

$$e^{tA} = [e^{-tA}]^{-1} : X \to X \quad \text{exists for each} \quad t \geq 0. \tag{6.9}$$

If D is the unit ball of X and A is an accretive operator on X, then the family $\{B_t\}_{t \geq 0} = \{e^{-tA}\}_{t \geq 0}$ forms a one-parameter semigroup of so-called proper contractions of D by the inequality

$$\|B_t\| < 1, \quad t > 0. \tag{6.10}$$

Obviously, this family satisfies the limit property

$$\lim_{t \to 0^+} B_t = I \tag{6.11}$$

with respect to the operator topology on X, *i.e.*,

$$\lim_{t \to 0^+} \|B_t - I\|_{L(X)} = 0. \tag{6.12}$$

Moreover, it is easy to see that, in fact,

$$\|B_t - I\|_{L(X)} \le t\|A\|e^{-t\|A\|_{L(X)}}. \tag{6.13}$$

Now it follows by the semigroup properties that

$$\lim_{t \to 0^+} \|B_{t+s} - B_s\|_{L(X)} = 0 \tag{6.14}$$

for all real t and s.

Therefore such a semigroup is called a uniformly continuous semigroup of linear operators.

Remark 6.1   *As a matter of fact, we will see below that each uniformly continuous semigroup (group) of linear bounded operators on X can be represented by the exponential form*

$$B_t = e^{-tA} \tag{6.15}$$

for some $A \in L(X)$.

In addition, it follows from the exponential series for e^{-tA} that

$$\lim_{t \to 0} \frac{1}{t}\left(I - B_t\right) = A, \tag{6.16}$$

where the limit again is taken with respect to the norm topology of $L(X)$. In view of the last property, the operator A is usually called the infinitesimal generator of the semigroup B_t (see a more general definition below).

Example 6.2   Let X be the space of all bounded uniformly continuous functions defined on $\mathbb{R} = (-\infty, \infty)$ with the topology of uniform convergence. Define the family $\{B_t\}_{t \in \mathbb{R}}$ of linear operators on X by the formula

$$[B_t(x)](y) = x(t + y), \quad t \in \mathbb{R}, \quad y \in \mathbb{R}, \tag{6.17}$$

where $x \in X$ is a bounded uniformly continuous function on $\mathbb{R} = (-\infty, \infty)$. It is clear that $[B_0(x)](y) = x(y)$, *i.e.*, $B_0 = I$ – the identity operator on X, and

$$\left[B_{t+s}(x)\right](y) = x(t + s + y) = x(t + (s + y)) = \left[B_t(B_s(x))\right](y), \tag{6.18}$$

i.e.,

$$B_{t+s} = B_t \circ B_s, \quad t, s \in \mathbb{R}. \tag{6.19}$$

So the family $\{B_t\}$ forms a semigroup of bounded linear operators on X. Actually this semigroup is a group because of

$$[B_t]^{-1} = B_{-t} \quad \text{for all} \quad t \geq 0. \tag{6.20}$$

However, regarding the continuity, one can assert only the pointwise convergence of this semigroup to the identity, ı.e.,

$$\lim_{t \to 0^+} (I - B_t)(x) = \lim_{t \to 0^+} \left\{ \sup_{y \in \mathbb{R}} |x(t + y) - x(y)| \right\} = 0 \tag{6.21}$$

for each $x \in X$. Moreover, the limit

$$\lim_{t \to 0^+} \frac{1}{t} (I - B_t)(x) \tag{6.22}$$

does not exist for all $x \in X$. Thus such a semigroup is not uniformly continuous on X.

Other examples of linear semigroups can be found, for instance, in [Hille and Phillips (1957)], [Daletskii and Krein (1970)] and [Krein (1971)].

Example 6.3 Let $X = \mathbb{C}$ be the complex plane, and let $D = \Delta$ be the open unit disk in $\mathbb{C}$. Let $\{F_t\}_{t \in \mathbb{R}}$ be the family of holomorphic self-mappings of Δ defined by

$$F_t(z) = \frac{z + \tanh t}{z \tanh t + 1}, \quad z \in \Delta. \tag{6.23}$$

It is easy to see that this family forms a one-parameter group of automorphisms of Δ which has the locally uniform continuity property

$$\lim_{t \to 0^+} F_t(z) = z, \tag{6.24}$$

uniformly on each compact subset of Δ. Moreover, there exists the locally uniform limit

$$\lim_{t \to 0^+} \frac{1}{t} (z - F_t(z)) = 1 - z^2, \tag{6.25}$$

which defines a holomorphic mapping on Δ.

Example 6.4 Another example of a semigroup on Δ which does not consist of automorphisms may be defined as follows:

$$F_t(z) = \frac{z}{e^t - z(e^t - 1)}, \quad t \geq 0. \tag{6.26}$$

Despite the fact that each F_t is a fractional linear transformation mapping Δ into itself, its inverse is not a self-mapping on Δ. So this semigroup cannot be extended to a group of self-mappings of Δ.

At the same time we again have that

$$\lim_{t \to 0^+} F_z(z) = z, \tag{6.27}$$

uniformly on each compact subset of Δ, and

$$\lim_{t \to 0^=} \frac{1}{t}\left(z - F_t(z)\right) = z - z^2 \tag{6.28}$$

exists for each $z \in \Delta$. Note that both the semigroups in Example 6.3 and 6.4 consist of mappings which are nonexpansive with respect to the Poincaré hyperbolic metric of Δ, *i.e.*,

$$\rho\big(F_t(z),\, F_t(w)\big) \le \rho(z, w) \tag{6.29}$$

for all pairs z and w in Δ and for each $t \ge 0$. Moreover, by the Schwarz–Pick Lemma, in Example 6.4 the strict inequality

$$\rho\big(F_t(z),\, F_t(w)\big) < \rho(z, w) \tag{6.30}$$

holds for all z and w in Δ, $z \ne \omega$, and $t > 0$, while in Example 6.3 we have equality

$$\rho\big(F_t(z),\, F_t(w)\big) = \rho(z, w) \quad z, w \in \Delta, \ \ t \ge 0. \tag{6.31}$$

In this case one says that the semigroup (group) $\{F_t\}_{t \ge 0}$ consists of ρ-isometries of the metric space (D, ρ).

As a matter of fact, all one-parameter semigroups (groups) of automorphisms of Δ can be described as follows.

Example 6.5   (see [Berkson *et al.* (1974)]).

(i) Each one-parameter group which consists of elliptic automorphisms of Δ is of the form

$$F_t(z) = \frac{(e^{i\varphi t} - |\alpha|^2)z + \alpha(1 - e^{i\varphi t})}{\alpha(e^{i\varphi t} - 1)z + 1 - |\alpha|^2 e^{i\varphi t}}, \quad t \ge 0, \tag{6.32}$$

where $\varphi \in \mathbb{R}$, $\alpha \in \Delta$ are given.

(ii) Each one-parameter group of hyperbolic automorphisms of Δ is of the form

$$F_t(z) = \frac{(\beta e^{\varphi t} - \alpha)z + \alpha\beta(1 - e^{\varphi t})}{(e^{\varphi t} - 1)z + \beta(1 - \alpha e^{\varphi t})}, \tag{6.33}$$

where $\varphi \in \mathbb{R}^+$, $\varphi \neq 0$, $\tau \in \partial\Delta$, $\alpha \in \partial\Delta$, $\beta \in \partial\Delta$, $\alpha \neq \beta$, are given numbers.

(iii) Each one-parameter semigroup (group) of parabolic automorphisms of Δ is of the form

$$F_t(z) = \frac{(1 - ict)z + ict\alpha}{-ic\bar{\alpha}tz + 1 + ict}, \quad t \geq 0, \tag{6.34}$$

where $c \in \mathbb{R}$, $c \neq 0$, $\alpha \in \partial\Delta$ are given.

Note that in (i) the point $\alpha \in \Delta$ is the unique common interior fixed point of the semigroup, *i.e.*,

$$F_t(\alpha) = \alpha, \quad t \geq 0. \tag{6.35}$$

However, this fixed point is not attractive, *i.e.*, $F_t(z)$ does not converge to α as $t \to \infty$ for all $z \in \Delta$, $z \neq \alpha$.

In cases (ii) and (iii), $\alpha \in \partial\Delta$ is a boundary locally uniformly attractive fixed point for $\{F_t\}_{t\geq 0}$, *i.e.*,

$$F_t(\alpha) = \alpha, \quad t \geq 0, \quad \text{and} \quad \lim_{t \to \infty} F_t(z) = \alpha, \tag{6.36}$$

uniformly on each subset strictly inside Δ.

Observe, however, that in case (ii) there is another (boundary) fixed point $\beta \in \partial\Delta$ of the semigroup $\{F_t\}_{t\geq 0}$, while in (iii), α is the unique fixed point in the whole plane.

These phenomena will be considered later in the context of the asymptotic behavior of more general semigroups.

It turns out that many important classes of linear semigroups are closely connected to corresponding classes of nonlinear semigroups. Moreover, if these connections are one-to-one, one may use different techniques to study both types.

An example of such a semigroup is the linear semigroup of isometries with respect to the so-called indefinite metric on Krein spaces. For simplicity we consider this construction just for the Cartesian product of complex planes although this scheme works for any product of Hilbert spaces (and sometimes even Banach spaces) (see, for example, [Vesentini (1987a-b); (1991); (1992); (1994a-b)] and [Khatskevich *et al.* (1995a)]).

Example 6.6 Let H be a complex Hilbert space with the inner product $\langle \cdot, \cdot \rangle$ and let $X = H \oplus \mathbb{C}$ be the Hilbert space of the direct sum of H and $\mathbb{C}$

with the inner product $(\cdot, \cdot)$ defined by

$$(x, y) = \langle u, v \rangle + z\bar{w}, \tag{6.37}$$

where $u, v \in H$, $z, w \in \mathbb{C}$, and $x = u + z$, $y = v + w$ are elements of X. Let $\alpha : X \to \mathbb{C}$ be a continuous hermitian sesquilinear form on X defined by

$$\alpha(x, y) = \langle u, v \rangle - z\bar{w}. \tag{6.38}$$

This form is called the indefinite metric on X. The space X endowed with this metric is variously called either a Krein or a Pontryagin space. If J denotes a self-adjoint operator on X defined by the block matrix

$$J = \begin{pmatrix} I & 0 \\ 0 & -1 \end{pmatrix}, \tag{6.39}$$

where I is the identity operator on H, then the indefinite metric $\alpha : X \to \mathbb{C}$ can be written as

$$\alpha(x, y) = (Jx, y). \tag{6.40}$$

A linear operator $A : X \to X$ is said to be an α-isometry if

$$\alpha(Ax, Ay) = \alpha(x, y). \tag{6.41}$$

In operator form this condition becomes

$$A^* J A = J, \tag{6.42}$$

where A^* is the adjoint operator of A. It turns out that a one-parameter semigroup of linear block operators on X, which are α-isometries, can be completely described by a (nonlinear) semigroup of fractional linear transformations on H, which preserve the infinitesimal Poincaré metric on the open unit ball $\mathbb{B}$ of H (see [Vesentini (1987a); Vesentini (1991)]).

For simplicity, we consider the case where $H = \mathbb{C}$ is the complex plane and $B = \Delta$ – the open unit disk in $\mathbb{C}$.

First observe that all groups of automorphisms of Δ of types (i)–(iii) described in Example 6.5 can be written in the general form

$$F_t(z) = e^{i\theta_t} \frac{z - \alpha_t}{1 - \bar{\alpha}_t z}, \quad |\alpha_t| < 1, \quad \theta_t \in \mathbb{R}, \quad t \in \mathbb{R}, \tag{6.43}$$

where the semigroup properties can be verified by the relations

$$e^{i(\theta_{t+s}-\theta_t-\theta_s)}\left(1+\alpha_s\bar{\alpha}_t e^{i\theta_s}\right)=1+\bar{\alpha}_s\alpha_t e^{-i\theta_s}$$

$$\alpha_{t+s}=\frac{\alpha_t e^{-i\theta_s}+\alpha_s}{1+\bar{\alpha}_s\alpha_t e^{-i\theta_s}},\quad \alpha_0=\theta_0=0. \tag{6.44}$$

(See, for example, [Berkson *et al.* (1974)].)

Let us define the families $\{a(t)\}_{t\geq 0}$ and $\{b(t)\}_{t\geq 0}$ by

$$a(t)=\frac{1}{\sqrt{1-|\alpha_t|^2}}\cdot e^{i\frac{\theta_t}{2}} \tag{6.45}$$

and

$$b(t)=-\alpha_t a(t). \tag{6.46}$$

Then

$$\frac{a(t)}{\overline{a(t)}}=e^{i\theta_t} \tag{6.47}$$

and

$$-\frac{b(t)}{a(t)}=\alpha_t. \tag{6.48}$$

Hence,

$$F_t(z)=\frac{a(t)}{\overline{a(t)}}\,\frac{z+\frac{b(t)}{a(t)}}{1+\frac{\overline{b(t)}}{a(t)}z}=\frac{a(t)z+b(t)}{\overline{b(t)}z+\overline{a(t)}}\,. \tag{6.49}$$

If we now define the family $\{A_t\}_t\geq 0$ of linear operators on $\mathbb{C}\oplus\mathbb{C}\ (=\mathbb{C}^2)$ by the matrix

$$A_t=\begin{pmatrix} a(t) & b(t) \\ \overline{b(t)} & \overline{a(t)} \end{pmatrix}, \tag{6.50}$$

then it is easy to verify that this family forms a one-parameter linear semigroup on $\mathbb{C}^2$. Moreover, clearly

$$A_t^* J A_t = J, \tag{6.51}$$

i.e., the family $\{A_t\}_{t\geq 0}$ is a semigroup of α-isometries on $\mathbb{C}^2$, where $\alpha:\mathbb{C}^2\to\mathbb{C}$ is defined by $\alpha(z,w)=z_1\bar{w}_1-z_2\bar{w}_2$, and $z=(z_1,z_2)$ and $w=$

(w_1, w_2) are elements of $\mathbb{C}^2$. In addition, for each $t \geq 0$, we have

$$\det A_t = |a(t)|^2 - |b(t)|^2 = \frac{1}{1 - |\alpha_t|^2} - |\alpha_t|^2 |a(t)|^2$$
$$= \frac{1 - |\alpha_t|^2}{1 - |\alpha_t|^2} = 1. \tag{6.52}$$

Now let $A : \mathbb{C}^2 \to \mathbb{C}^2$ be any matrix operator on $X = \mathbb{C}^2$,

$$A = \begin{pmatrix} a & b \\ c & d \end{pmatrix}, \tag{6.53}$$

preserving the α-metric on X and normalized by the condition $\det A = 1$.

It then follows by the condition

$$A^* J A = J \tag{6.54}$$

that $c = \bar{b}$ and $d = \bar{a}$, *i.e.*, A has the form

$$\begin{pmatrix} a & b \\ \bar{b} & \bar{a} \end{pmatrix} \tag{6.55}$$

with $|a|^2 - |b|^2 = 1$.

If we define the fractional linear transformation

$$F(z) = \frac{az + b}{\bar{b}z + \bar{a}}, \tag{6.56}$$

then we have

$$F(z) = \frac{a}{\bar{a}} \frac{z + \frac{b}{a}}{1 + \frac{\bar{b}}{\bar{a}} z}. \tag{6.57}$$

Since $\left|\frac{a}{\bar{a}}\right| = 1$ and $\left|\frac{b}{a}\right| < 1$, we obtain that F is an automorphism of Δ.

Another useful class of one-parameter semigroups of linear operators is the so-called one-parameter semigroup of composition operators defined on Hardy spaces.

Example 6.7   Let $F : \Delta \to \Delta$ be a holomorphic self-mapping on Δ. If $0 < p \leq \infty$, then the operator $C : H^p(\Delta) \to H^p(\Delta)$ defined by

$$C_F f = f \circ F \tag{6.58}$$

is a well defined linear operator on $H^p(\Delta)$. It is called the composition operator on $H^p(\Delta)$ (see, for example, [Shapiro (1993)] and [Cowen and

MacCluer (1995)]). If $S = \{F_t\}_{t\geq 0}$ is a semigroup of holomorphic self-mappings on Δ, then the family $\{A_t\}_{t\geq 0}$ defined by

$$A_t f = f \circ F_t, \quad t \geq 0, \tag{6.59}$$

defines a semigroup of linear operators on $H^p(\Delta)$.

In addition, if $S = \{F_t\}_{t\geq 0}$ is a continuous semigroup, *i.e.*,

$$\lim_{t\to 0^+} F_t(z) = z, \quad z \in \Delta, \tag{6.60}$$

then $\{A_t\}_{t\geq 0}$ is a continuous semigroup on $H^p(\Delta)$ in the sense of the strong topology in this space, *i.e.*,

$$\lim_{t\to 0^+} A_t f = f \tag{6.61}$$

for all $f \in H^p(\Delta)$. Note that this convergence is not uniform on $H^p(\Delta)$. The converse is also true: *i.e.*, for each strongly continuous semigroup $\{A_t\}_{t\geq 0}$ of composition operators the corresponding semigroup $\{F_t\}$ is continuous on the unit disk Δ in the topology of compact convergence on Δ (see [Berkson and Porta (1978)]). Of course, a subject of interest is the question when A_t is an isometry with respect to the norm topology on $H^p(\Delta)$. For $p \neq 2$ the answer was given by F. Forelli [Forelli (1964)].

Namely, each one-parameter semigroup, (actually, group) $\{A_t\}$ of composition operators on $H^p(\Delta)$, $p \neq 2$, $p > 1$, is given by

$$(A_t f)(z) = e^{iwt}[F_t'(z)]^{\frac{1}{p}} f(F_t(z)), \quad t \in \mathbb{R}, \quad w \in \mathbb{R}, \tag{6.62}$$

where F_t is an automorphism of Δ. On the other hand, for each $1 \leq p < \infty$, such an operator A_t is an isometry on $H^p(\Delta)$.

6.2 Linear semigroups

Let D be a domain in a Banach space X. A semigroup $S = \{F_t\}_{t\in[0,T)}$ is linear if for each $t \in [0, T)$ the mapping $F_t : D \to D$ is the restriction of a linear operator A_t acting on X. We will mostly be interested in the cases where D is either all of X or the open unit ball in X.

So, let us first assume that $D = X$, and let $S = \{B_t\}_{t\geq 0}$ be a continuous semigroup of linear bounded operators on X, *i.e.*,

$$(i) \;\; B_{t+s} = B_t \circ B_s, \quad t, s \geq 0, \tag{6.63}$$

$$(ii) \;\; B_0 = \lim_{t\to 0^+} B_t = I \; (\lim_{t\to 0^+} B_t x = x, \; x \in X), \tag{6.64}$$

where the limit in (ii) is taken with respect to the strong topology on X. Usually, such a semigroup is known as a C_0-semigroup. It is easy to see by the semigroup property (i) that condition (ii) implies continuity at each point $t_0 > 0$, *i.e.*,

$$\lim_{t \to t_0} B_t = B_{t_0}. \tag{6.65}$$

Again we understand the last equality as

$$\lim_{t \to t_0} B_t x = B_{t_0} x \tag{6.66}$$

for each $x \in X$.

For each $t > 0$, define $A_t = (I - B_t)/t$. Clearly, A_t belongs to $L(X)$, the space of bounded linear operators on X, for each $t > 0$.

Theorem 6.1 *The set* $\Omega = \left\{ x \in X : \lim_{t \to 0^+} A_t x \left(= \lim_{t \to 0^+} \frac{1}{t} (I - B_t) x \right) \text{ exists} \right\}$ *in dense* X.

Proof. Let X^* denote (as usual) the dual of X, and let $f(t)$ be the continuous scalar–valued function on $\mathbb{R}^+ = [0, \infty)$ defined by

$$f(t) = \ell(B_t x), \tag{6.67}$$

where x is a fixed element of X and $\ell \in X^*$. Now it follows by the mean value theorem that

$$f(t) = \lim_{s \to 0^+} \frac{1}{s} \int_t^{t+s} f(\theta) d\theta. \tag{6.68}$$

Since $\ell \in X^*$ is arbitrary, we see that for each $t \geq 0$ and each $x \in X$,

$$B_t x = \lim_{s \to 0^+} \frac{1}{s} \int_t^{t+s} B_\theta x d\theta. \tag{6.69}$$

Define now

$$y = \int_0^t B_\theta x d\theta, \quad x \in X, \quad t > 0. \tag{6.70}$$

Then

$$\frac{1}{s} [I - B_s] y = \frac{1}{s} \int_0^s B_\theta x d\theta - \frac{1}{s} \int_t^{t+s} B_\theta x d\theta. \tag{6.71}$$

Letting here $s \to 0^+$ we obtain from (6.69) that

$$\lim_{s \to 0^+} \frac{1}{s}(I - B_s)y = x - B_t x \tag{6.72}$$

exists. In other words, $y \in \Omega$.

Let now x be an arbitrary element of X. Define

$$x_n = n \int_0^{\frac{1}{n}} B_\theta x \, d\theta. \tag{6.73}$$

As we have seen, $x_n \in \Omega$ for all $n = 1, 2, 3 \dots$. On the other hand, we can represent x_n in the form

$$x_n = \int_0^1 B_{\frac{\theta}{n}} x \, d\theta. \tag{6.74}$$

Letting $n \to \infty$, we see that $x_n \to x$ strongly. Thus the closure of Ω equals X, as claimed. $\qquad\square$

Definition 6.3 The mapping $A : \Omega \to X$ defined by

$$Ax = \lim_{t \to 0^+} \frac{1}{t}(x - B_t x) \tag{6.75}$$

is called the infinitesimal generator of the (strongly) continuous semigroup $S = \{B_t\}_{t \geq 0}$. Obviously, the mapping $A : \Omega \to X$ is linear, but not necessarily bounded unless Ω is all of X.

The main property of the infinitesimal generator (which actually is an explanation of the name) is that a given semigroup and its generator are related by the Cauchy problem with an arbitrary initial data in Ω.

Theorem 6.2 *Let $S = \{B_t\}_{t \geq 0}$ be a strongly continuous semigroup and let*

$$\Omega = \left\{ x \in X : Ax := \lim_{t \to 0^+} \frac{1}{t}(I - B_t)x \ \ \text{exists} \right\}. \tag{6.76}$$

Then for each $x \in \Omega$, $B_t x \in \Omega$ and

$$\frac{d}{dt}(B_t x) + AB_t x = 0. \tag{6.77}$$

Proof. First we note that by the semigroup property, for t and $s \geq 0$ the operators B_t and B_s commute. By definition, the right derivative of B_t at

$t \geq 0$ is given by

$$\frac{d^+}{dt}\,(B_t x) = \lim_{s \to 0^+} \frac{1}{s}\,(B_{t+s}x - B_t x) = \lim_{s \to 0^+} B_t \frac{1}{s}\,(B_s x - x)$$

$$= \lim_{s \to 0^+} \frac{1}{s}\,(B_s - I)B_t x = -A B_t x, \quad x \in \Omega. \qquad (6.78)$$

Also

$$\frac{d^+}{dt}\,B_t x = B_t A x, \quad x \in \Omega. \qquad (6.79)$$

Thus

$$A B_t x = B_t A x, \quad x \in \Omega, \ \ t \geq 0, \qquad (6.80)$$

which means that Ω is B_t-invariant for each $t \geq 0$. If $t > 0$, then to obtain the left derivative it suffices to verify that for $0 < s \leq t$,

$$\lim_{s \to 0^+} \left\| \frac{1}{s}\,(B_t x - B_{t-s}x) + B_t A x \right\| = 0. \qquad (6.81)$$

To do this, we first observe that it can be shown by using the uniform boundedness principle and the semigroup property that for each $t \geq 0$, there is a number

$$1 \leq M < \infty \quad \text{and} \quad \omega \in (-\infty, \infty) \qquad (6.82)$$

such that

$$\|B_t\| \leq M e^{\omega t}. \qquad (6.83)$$

Now we have, for $x \in \Omega$,

$$\left\| \frac{1}{s}\,(B_t x - B_{t-s}x) + B_t A x \right\|$$

$$= \left\| B_{t-s}\!\left(\frac{1}{s}\,(B_s x - x) \right) + B_t A x \right\|$$

$$= \left\| B_{t-s}\!\left(\left[\frac{1}{s}\,(B_s x - x) + (B_s - I)A x \right] \right) \right\|$$

$$\leq M e^{\omega(t-s)} \left\{ \left\| \frac{1}{s}\,(B_s x - x) + A x \right\| \right.$$

$$\left. + \|(B_s - I)A x\| \right\} \to 0, \quad s \to 0^+. \qquad (6.84)$$

Hence, for $t > 0$ we obtain

$$\frac{d}{dt}(B_t x) = -AB_t x. \tag{6.85}$$

$\square$

Corollary 6.1 *Each semigroup generator generates a unique semigroup.*

Examples of strongly continuous semigroups are given in examples 6.1, 6.2 and 6.6 of the previous section.

Note, however, that the semigroup $S = \{B_t = e^{-tA}\}_{t \geq 0}$ defined in Example 6.1 satisfies the so-called property of uniform continuity.

Definition 6.4 A semigroup $S = \{B_t\}_{t \geq 0}$ is called uniformly continuous if

$$\lim_{t \to 0^+} \|I - B_t\| = 0, \tag{6.86}$$

where the limit is taken with respect to the operator topology on X.

As we have already seen, for each bounded linear operator $A : X \to X$ the semigroup $S = \{B_t\}$, defined by $B_t = e^{-tA}$, is a uniformly continuous semigroup. In this case

$$A = \lim_{t \to 0^+} \frac{1}{t}(I - B_t) \tag{6.87}$$

is its infinitesimal generator. Moreover, we have

$$\frac{d}{dt}(e^{-tA}) = -Ae^{-tA} \tag{6.88}$$

uniformly on X.

As a matter of fact, each uniformly continuous semigroup S can be represented in the form $S = \{e^{-tA}\}_{t \geq 0}$ for some $A \in L(X)$.

To show this we just need to prove the following assertion.

Theorem 6.3 *Let $S = \{B_t\}_{t \geq 0}$ be a uniformly continuous semigroup, and let $A : \Omega \to X$ be its infinitesimal generator. Then $A \in L(X)$.*

Proof. It follows by the continuity in the uniform operator topology that

$$\lim_{t \to 0^+} \left\| I - \frac{1}{t} \int_0^t B_\tau d\tau \right\| = 0. \tag{6.89}$$

Hence there exists $t_0 > 0$ such that

$$\left\| I - \frac{1}{t_0} \int_0^t B_\tau d\tau \right\| < 1. \tag{6.90}$$

Now if we define $A_0 \in L(X)$ by

$$A_0 = \frac{1}{t_0} \int_0^{t_0} B_\tau d\tau, \tag{6.91}$$

then A_0 is invertible and $A_0^{-1} = \left(I - (I - A_0)\right)^{-1}$. In addition,

$$\frac{1}{t}(I - B_t)B_0 = \frac{1}{t_0 t}\left\{\int_0^{t_0} B_\tau d\tau - \int_0^{t_0} B_{t+\tau}d\tau\right\}$$
$$= \frac{1}{t_0}\left\{\frac{1}{t}\int_0^t B_\tau d\tau - \frac{1}{t}\int_{t_0}^{t+t_0} B_\tau d\tau\right\}. \tag{6.92}$$

Letting $t \to 0^+$, we see that the expression on the right-hand side converges to the limit $\frac{1}{t_0}(I - B_{t_0})$ which belongs to $L(X)$. Hence, $AB_0 = \frac{1}{t_0}(I - B_{t_0})$ and $A = \frac{1}{t_0}(I - B_{t_0})B_0^{-1} \in L(X)$, and we are done. $\qquad\square$

Despite the fact that the class of uniformly continuous semigroups is quite narrow, this class has a large variety of applications to various problems. In particular, each strongly continuous semigroup in a finite dimensional space is uniformly continuous and is generated by a matrix operator.

Another useful example of a uniformly continuous semigroup can be given by the integro-differential equation

$$\frac{\partial f}{\partial t} + \int_D k(x,y)f(t,y)dy = 0, \tag{6.93}$$

where D is a domain in $\mathbb{R}^n$ and $k(x,y)$ is a $L_2(D \times D)$ kernel. If we define a linear operator $A : L_2(D) \to L_2(D)$ by

$$(Ag)(x) = \int_D k(x,y)g(y)dy, \quad x \in \mathbb{R}^n, \tag{6.94}$$

we have that the solution of the Cauchy problem

$$\begin{cases} \dfrac{\partial f}{\partial t} + \displaystyle\int_D k(x,y)f(t,y)dy = 0 \\ f(0,x) = g(x). \end{cases} \tag{6.95}$$

is given by the exponent

$$f(t,x) = e^{-tA}x. \tag{6.96}$$

Hence we obtan a uniformly continuous semigroup on $X = L_2(D)$.

An important question in the general semigroup theory is when a given semigroup leaves a given domain D invariant.

For linear semigroups this question mostly arises when D is the open unit ball in X. In this case the semigroup is said to be a semigroup of contractions on X. In other words, the problem is to find the properties of the infinitesimal generator A of a semigroup $\{B_t\}_{t\geq 0}$ with

$$\|B_t\| \leq 1 \tag{6.97}$$

for all $t \geq 0$.

Basic results in this connection are given in the famous Hille–Yosida and Lumer–Phillips theorems. Although these theorems are known for strongly continuous semigroup in locally convex spaces, it is enough for our further purposes to formulate them for uniformly continuous semigroups in Banach spaces.

Definition 6.5 Let X be a Banach space and let $A \in L(X)$. The set $\rho(A) = \{\lambda \in \mathbb{C} : (\lambda I - A)^{-1} \in L(X)\}$ is called the resolvent set of the operator A. If $\rho(A)$ is nonempty, then the operator–valued function $R = (\lambda I - A)^{-1}$ is called the resolvent of the operator A.

Now we formulate a version of the Hille–Yosida theorem (see [Yosida (1974)]) for a uniformly continuous semigroup of contractions on X.

Theorem 6.4 *Let $A \in L(X)$. Then A is an (infinitesimal) generator of a semigroup of proper contractions $S = \{B_t\}$, $\|B_t\| < 1$, if and only if*

(i) $\rho(A) \supset (-\infty, 0)$;

(ii) $\lambda R(\lambda, -A) = \left(I + \frac{1}{\lambda} A\right)^{-1}$ *is a contraction for all $\lambda > 0$.*

Moreover, the following exponential formula holds:

$$B_t = e^{-tA} = \lim_{n\to\infty} \left[\left(I + \frac{1}{n} A\right)^{-1}\right]^n. \tag{6.98}$$

Remark 6.2 *It can be shown that, actually, the spectrum $\sigma(A)$ of the operator A lies in the closed right-half plane. This, of course, implies condition (i).*

Definition 6.6 A linear operator $A \subset L(X)$ is said to be **accretive** (respectively, **dissipative**) if for each $x \in X$, there is $x^* \in J(x)$ such that

$$\mathrm{Re}\langle Ax, x^*\rangle \geq 0 \tag{6.99}$$

(respectively,

$$\mathrm{Re}(\langle x, x^*\rangle \leq 0). \tag{6.100}$$

A very useful characterization of accretive (respectively, dissipative) operators can be given as follows.

Theorem 6.5   *A linear operator $A \in L(X)$ is accretive if and only if*

$$\|(\lambda I + A)x\| \geq \lambda \|x\| \tag{6.101}$$

(respectively,

$$\|(\lambda I - A)x\| \geq \lambda \|x\| \,) \tag{6.102}$$

for all $x \in X$ and $\lambda > 0$.

This result and the Hille–Yoside theorem imply the following important assertion known as the Lumer–Phillips theorem (see [Yosida (1974)]).

Theorem 6.6   *Let $A \in L(X)$. If A is accretive and for some $\lambda_0 \in \mathbb{R}_0^+ = (0, \infty)$, $(\lambda_0 I + A)^{-1} \in L(X)$, then A is the generator of a one-parameter semigroup of proper contractions on X, i.e,*

$$\|e^{-tA}\| < 1, \quad t \geq 0. \tag{6.103}$$

Conversely, if for some $A \in L(X)$

$$\|e^{-tA}\| \leq 1, \quad t \geq 0, \tag{6.104}$$

then A is accretive (or – which is the same – A is dissipative).

Definition 6.7   An operator $A \in L(X)$ is said to be **conservative** (or hermitian) if

$$\mathrm{Re}\langle Ax, x^* \rangle = 0 \tag{6.105}$$

for all $x \in X$ and $x^* \in X^*$. A conservative operator A is simultaneously accretive and dissipative.

The following assertion holds.

Theorem 6.7   *A linear operator $A \in L(X)$ is conservative if and only if its spectrum lies on the imaginary axis of the complex plane and it is a generator of a group $\left\{ B_t = e^{-tA} : t \in (-\infty, \infty) \right\}$ of linear isometrics, i.e.,*

$$\|e^{-tA}x\| = \|x\| \tag{6.106}$$

for all $t \in \mathbb{R}$ and $x \in X$.

6.3 Generated Semigroups of Nonexpansive and Holomorphic Mappings

Now let D be a domain (open, connected subset) in a Banach space X with the topology induced by the norm of X.

Definition 6.8 A semigroup $S = \{F_t : t \in (0, T)\}$, $T > 0$, on D is said to be generated if for each $x \in D$, there exists the strong limit

$$f(x) = \lim_{t \to 0^+} \frac{1}{t} \left(x - F_t(x) \right). \tag{6.107}$$

In this case the mapping $f : D \to X$ is called the **(infinitesimal) generator** of S.

One of the most important problems in the context of semigroup theory is to find out when a continuous semigroup of nonexpansive mappings in the norm or metric sense (in particular, ρ-nonexpansive or holomorphic mappings) has a generator.

To trace an analogy with the classical linear case, we note that each semigroup of bounded linear operators which is continuous in the operator topology is differentiable at zero, and its generator is also a bounded linear operator. And conversely, if f is a linear holomorphic mapping, then it is bounded by definition, and we obtain the simplest case: the semigroup S generated by f is a uniformly continuous linear semigroup $F_t = e^{-tf}$. In addition, we have the exponential formula

$$e^{-tf} = \lim_{n \to \infty} \left(I + tf/n \right)^{-n}. \tag{6.108}$$

For the nonlinear case the analogous facts are not trivial. For nonlinear semigroups of norm nonexpansive mappings this problem has been considered by many mathematicians, mostly in the framework of the study of the asymptotic behavior of semigroups as well as the structure of their stationary point sets. For semigroups of holomorphic mappings it was also studied in connection with the theory of branching processes. In the one-dimensional case, the differentiability with respect to the parameter of nonlinear semigroups of holomorphic mappings was proved by E. Berkson and H. Porta [Berkson and Porta (1978)] in their study of linear semigroups of composition operators on Hardy spaces. This nice result was extended by M. Abate [Abate (1992)] to the case of $\mathbb{C}^n$. However, it is no longer true in the infinite dimensional case. E. Vesentini has investigated semigroups of those fractional linear transformations which are isometries with respect to the infinitesimal hyperbolic metric on the unit ball of a Banach space. He used this approach to study several important problems in the theory of linear operators on indefinite metric spaces. Observe that in general, such semigroups are not everywhere differentiable.

As a matter of fact, it turns out that *a continuous one-parameter semigroup of holomorphic self-mappings of a domain in a Banach space is differentiable with respect to the parameter if and only if it is right locally uniformly continuous with respect to the parameter at zero.*

Let us recall that *a net* $\{f_j\}_{j\in\mathcal{A}} \subset \mathrm{Hol}(D,X)$ *is said to converge to a mapping* $f \in \mathrm{Hol}(D,X)$ *in the topology of locally uniform convergence over* D *(or, briefly, T-converge) if for every ball* $B \subset\subset D$ *(B is strictly inside* D*),*

$$\limsup_{\substack{j\in\mathcal{A}\ x\in B}} \left\| f_j(x) - f(x) \right\| = 0. \tag{6.109}$$

We write in this case $f = T- \lim_{j\in\mathcal{A}} f_j$. For the finite dimensional case this topology coincides with the compact open topology on D.

Definition 6.9   A family $S = \{F_t\}_{t>0} \subset \mathrm{Hol}(D)$ is said to be a locally uniformly continuous one-parameter semigroup (or, briefly, a T-continuous semigroup) if it satisfies the semigroup property

(i) $F_{s+t} = F_s \circ F_t, \quad s,t > 0,$
     and

(ii) $T- \lim_{t\to 0^+} F_t = I|_D.$

Theorem 6.8 ([Reich and Shoikhet (1998b)]) *Let D be a bounded domain in X, and let $S = \{F_t\}_{t>0} \subset \mathrm{Hol}(D)$ be a strongly continuous semigroup. The following conditions are equivalent:*

(a) S is a T-continuous semigroup;
(b) The differences

$$f_t = \frac{1}{t}\left(I - F_t\right) \tag{6.110}$$

are uniformly bounded on each subset strictly inside D;
(c) For each $x \in D$, there exists the strong limit

$$\lim_{t\to 0^+} \frac{1}{t}\left(I - F_t\right)(x) = f(x), \tag{6.111}$$

which is bounded on each subset strictly inside D.

Remark 6.3 *It is remarkable that actually*

$$f = T-\lim_{t\to 0^+} \frac{1}{t}(I - F_t), \tag{6.112}$$

i.e., the convergence in (c) is actually locally uniform convergence over D (see [Reich and Shoikhet (1998a)]). It is clear that if (c) holds, then the infinitesimal generator f belongs to $\mathrm{Hol}(D, X)$.

Since for the finite dimensional case a continuous semigroup of holomorphic self-mappings is T-continuous, it follows that such a semigroup always has a holomorphic infinitesimal generator (see also [Abate (1992)]).

We now prove the above assertions in a more general setting.

Definition 6.10 Let X be an arbitrary Banach space and let D be a in X. We say that a family $\{G_s : s \in (0,T)\}$, $T > 0$, of self-mappings of domain D satisfies the **approximate semigroup property** if for each subset $\tilde{D}$ strictly inside D the following conditions hold:

(i) for each $\varepsilon > 0$, there is a positive $\delta = \delta(\tilde{D}, \varepsilon) \leq T$ such that

$$\sup_{x\in\tilde{D}} \left\| G_s(x) - G^p_{s/p} \right\| < \varepsilon s \tag{6.113}$$

for all positive integers p and all $s \in (0, \delta)$;

(ii) for each pair $s, t \in (0, T)$, $s + t < T$, there exists $L = L(\tilde{D})$ such that

$$\sup_{x\in\tilde{D}} \left\| G_{s+t}(x) - G_s(G_t(x)) \right\| < L\sqrt{st}. \tag{6.114}$$

Theorem 6.9 *Let D be a domain in a complex Banach space X, and let $\{G_s : s \in (0, T)\}$ be a family of holomorphic self-mapping of D which satisfies the approximate semigroup property. Suppose that G_s converges to the identity as $s \to 0^+$, uniformly on each subset $\tilde{D}$ strictly inside D, i.e.,*

$$\limsup_{s\to 0^+\ x\in\tilde{D}} \|G_s(x) - x\| = 0. \tag{6.115}$$

Then the strong limit

$$\lim_{s\to 0^+} \frac{1}{s}(I - G_s) = f \tag{6.116}$$

exists and is a holomorphic mapping from D into X, which is bounded on each subset strictly inside D.

To prove our theorem we need two lemmas.

Lemma 6.1 *Let D be a domain in a complex Banach space and let $\phi \in$ Hol(D). Suppose that for some subset $D_1 \subset D$ with dist$(D_1, \partial D) > 0$ there are two numbers μ and d, $0 < \mu < d$, an integer $\rho \geq 1$, and a domain D_2, $D_1 \subset D_2 \subset D$ with dist$(D_1, \partial D_2) \geq d$ such that*

$$\sup_{x \in D_2} \|x - \phi^k(x)\| < \mu \tag{6.117}$$

for all $k = 0, 1, 2, \ldots, \rho - 1$. Then for $x \in D_1$ the following inequality holds:

$$\left\| x - \phi^\rho(x) - \rho(x - \phi(x)) \right\| \leq \frac{\mu}{d - \mu}(\rho - 1)\|x - \phi(x)\|. \tag{6.118}$$

Proof. Let $x \in D_1$ and $z \in D_2$ be such that $\|z - x\| \leq \mu$. Then the ball $B_{d-\mu}(z)$ with its center at z and radius $d - \mu$ lies in D_2. Hence it follows from (6.117) and the Cauchy inequality that

$$\left\| (I - \phi^k)'(z) \right\| \leq \frac{\mu}{d - \mu}. \tag{6.119}$$

Therefore, for $x \in D_1$ and $y \in D$ such that $\|x - y\| < \mu$ we have, by (6.119),

$$\left\| x - \phi^k(x) - (y - \phi^k(y)) \right\| \leq \frac{\mu}{d - \mu}\|x - y\|. \tag{6.120}$$

Now setting $y = \phi(x)$ and using (6.117) and (6.120) we obtain by the triangle inequality

$$\left\| x - \phi^p(x) - p(x - \phi(x)) \right\| = \left\| \sum_{k=0}^{p-1} [\phi^k(x) - \phi^{(k+1)}(x) - x + \phi(x)] \right\|$$

$$\leq \sum_{k=1}^{p-1} \left\| \phi^k(x) - x - [\phi^k(\phi(x)) - \phi(x)] \right\|$$

$$\leq \frac{\mu}{d - \mu}(p - 1)\|x - \phi(x)\|, \tag{6.121}$$

and we are done. $\square$

Lemma 6.2 *Let D be a domain in a complex Banach space and let a family $\{G_s : 0 < s < T\}$, $G_s \in$ Hol(D), satisfy the approximate semigroup property. Suppose that G_s converges to the identity as $s \to 0^+$, uniformly on each subset strictly inside D. Then for $s > 0$ small enough the net*

$$f_s = \frac{1}{s}(I - G_s) \tag{6.122}$$

is uniformly bounded on each subset strictly inside D.

Proof. Let D_1 be a subset strictly inside D and let $0 < d < dist(\partial D, D_1)$. Take any domain $D_2 \subset\subset D$ such that $D_1 \subset\subset D_2$ and $dist(D_1, \partial D_2) > d$. Choose μ, $0 < \mu < d$, such that $\mu(d - \mu)^{-1} < \frac{1}{2}$, and choose $\sigma, 0 < \sigma \le T$, such that for all $\tau \in (0, \sigma]$,

$$\sup_{x \in D_2} \|x - G_\tau(x)\| < \frac{\mu}{2}. \tag{6.123}$$

In addition, it follows by the approximate semigroup property (i) that there exists $0 < \delta < \frac{\sigma}{2}$ such that

$$\left\|G_s^k(x) - G_{sk}(x)\right\| < \frac{\mu}{2} \tag{6.124}$$

for all $x \in D_2$ and each $k = 1, 2, \ldots$. Now set $n = \left[\frac{\sigma}{s}\right]$. For $s \in (0, \delta)$, we have $n \ge 2$, $ns \ge \frac{\sigma}{2}$, and $ks \le \sigma$ for all $k = 1, 2, \ldots, h$. Hence it follows from (6.123) and (6.124) that

$$\sup_{x \in D_2} \left\|x - G_s^k(x)\right\| < \mu, \quad s \in (0, \delta). \tag{6.125}$$

Now for all $x \in D_1$ we get, by Lemma 6.1,

$$n\|x - G_s(x)\| - \|x - G_s^n(x)\| \le \left\|n(x - G_s(x)) - (x - G_s^n(x))\right\|$$

$$\le \frac{1}{2} n \left\|x - G_s(x)\right\|, \tag{6.126}$$

or

$$\|x - G_s(x)\| \le \frac{2}{n} \|x - G_s^n(x)\|. \tag{6.127}$$

Therefore, by (6.122)–(6.127), we obtain

$$\|f_x(x)\| \le \frac{2}{ns} \left(\|x - G_{ns}(x)\| + \|G_{ns}(x) - G_s^n(x)\|\right)$$

$$\le \frac{2}{ns} \left(\frac{\mu}{2} + \frac{\mu}{2}\right) \le \frac{4\mu}{\sigma} = \mathcal{L} < \infty, \tag{6.128}$$

whenever $s \in (0, \delta)$. The lemma is proved. $\qquad\square$

Proof of Theorem 6.9. Let D be a domain in a complex Banach space X, and let $\{G_s : s \in (0, T)\}$ be a family of holomorphic self-mappings of D which satisfies the approximate semigroup property. Suppose that G_s

converges to the identity uniformly on each subset strictly inside D. To show that the net

$$f_s = \frac{1}{s}\left(I - G_s\right) \tag{6.129}$$

is a Cauchy net, as $s \to 0^+$, on each subset D_1 strictly inside D, assume that $\varepsilon > 0$ has been given. Choose $0 < d < dist(D_1, \partial D)$, $0 < \mu < d$, such that

$$\frac{\mu}{d - \mu} < \varepsilon, \tag{6.130}$$

and choose $0 < \omega \leq T$ such that

$$\sup_{x \in D_1} \|x - G_\tau(x)\| < \frac{\mu}{2} \tag{6.131}$$

for all $\tau \in (0, \omega)$.

Let $0 < \delta = \delta(D_1, \varepsilon) \leq \omega$ be such that condition (i) of the definition is satisfied, *i.e.*,

$$\left\|G_\tau(x) - G^p_{\frac{\tau}{p}}(x)\right\| \leq \varepsilon\tau \tag{6.132}$$

whenever $x \in D_1$ and $\tau \in (0, \delta)$, $p = 1, 2, \ldots$. Now choose an integer $N > 0$ such that $N^{-1} < \delta$ and $\varepsilon \cdot N^{-1} < \frac{1}{2}\mu$. Then, for all integers $m, n \geq N$ and all $k = 0, 1, 2, \ldots, p = \max\{m, n\}$, we have, by (6.131) and (6.132),

$$\sup_{x \in D_1} \left\|x - G^k_{\frac{1}{m \cdot n}}(x)\right\| \leq \sup_{x \in D_1} \left\|G^k_{\frac{k}{mn}}(x) - G_{\frac{k}{mn}}(x)\right\|$$

$$+ \sup_{x \in D_1} \left\|x - G_{\frac{k}{mn}}(x)\right\| < \varepsilon\frac{k}{mn} + \frac{\mu}{2} < \mu. \tag{6.133}$$

Therefore, by (i) and Lemma 6.1, setting in this lemma $\phi = G_{\frac{1}{nm}}$ and $p = m$ we get for all $x \in D_1$,

$$\left\|x - G_{\frac{1}{n}}(x) - m\big(x - G_{\frac{1}{nm}}(x)\big)\right\| \leq \left\|x - G^m_{\frac{1}{nm}}(x) - m\big(x - G_{\frac{1}{nm}}(x)\big)\right\|$$

$$+ \left\|G^m_{\frac{1}{nm}}(x) - G_{\frac{1}{n}}(x)\right\| \leq \frac{\mu}{d - \mu}m\left\|x - G_{\frac{1}{nm}}(x)\right\| + \varepsilon \cdot \frac{1}{n}. \tag{6.134}$$

Multiplying this inequality by n and using (6.129) and (6.130) we obtain, for $x \in D_1$,

$$\left\|f_{\frac{1}{n}}(x) - f_{\frac{1}{nm}}(x)\right\| \leq \varepsilon\left(\left\|f_{\frac{1}{nm}}(x)\right\| + 1\right). \tag{6.135}$$

Now it follows, by Lemma 6.2, that there is $\mathcal{L} = \mathcal{L}(D_1)$ such that

$$\left\| f_{\frac{1}{nm}}(x) \right\| < \mathcal{L} \tag{6.136}$$

for all $x \in D_1$, whenever N (and therefore $n \cdot m$) is big enough. So, by (6.131) we have

$$\left\| f_{\frac{1}{n}}(x) - f_{\frac{1}{nm}}(x) \right\| \le \varepsilon(\mathcal{L} + 1). \tag{6.137}$$

In a similar way we can get

$$\left\| f_{\frac{1}{n}}(x) - f_{\frac{1}{nm}}(x) \right\| \le \varepsilon(\mathcal{L} + 1) \tag{6.138}$$

for all x in D_1 and $n, m \ge N$, and hence

$$\left\| f_{\frac{1}{n}}(x) - f_{\frac{1}{m}}(x) \right\| \le 2\varepsilon(\mathcal{L} + 1) \tag{6.139}$$

for all $x \in D_1$ whenever $n, m \ge N$. This inequality means that the sequence $\left\{ f_{\frac{1}{n}} \right\}_{n=N}^{\infty}$ converges as $n \to \infty$ uniformly on each subset D_1 strictly inside D. In particular, it converges uniformly on each ball strictly inside D and is uniformly bounded on such a ball. Therefore, its limit

$$f = \lim_{n \to \infty} f_{\frac{1}{n}} \tag{6.140}$$

is a holomorphic mapping from D into X. Now we show that the net $\{f_s\}_{s \in (0,T)}$ converges to f uniformly on each subset D_1 strictly inside D. This will conclude the proof of our theorem.

For given $\varepsilon > 0$ and $x \in D_1$, setting $n = \left[\frac{1}{s^2} \right]$, we can choose s so small that

$$\left\| f_{\frac{1}{n}}(x) - f(x) \right\| < \varepsilon. \tag{6.141}$$

In addition, for such s and n we have

$$\begin{aligned}
f_s - f_{\frac{1}{n}} &= \frac{1}{s}\left(I - G_s\right) - n\left(I - G_{\frac{1}{n}}\right) \\
&= \frac{1}{s}\left(G_{\frac{1}{n}}^{[sn]} - G_{\frac{[sn]}{n}}\right) + \frac{1}{s}\left(G_{\frac{[sn]}{n}} - G_s\right) \\
&\quad + \frac{1}{s}\left[\left(I - G_{\frac{1}{n}}^{[sn]}\right) - ns\left(I - G_{\frac{1}{n}}\right)\right].
\end{aligned} \tag{6.142}$$

Observe that in our setting $n = \left[\frac{1}{s^2} \right]$, so that we have $ns \to \infty$ and $\frac{[ns]}{ns} \to 1$ as $s \to 0$. Thus we can find $\delta > 0$ such that $1 - \frac{[ns]}{ns} < \varepsilon$ and $G_{\frac{[sn]}{n}}(x) \in D_2 \subset\subset D$ whenever $s \in (0, \delta)$ and $x \in D_1$. Using the approximate semigroup property (ii) (see the Definition) we get for such s and all $x \in D_1$,

$$\frac{1}{s}\left\|G_{\frac{[sn]}{n}}(x) - G_s(x)\right\| \leq \frac{1}{s}\left\|G_{\frac{[sn]}{n}}(x) - G_{\frac{[sn]}{n}} \circ G_{s - \frac{[sn]}{n}}(x)\right\|$$

$$+ \left\|G_{\frac{[sn]}{n}} \circ G_{s - \frac{[sn]}{n}}(x) - G_s(x)\right\|$$

$$\leq \frac{1}{s}\left[M\left\|x - G_{s - \frac{[ns]}{n}}\right\|\right.$$

$$\left. + L\sqrt{\left(s - \frac{[sn]}{n}\right)\frac{[sn]}{n}}\,\right], \tag{6.143}$$

where $M = \displaystyle\sup_{x \in D_2,\, s \in (0,\delta)} \|(G_s)'(x)\|$. Once again, using Lemma 6.2, we have

$$\left\|x - G_{s - \frac{[ns]}{n}}\right\| \leq \mathcal{L}\left(s - \frac{[ns]}{n}\right), \tag{6.144}$$

and therefore (6.143) implies that

$$\frac{1}{s}\left\|G_{\frac{[sn]}{n}}(x) - G_s(x)\right\| \leq \varepsilon(M\mathcal{L} + L). \tag{6.145}$$

Now condition (i) of the definition implies

$$\frac{1}{s}\left\|G_{\frac{1}{n}}^{[sn]}(x)\right\| \leq \varepsilon\frac{[sn]}{sn} < \varepsilon. \tag{6.146}$$

Finally, by Lemmas 6.1 and 6.2, we obtain for $x \in D_1$ and $s \in (0,\delta)$,

$$\frac{1}{s}\left\|x - G_{\frac{1}{n}}^{[sn]}(x) - ns\left(x - G_{\frac{1}{n}}(x)\right)\right\|$$

$$\leq \frac{1}{s}\left\|x - G_{\frac{1}{n}}^{[sn]}(x) - [ns]\left(x - G_{\frac{1}{n}}(x)\right)\right\| + \frac{1}{s}\left|[ns] - ns\right|\left\|x - G_{\frac{1}{n}}(x)\right\|$$

$$\leq \left(\varepsilon\frac{1}{s}[ns] + \frac{1}{s}\left|[ns] - ns\right|\right)\left\|x - G_{\frac{1}{n}}(x)\right\|$$

$$\leq \left(\varepsilon\frac{[ns]}{ns} + \left|\frac{[ns]}{ns} - 1\right|\right)\mathcal{L} \leq 2\mathcal{L}\varepsilon. \tag{6.147}$$

Thus for $x \in D_1$ and $s \in (0,\delta)$ we get from (6.141)–(6.147)

$$\|f_s(x) - f(x)\| \leq \left\|f_{\frac{1}{n}}(x) - f(x)\right\| + \left\|f_s(x) - f_{\frac{1}{n}}(x)\right\|$$

$$\leq \varepsilon(2 + M\mathcal{L} + L + 2\mathcal{L}), \tag{6.148}$$

and we are done. $\qquad\square$

Proof of Theorem 6.8. It follows by Theorem 6.2 that condition (a) implies condition (c). By Lemma 6.2, conditions (a) and (b) are equivalent. The implication (c)$\Longrightarrow$(b) is obvious. $\qquad\square$

6.4 The Cauchy Problem and the Product Formula

We know already that a T-continuous semigroup of holomorphic mappings is right-differentiable with respect to the parameter t at zero. As a matter of fact, this implies, in turn, that F_t is differentiable at each point $t \in \mathbb{R}^+$ and that the function $u(t, x) = F_t(x)$ is the solution of the Cauchy problem $u'_t(t, x) = -f(u(t, x))$, $u(0, x) = x$ [Cartan (1967)]. In the context of the Hille–Yosida theory the following question is also of interest. If in the infinite dimensional case we have a family of holomorphic mappings which satisfies the approximate semigroup property and converges to the identity uniformly on each subset strictly inside D, is this family differentiable with respect to the parameter and does its derivative generate a semigroup which may be represented by the so-called product or exponential formula?

To answer these questions we first formulate the following general assertion.

Theorem 6.10 *Let (D, ρ) be a complete metric space, and let $\{G_s : 0 < s < T\}$ be a family of ρ-nonexpnsive mappings on (D, ρ) with the following properties:*

(i) For each ρ-ball $\mathcal{B} \subset (D, \rho)$ and each $\varepsilon > 0$, there is a positive $\delta = \delta(\mathcal{B}, \varepsilon) \leq T$ such that

$$\rho\left(G_s(x),\, G^p_{\frac{s}{p}}(x)\right) < \varepsilon \cdot s \qquad (6.149)$$

for all $x \in \mathcal{B}$ and for all integers p whenever $s \in (0, \delta)$;

(ii) For each ρ-ball $\mathcal{B} \subset (D, \rho)$ there exist $\mu = \mu(\mathcal{B}) > 0$ and $\mathcal{L} = \mathcal{L}(\mathcal{B})$ such that

$$\rho(G_s(x),\, x) \leq \mathcal{L} \cdot s \qquad (6.150)$$

for all $x \in \mathcal{B}$, whenever $s \in (0, \mu)$.

Then for each pair $s \in (0, T)$ and $t > 0$ and each sequence of integers $\{t_n\}$

such that

$$\frac{t_n s}{nt} \to 1, \quad \text{as } n \to \infty, \tag{6.151}$$

there exists the limit

$$\lim_{n \to \infty} G^{t_n}_{\frac{s}{n}} = F_t \tag{6.152}$$

uniformly on each ρ-ball in (D, ρ). This limit does not depend on the sequence $\{t_n\}$ and it is a locally uniformly continuous one-parameter semigroup with respect to $t > 0$.

Proof. First we establish two simple inequalities. For each $\tau \in (0, T)$ and each integer ℓ, we have

$$\rho\big(G^\ell_\tau(x), x\big) \leq \sum_{j=0}^{\ell-1} \rho\big(G^{j+1}_\tau(x), G^j_\tau(x)\big) \leq \ell\rho\big(G_\tau(x), x\big). \tag{6.153}$$

Now if ℓ_1 and ℓ_2 are two arbitrary integers, (6.153) implies

$$\rho\big(G^{\ell_1}_\tau(x), \ G^{\ell_2}_\tau(x)\big) = \rho\left(G^{\min(\ell_1, \ell_2)}_\tau\big(G^{|\ell_1 - \ell_2|}_\tau(x)\big), \ G^{\min(\ell_1, \ell_2)}_\tau(x)\right)$$

$$\leq \rho\big(G^{|\ell_1 - \ell_2|}_\tau(x), x\big) \leq |\ell_1 - \ell_2|\rho\big(G_\tau(x), x\big) \tag{6.154}$$

for each $\tau \in (0, T)$.

Take a ρ-ball $B \subset (D, \rho)$ and choose $\mu > 0$ so that condition (ii) holds. Then, for each $\tau \in (0, \mu)$ we have , by (6.153) and (6.154),

$$\rho\big(G^\ell_\tau(x), x\big) \leq \ell \cdot \mathcal{L}\tau \tag{6.155}$$

and

$$\rho\big(G^{\ell_1}_\tau(x), \ G^{\ell_2}_\tau(x)\big) \leq |\ell_1 - \ell_2|\mathcal{L}\tau \tag{6.156}$$

for all $x \in B$ and for all integers ℓ, ℓ_1, ℓ_2.

For a given $s \in (0, T)$ and $t > 0$, consider the sequence of mappings $\left\{G^{t_n}_{\frac{s}{n}}\right\}_1^\infty$ on B, where $\{t_n\}$ is a sequence of integers which satisfies (6.151). Taking an integer N so that $s/N < \mu$ we get, by (6.155),

$$\rho\left(G^{t_n}_{\frac{s}{n}}(x), x\right) < \frac{t_n \cdot s}{n}\mathcal{L} < \infty \tag{6.157}$$

for all $n \geq N$ and all $x \in B$. In addition, for each $j = 1, 2, \ldots, t_n$ and $m = 1, 2, \ldots,$

$$\rho\left(G^{mj}_{\frac{s}{nm}}(x), x\right) \leq mj \cdot \frac{s}{nm}\mathcal{L} < \infty \tag{6.158}$$

whenever $n \geq N$ This means that there exists a ρ-ball $B_1 \subset (D, \rho)$ such that the sequences $\left\{ G_{\frac{s}{n}}^{t_n}(x) \right\}_N^\infty$ and $\left\{ G_{\frac{s}{nm}}^{mj}(x) \right\}_N^\infty$ are in B_1 for all $x \in B$, $j = 1, 2, \ldots, t_n$, $m = 1, 2, \ldots$. Now for a given $\varepsilon > 0$ we can choose, by (i),
$\delta = \delta(\varepsilon, B_1) \leq T$ such that

$$\rho\left(G_\tau(z), \ G_{\frac{\tau}{m}}^m(z) \right) < \varepsilon \tau \tag{6.159}$$

for all $z \in B_1$ and all $m = 1, 2, \ldots$, whenever $0 < \tau < \delta$. Taking N so large that $s/N < \min\{\mu, \delta\}$ and setting $z = G_{\frac{s}{nm}}^{jm}(x)$, $x \in B$, $m = 1, 2, \ldots$, $j = 1, 2, \ldots, t_n$, $n \geq N$ and $\tau = \frac{s}{n}$, we obtain, by the triangle inequality, the nonexpansiveness of G_s and inequality (6.159), that

$$\rho\left(G_{\frac{s}{n}}^{t_n}(x), \ G_{\frac{s}{nm}}^{t_n \cdot m}(x) \right) \leq \sum_{j=0}^{t_n-1} \rho\left(G_{\frac{s}{n}}^{t_n-j}\left(G_{\frac{s}{nm}}^{j \cdot m}(x) \right), \ G_{\frac{s}{n}}^{t_n-j-1}\left(G_{\frac{s}{nm}}^{(j+1)m}(x) \right) \right)$$

$$= \sum_{j=0}^{t_n-1} \rho\Big(G_{\frac{s}{n}}^{t_n-j-1}\left(G_{\frac{s}{n}}\left(G_{\frac{s}{nm}}^{j \cdot m}(x) \right) \right),$$
$$G_{\frac{s}{n}}^{t_n-j-1}\left(G_{\frac{s}{nm}}^{(j+1)m}(x) \right) \Big)$$

$$\leq \sum_{j=0}^{t_n-1} \rho\left(G_{\frac{s}{n}}\left(G_{\frac{s}{nm}}^{jm}(x) \right), \ \left(G_{\frac{s}{nm}}^{(j+1)m}(x) \right) \right)$$

$$= \sum_{j=0}^{t_n-1} \rho\left(G_{\frac{s}{n}}\left(G_{\frac{s}{nm}}^{jm}(x) \right), \ G_{\frac{s}{nm}}^m\left(G_{\frac{s}{nm}}^{jm}(x) \right) \right)$$

$$\leq t_n \cdot \varepsilon \cdot \frac{s}{n} < a \cdot \varepsilon, \tag{6.160}$$

where $a = \sup\left\{ \frac{t_n \cdot s}{n} \right\} < \infty$ because of (6.151). In the same way and for the same $\varepsilon > 0$ we obtain the inequality

$$\rho\left(G_{\frac{s}{m}}^{t_m}(x), \ G_{\frac{s}{nm}}^{t_m \cdot n}(x) \right) < \varepsilon \cdot a \tag{6.161}$$

for all $x \in B$ whenever $m \geq N$, $n = 1, 2, \ldots$. In addition, it follows by (6.156) that

$$\rho\left(G_{\frac{s}{nm}}^{t_n \cdot m}(x), \ G_{\frac{s}{nm}}^{t_{nm}}(x) \right) \leq \left| \frac{st_n}{n} - \frac{st_{nm}}{nm} \right| \mathcal{L} < \varepsilon \cdot a \tag{6.162}$$

and

$$\rho\left(G_{\frac{s}{nm}}^{t_m \cdot n}(x), \ G_{\frac{s}{nm}}^{t_{nm}}(x) \right) \leq \left| \frac{st_m}{m} - \frac{st_{nm}}{nm} \right| \mathcal{L} < \varepsilon \cdot a \tag{6.163}$$

for $n, m \geq N$, where N is large enough.

Thus, by the triangle inequality, (6.160)–(6.163) imply that for a given ε there is $N > 0$ such that

$$\rho\left(G^{t_n}_{\frac{s}{n}}(x),\ G^{t_m}_{\frac{s}{m}}(x)\right) < 4a\varepsilon \tag{6.164}$$

whenever $n, m > N$. This means that $\left\{G^{t_n}_{\frac{s}{n}}\right\}$ is a Cauchy sequence uniformly on each ρ-ball $B \subset (D, \rho)$, and since (D, ρ) is complete, its limit exists and is a ρ-nonexpansive mapping on (D, ρ).

Once again it follows from (6.156) that it $\{r_n\}$ is another sequence of integers such that

$$\frac{sr_n}{n} \to t \tag{6.165}$$

then, for a given $\varepsilon > 0$ and $x \in B$,

$$\rho\left(G^{t_n}_{\frac{s}{n}}(x),\ G^{r_n}_{\frac{s}{n}}(x)\right) \leq \left|\frac{st_n}{n} - \frac{st_n}{n}\right| \mathcal{L} < \varepsilon \tag{6.166}$$

whenever n is large enough. This means that

$$F_t = \lim_{n \to \infty} G^{t_n}_{\frac{s}{n}} \tag{6.167}$$

does not depend on the sequence $\{t_n\}$ satisfying (6.151).

Now let $s \in (0, T)$, $t > 0$ and $r > 0$ be given numbers. let $\{t_n\}_1^\infty$ and $\{r_n\}_1^\infty$ be two sequences of integers such that $\frac{t_n \cdot s}{n} \to t$ and $\frac{r_n \cdot s}{n} \to r$ as $n \to \infty$. Then, for a given $\varepsilon > 0$ and $x \in B$,

$$
\begin{aligned}
\rho\big(F_t(F_r(x)),\ F_{t+r}(x)\big) &\leq \rho\left(F_t(F_r(x)),\ G^{t_n+r_n}_{\frac{s}{n}}(x)\right) \\
&\quad + \rho\left(F^{t_n+r_n}_{\frac{s}{n}}(x),\ T_{t+r}(x)\right) \\
&\leq \rho\left(F_t(F_r(x)),\ G^{t_n}_{\frac{s}{n}}\left(G^{r_n}_{\frac{s}{n}}(x)\right)\right) + \varepsilon \\
&\leq \rho\left(F_t(F_r(x)),\ G^{t_n}_{\frac{s}{n}}(F_r(x))\right) \\
&\quad + \rho\left(G^{t_n}_{\frac{s}{n}}(F_r(x)),\ G^{t_n}_{\frac{s}{n}}\left(G^{r_n}_{\frac{s}{n}}(x)\right)\right) + \varepsilon \\
&\leq \rho\left(F_r(x),\ G^{r_n}_{\frac{s}{n}}(x)\right) + 2\varepsilon \leq 3\varepsilon \tag{6.168}
\end{aligned}
$$

whenever n is big enough. Since $\varepsilon > 0$ is arbitrary, we have

$$F_{t+r} = F_t \cdot F_r. \tag{6.169}$$

Thus, $F_r : R^+ \to (D, \rho)$ is a one-parameter semigroup which is uniformly continuous on each ρ-ball in (D, ρ) with respect to $t > 0$. The theorem is proved. $\qquad\square$

A consequence of this theorem is the following important assertion.

Theorem 6.11 *Let D be a metric domain in X with a metric $\rho \in (SPS)$ and let $\{G_s : s \in (0, T)\}$ be a family of holomorphic self-mapping of D which satisfies the approximate semigroup property. Suppose that G_s converges to the identity as $s \to 0^+$, uniformly on each subset $\widetilde{D}$ strictly inside D, i.e.,*

$$\limsup_{s \to 0^+ \ x \in \tilde{D}} \|G_s(x) - x\| = 0. \tag{6.170}$$

Then, for each pair s and $t, s \in (0, T)$, $t > 0$, and each sequence of integers $\{t_n\}$ such that

$$\frac{t_n s}{nt} \to 1 \quad \text{as} \quad n \to \infty, \tag{6.171}$$

there exists the strong limit

$$\lim_{n \to \infty} G_{\frac{s}{n}}^{t_n} = F_t, \tag{6.172}$$

uniformly on each subset strictly inside D. This limit does not depend on $\{t_n\}$ and s in (6.171), and the family $\{F_t : 0 < t < \infty\}$ is a one-parameter semigroup of holomorphic self-mappings of D generated by

$$f = \lim_{s \to 0^+} \frac{1}{s} (I - G_s). \tag{6.173}$$

For $x \in D$ the mapping $u(t, x) = F_t(x)$ defined by (6.172) is the solution of the Cauchy problem

$$\begin{cases} \dfrac{\partial u(t, x)}{\partial t} + f(u(t, x)) = 0 \\[2mm] \lim_{t \to 0^+} u(t, x) = x \end{cases}, \tag{6.174}$$

where f is defined by (6.173).

Proof. Let D be a metric domain in X with some metric $\rho \in (SPS)$. Then it follows that conditions (i) and (ii) of the above theorem are satisfied. Therefore, by this theorem, for each pair s and t, $s \in (0, T)$, $t > 0$, and each sequence of integers $\{t_n\}$ such that $\frac{t_n s}{nt} \to 1$, there exists the strong limit $F_t = \lim_{n \to \infty} G_{\frac{s}{n}}^{t_n}$ uniformly on each subset strictly inside D. This limit, $F_t : R^+ \to \mathrm{Hol}(D)$, is a one-parameter semigroup of holomorphic self-mappings of D.

Now we already know that the limit

$$\lim_{t \to 0^+} \frac{I - G_t}{t} =: f \tag{6.175}$$

exists and is a holomorphic mapping on D. We want to show that this mapping $f : D \to X$ generates the semigroup $\{F_t\}$, *i.e.*, that it also satisfies the condition

$$f = \lim_{t \to 0^+} \frac{I - F_t}{t}, \tag{6.176}$$

where the convergence in (6.176) is uniform on each $D_1 \subset\subset D$. Indeed, for given $\varepsilon > 0$ and $D_1 \subset\subset D$, we can find a small enough $\delta > 0$ such that

$$\left\| f(x) - \frac{x - G_t(x)}{t} \right\| < \varepsilon \tag{6.177}$$

for all $x \in D_1$ and all $t \in (0, \delta)$. In addition, setting $s = t$ and $t_n = n$ we can find $\delta_1 < \delta$ such that

$$\frac{1}{t} \left\| G_{\frac{t}{n}}^m (x) - F_t(x) \right\| \le \frac{1}{t} \varepsilon t = \varepsilon \tag{6.178}$$

(see (6.160)) for all $t \in (0, \delta)$. Once again, using (6.175), we take $\delta_2 < \delta$ such that

$$\frac{1}{t} \left\| G_t(x) - G_{\frac{t}{n}}^n (x) \right\| < \varepsilon \quad \text{for } 0 < t < \delta_2. \tag{6.179}$$

Thus we get for $t \in (0, \delta_2)$,

$$\left\| f(x) - \frac{x - F_t(x)}{t} \right\| \le \left\| f(x) - \frac{x - G_t(x)}{t} \right\|$$
$$+ \frac{1}{t} \left\| G_t - G_{\frac{t}{n}}^n (x) \right\| + \frac{1}{t} \left\| G_{\frac{t}{n}}^n (x) - F_t(x) \right\| \le 3\varepsilon, \tag{6.180}$$

and we are done.

Finally, it follows by the semigroup property and (6.176) that for each $x \in D$ the mapping $u(t, x) = F_t(x)$ is a solution of the Cauchy problem (6.174). This concludes the proof of the theorem. $\qquad\square$

Corollary 6.2　*Let D be a metric domain in X with a metric $\rho \in (SPS)$ and let $\{F_t : t \in (0, T)\}$, $T > 0$, be a one-parameter semigroup of holomor-*

phic self-mappings of D such that

$$\lim_{t \to 0^+} F_t = I, \tag{6.181}$$

uniformly on each subset strictly inside D. Then this semigroup can be continuously extended to a flow $\{F_t : 0 < t < \infty\}$ on all of $\mathbb{R}^+$.

Remark 6.4 *Formula (6.172) is called the product formula. The crucial point in establishing this formula is that the family $\{G_s : s \in (0,T)\}$ of holomorphic mappings in D satisfies the approximate semigroup property and has a right-hand derivative at $s = 0$ which is equal to f.*

The question is what happens when we have an arbitrary continuous family $\{G_s\}_{s>0} \subset \mathrm{Hol}(D)$ which is differentiable at $s = 0^+$. Actually, we will see below that for a bounded convex domain, the above theorem and the so-called resolvent method and range condition imply a somewhat more general assertion.

Theorem 6.12 *Let D be a bounded convex domain in a complex Banach space X, and let $\{G_s\}_{s \in (0,T)}$ be an arbitrary family of holomorphic self-mappings of D such that*

$$\lim_{s \to 0^+} \frac{x - G_s(x)}{s} = f(x) \tag{6.182}$$

exists uniformly on each subset strictly inside D and is bounded on such subsets. Then

(1) the Cauchy problem (6.174) has a global solution $u(\cdot, \cdot)$ defined on $R^+ \times D$;

(2) this solution can be obtained by the following product formula:

$$u(t, \cdot) = \lim_{n \to \infty} G_{\frac{t}{n}}^n, \tag{6.183}$$

where the limit is uniform on each subset strictly inside D.

An immediate but very important consequence of this theorem is the following assertion.

Corollary 6.3 *Let D be as in Theorem 6.10, and let f and g be two holomorphic generators of one-parameter semigroups on D, i.e., $\{F_t\}_{t>0}$ and $\{G_t\}_{t>0}$, respectively. Then the mapping $h = f + g$ is also a generator and the semigroup H_t generated by it can be obtained by the formula*

$$H_t = \lim_{n \to \infty} \left[F_{\frac{t}{n}} \cdot G_{\frac{t}{n}} \right]^n, \tag{6.184}$$

where the limit is uniform on each subset strictly inside D.

Corollary 6.4   *The set of holomorphic generators on a bounded convex domain is a real cone.*

6.5   Nonlinear Resolvents, the Range Condition and Exponential Formulas

Definition 6.11   We will say that a mapping $f : D \to X$ satisfies the **range condition** if there exists a positive $T > 0$ such that for each $s \in (0, T)$,

$$(I + sf)(D) \supseteq D \tag{6.185}$$

and $T_s = (I + sf)^{-1}$ is a well-defined self-mapping of D. This mapping T_s is called the **(nonlinear) resolvent** of f.

It turns out that similarly as in the linear Hille–Yosida theory, the resolvent method plays an important role in the study of nonlinear generators of semigroups, as well as in the study of the structure of the fixed point sets of self-mappings and null point sets of vector fields.

In this context it is natural to look for geometrical conditions which will ensure that any semigroup of holomorphic mappings can be represented by exponential formulas or, in other words, to find out when the range condition holds for each holomorphic generator. To answer this query we need the following formula, which is usually called the resolvent identity.

Lemma 6.3   *Let D be a convex domain in a Banach space X, and let $f : D \to X$ be a mapping which satisfies the range condition. Then for $0 \leq s \leq t < T$ the following **resolvent identity** holds:*

$$T_t = T_s\left(\frac{s}{t} I + \left(1 - \frac{s}{t}\right) T_t\right). \tag{6.186}$$

Proof.   For each $x \in D$ the element $y = \frac{s}{t} x + \left(1 - \frac{s}{t}\right) T_t(x)$ belongs to D, by the convexity of D. It follows by the definition of the resolvent that $I - T_t = tf(T_t)$ for $t \in [0, T)$. Thus,

$$y = T_t(x) + \frac{s}{t}\left(x - T_t(x)\right) = T_t(x) + sf(T_t(x)) = (I + sf)T_t(x). \tag{6.187}$$

Hence $T_s(y) = (I + sf)^{-1}(y) = T_t(x)$, and we are done. $\square$

Lemma 6.4 *Let D be a convex metric domain with a metric $\rho \in (SPS)$. If $f \in \mathrm{Hol}(D, X)$ is bounded on each subset strictly inside D and satisfies the range condition, then the family $\{T_s = (I + sf)^{-1} : s \in (0, T)\}$ converges to the identity, uniformly on each subset strictly inside D and satisfies the approximate semigroup property.*

Proof. Indeed, let D be a metric domain with a metric $\rho \in (SPS)$, and let D_1 be a subset of D such that $\mathrm{dist}(D_1, \partial D) > 0$.

Denote $M_1 = \sup\{\|f(y)\| : y \in D_1\}$ and $\delta = \min\{\frac{d}{M}, T\}$, where $0 < d < \mathrm{dist}(D_1, \partial D)$. Setting $y = x + sf(x)$ for $x \in D$ and $s \in (0, \delta)$, we have $T_s(y) = x$, $\|x - y\| < d$, and

$$\rho(T_s(x), x) = \rho(T_s(x), T_s(y)) \leq \rho(x, y) \leq L\|x - y\| \leq LM_1 s. \quad (6.188)$$

Since for each $\tau \in (0, T)$ and each integer ℓ,

$$\rho(T_\tau^\ell(x), x) \leq \sum_{j=0}^{\ell-1} \rho(T_\tau^{j+1}(x), T_\tau^j(x)) \leq \ell\rho(T_\tau(x), x), \quad (6.189)$$

we have that for $x \in D$, $n = 1, 2, \ldots$, and $k = 1, 2, \ldots, n$,

$$\left\| T_{\frac{s}{n}}^k(x) - x \right\| \leq \frac{1}{m}\, \rho\left(T_{\frac{s}{n}}^k(x), x\right) \leq \mathcal{L} \cdot s < \frac{d}{2}, \quad (6.190)$$

whenever $0 < s < \delta_1 = \min\{\delta, \frac{d}{2\mathcal{L}}\}$.

Firstly, (6.188) and (6.190) mean that the subset $D_2 = D_1 \bigcup_{x \in D_1} B_d(x)$, which lies strictly inside D, where $B_d(x)$ is the closed ball with its center at x and radius d, contains the sets $\{T_s(x)\}_{s \in (0, \delta_1)}$ and $\{T_{\frac{s}{n}}^k\}$, $s \in (0, \delta)$, $n = 1, 2, \ldots$, and $k = 1, \ldots, n$, where $x \in D_1$.

Secondly, it follows from (6.188) that the net $\{T_s\}_{s \in (0, \delta)}$ converges to the identity uniformly on D_1. Now denote $M_2 = \sup_{x \in D_2}\{\|f(x)\|\}$. It follows from the Cauchy inequalities that for each $x \in D_1$ and $y \in D$ such that $\|x - y\| < \frac{d}{2}$, $\|f'(y)\| < 2M/d$, and hence, for such x and y we have

$$\|f(x) - f(y)\| \leq \frac{2M_2}{d}\, \|x - y\|. \quad (6.191)$$

Now, because of the identity

$$x - T_s(x) = sf(T_s(x)), \quad x \in D, \quad (6.192)$$

we have that

$$f(x) = \lim_{s \to 0^+} \frac{x - \mathcal{T}_s(x)}{s}, \tag{6.193}$$

uniformly on D_1.

In addition, (6.190)–(6.192) imply that

$$\left\| sf(x) - x + \mathcal{T}_{\frac{s}{n}}^n(x) \right\| \le \sum_{k=1}^{n} \left\| \frac{s}{n} f(x) + \mathcal{T}_{\frac{s}{n}}\left(\mathcal{T}_{\frac{s}{n}}^{k-1}(x) \right) - \left(\mathcal{T}_{\frac{s}{n}}^{k-1}(x) \right) \right\|$$

$$= \sum_{k=1}^{n} \frac{s}{n} \left\| f(x) - f\left(\mathcal{T}_{\frac{s}{n}}^{k}(x) \right) \right\|$$

$$\le \mathcal{L} \cdot \frac{2M_2}{d} \cdot s^2, \ x \in D. \tag{6.194}$$

Thus we obtain, by (6.188), (6.192) and (6.194),

$$\left\| \mathcal{T}_x(x) - \mathcal{T}_{\frac{s}{n}}^n \right\| \le \left\| x - \mathcal{T}_{\frac{s}{n}}^n(x) - sf(x) \right\| + \left\| sf(x) - x + \mathcal{T}_s(x) \right\|$$

$$\le \mathcal{L}\frac{2M_2}{d} s^2 + s\left\| f(x) - f(\mathcal{T}_s(x)) \right\|$$

$$\le \mathcal{L}\frac{4M_2}{d} \cdot s^2 \text{ for all } x \in D_1, \ s \in (0, \delta_1), \ n = 1, 2, \tag{6.195}$$

This inequality shows that condition (i) of the definition of the approximate semigroup property is satisfied.

Now take positive s, t such that $s + t < \delta_1$. Then it follows by the resolvent identity and (6.188) that

$$\left\| \mathcal{T}_{s+t}(x) - \mathcal{T}_t(x) \right\| \le \frac{1}{m} \rho\big(\mathcal{T}_{s+t}(x), \ \mathcal{T}_t(x) \big)$$

$$\le \frac{1}{m} \rho\left(\mathcal{T}_t\Big(\frac{t}{s+t} x + \frac{s}{s+t} \mathcal{T}_{t+s}(x) \Big), \ \mathcal{T}_t(x) \right)$$

$$\le \frac{1}{m} \rho\Big(\frac{t}{s+t} x + \frac{s}{s+t} \mathcal{T}_{t+s}(x), \ x \Big)$$

$$\le \frac{L}{M} \left\| x - \mathcal{T}_{s+t}(x) \right\| \frac{s}{s+t} \le \frac{L}{m} \cdot \mathcal{L} \cdot s. \tag{6.196}$$

Thus we have

$$\left\| T_{s+t}(x) - T_s(T_t(x)) \right\| \leq \frac{1}{m} \rho\big(T_{s+t}(x),\, T_s T_t(x)\big)$$

$$\leq \frac{1}{m} \rho\left(T_s\Big(\frac{s}{s+t}\,x + \frac{t}{s+t}\,T_{s+t}(x)\Big),\, T_s(T_t(x)) \right)$$

$$\leq \frac{L}{m} \left\| \frac{s}{s+t}\,x + \frac{t}{s+t}\,T_{s+t}(x) - T_t(x) \right\|$$

$$\leq \frac{L}{m} \left[\frac{s}{s+t}\,\|x - T_t(x)\| + \frac{t}{s+t}\,\|T_{s+t}(x) - T_t(x)\| \right]$$

$$\leq \frac{L}{m}\,\frac{st}{s+t}\left[\mathcal{L} + \frac{L}{m}\,\mathcal{L} \right] \leq \frac{1}{2}\, L_1 \sqrt{st}. \tag{6.197}$$

This proves condition (ii) of the definition of the approximate semigroup property, and we are done. $\qquad\qquad\square$

Thus we have proved the main theorem of this section.

Theorem 6.13 *Let D be a convex metric domain in a complex Banach space X with a metric $\rho \in (SPS)$, and let $f \in \mathrm{Hol}(D, X)$ be bounded on each subset strictly inside D and satisfy the range condition. Then f is the infinitesimal generator of the one-parameter semigroup of holomorphic self-mappings of D, which can be defined by the following analogs of the exponential formula:*

$$F_t = \lim_{n\to\infty} \left(I + \frac{t}{n}\,f \right)^{-n} \tag{6.198}$$

or

$$F_t = \lim_{n\to\infty} \left(I + \frac{1}{n}\,f \right)^{[-tn]}, \tag{6.199}$$

where the convergence in (6.198) and (6.199) is uniform on each subset strictly inside D.

As a matter of fact, for bounded convex domains the converse also holds. We prove this assertion in a more general setting.

Let D be a metric convex domain in a Banach space X with a corresponding metric ρ on D. We recall that the metric ρ is compatible with the convex structure of D if the following conditions hold:

(i) $\quad \rho\big(sx + (1-s)y,\, sw + (1-s)z\big) \leq \max\big[\rho(x, w),\, \rho(y, z)\big];$ $\qquad$ (6.200)

(ii) there is a real function $\varphi : [0, 1) \to [0, 1)$ such that

$$\limsup_{s \to 1^-} \frac{1 - s}{1 - \varphi(s)} < \infty, \tag{6.201}$$

and for each three elements $x, y, z \in D$ and $s \in [0, 1]$, the following inequality holds:

$$\rho\big(sx + (1 - s)z,\ sy + (1 - s)z\big) \le \varphi(s)\rho(x, y). \tag{6.202}$$

Of course, each convex bounded domain in a complex Banach space is a compatible metric domain with the hyperbolic metric ρ on D.

Theorem 6.14 *Let D be a metric convex domain in a Banach space X with a corresponding metric ρ on D, and let the metric ρ be compatible with the convex structure of D. Suppose that $\{F_t : 0 < t < T\}$ is a family of ρ-nonexpansive self-mappings of D, i.e.,*

$$\rho\big(F_t(x),\ F_t(y)\big) \le \rho(x, y) \tag{6.203}$$

for each $0 < t < T$ and for all x, y in D. If the limit

$$\lim_{t \to 0^+} \frac{I - F_t}{t} =: f \tag{6.204}$$

exists uniformly on each ρ-ball in D and $f : D \to X$ is a continuous mapping, then f satisfies the range condition. Moreover, the resolvent $T_s = (I + sf)^{-1} : D \to D$ is defined for each $s > 0$ and is a ρ-nonexpansive self-mappings of D.

Proof. Suppose that the conditions of the theorem are satisfied. For a fixed $z \in D$ and $s, t > 0$, we consider the equation

$$x = \frac{s}{s + t}\, F_t(x) + \frac{t}{s + t}\, z = G_{s,t}(x). \tag{6.205}$$

Since ρ is compatible with the convex structure of D we have, by definition, that

$$\rho\big(G_{s,t}(x), G_{s,t}(y)\big) \le \varphi\Big(\frac{s}{s + t}\Big)\rho\big(F_t(x),\ F_t(y)\big)$$

$$\le \varphi\Big(\frac{s}{s + t}\Big)\rho(x, y), \tag{6.206}$$

where $0 < \varphi\left(\frac{s}{s+t}\right) < 1$, and

$$\limsup_{t \to 0^+} \frac{t}{1 - \varphi\left(\frac{s}{s+t}\right)} < \infty. \tag{6.207}$$

So, by the Banach fixed point theorem, it follows that equation (6.205) has a unique solution in $D, x = x_{s,t}$ for each $s, t > 0$. Since the equation

$$x + sf_t(x) = z, \quad z \in D, \tag{6.208}$$

where $f_t = \frac{1}{t}\left(I - F_t\right)$ is equivalent to (6.205), this implies that the mapping $T_{s,t} = (I + sf_t)^{-1}$ is a well-defined self-mapping of D and $T_{s,t}(z) = x_{s,t}$. In addition, $T_{s,t}(z)$ can be obtained by the approximation method

$$x_{s,t}^{(n)}(z) = \frac{s}{s+t} \, F_t\left(x_{s,t}^{(n-1)}(z)\right) + \frac{t}{s+t} \, z, \tag{6.209}$$

where $n = 1, 2, \ldots,$ and $x_{s,t}^0(z) = y$ is an arbitrary element of D. Thus it follows by induction that

$$\rho\left(x_{s,t}^{(n)}(z), \, x_{s,t}^{(n)}(w)\right) \leq \max\left\{\rho\left(x_{s,t}^{(n-1)}(z), \, x_{s,t}^{(n-1)}(w), \, \rho(z,w)\right\}\right.$$
$$\leq \rho(z,w), \tag{6.210}$$

because F_t is a ρ-nonexpansive mapping on D. Hence $T_{s,t} : D \to D$ is also ρ-nonexpansive on D. Now we want to show that the net $\{T_{s,t}\}_{s>o}$ converges to a mapping $T_s : D \to D$, as $t \to 0^+$. If this holds it is clear that $T_s = (I + sf)^{-1}$ is a ρ-nonexpansive mapping of D, and this will conclude our proof.

First we show that for each $z \in D$ and each $s > 0$, the net $\{T_{s,t}(z)\}(= \{x_{s,t}\})$ is strictly inside D for t small enough. Indeed,

$$\rho\big(T_{s,t}(z), \, z\big) = \rho\left(\frac{s}{s+t} \, F_t(x_{s,t}) + \frac{t}{s+t} \, z, z\right)$$
$$\leq \varphi\left(\frac{s}{s+t}\right)\rho\big(F_t(x_{s,t}), \, z\big)$$
$$\leq \varphi\left(\frac{s}{s+t}\right)\left[\rho\big(F_t(x_{s,t}, \, F_t(z)\big) + \rho\big(F_t(z), \, z\big)\right]$$
$$\leq \varphi\left(\frac{s}{s+t}\right)\left[\rho(x_{s,t}, \, z) + \rho(F_t(z), \, z)\right]. \tag{6.211}$$

Thus it follows by the approximate semigroup property of the resolvent and by (6.207) that

$$\limsup_{t\to 0^+} \rho(x_{s,t},\, z) \le \limsup_{t\to 0^+} \frac{L \cdot t}{1 - \varphi\left(\frac{s}{s+t}\right)} \frac{\|z - F_t(z)\|}{t} \le C < \infty.$$

$$(6.212)$$

So, there is a ρ-ball in D which contains the set $\{x_{s,t}(z)\}$ whenever t is small enough. Now we want to show that for a fixed $s > 0$ and a given $z \in D$ this net is a Cauchy net as $t \to 0^+$. Indeed, if we denote, as before, $x_{s,t} = T_{s,t}(z)$ and $z_{t,r} = (I + sf_r)(x_{s,t})$, we have $z = (I + sf_t)(x_{s,t})$, and $\{z_{t,r}\}$ converges to z as $t, r \to 0^+$, because $\{f_t\}$ is a Cauchy net. But $T_{s,r}(z_{t,r}) = x_{s,t}$ and we get

$$\rho\big(T_{s,t}(z),\, T_{s,r}(z)\big) \le \rho\big(T_{s,t}(z),\, T_{s,r}(z_{t,r})\big) + \rho\big(T_{s,r}(z_{t,r}),\, T_{s,r}(z)\big)$$
$$\le \rho(z, z_{t,r}) \to 0, \qquad\qquad (6.213)$$

as $t, r \to 0^+$. The proof is complete. $\qquad\square$

We denote by $RN_\rho(D)$ the class of all mappings $f : D \to X$ for which the resolvent $(I + rf)^{-1}$ is a well-defined ρ-nonexpansive self-mapping of D for each positive r.

We say that a mapping $f : D \to X$ is an infinitesimal ρ-generator if for some $T > 0$, there exists a one parameter semigroup $S = \{S_t\}_{t\in(0,T)}$ of self-mappings $S_t : D \to D$ $\big(S_{t+r}(x) = S_t(S_r(x)),\ x \in D,\ 0 < t + r < T\big)$ which are ρ-nonexpansive for each $t \in (0, T)$:

$$\rho\big(S_t(x),\, S_t(y)\big) \le \rho(x, y), \quad x, y \in D, \qquad (6.214)$$

and such that

$$\lim_{t\to 0^+} \frac{1}{t}\, \rho\big(x - tf(x),\, S_t(x)\big) = 0 \qquad (6.215)$$

exists for all $x \in D$ uniformly on each ρ-ball in D.

In this case we will also say that f is a ρ-generator on $(0, T)$. if $T = \infty$ we will write $f \in GN_\rho(D)$.

So if, in particular, D is a bounded convex domain in a complex Banach space and ρ is its hyperbolic metric, then a continuous ρ-generator on some interval $(0, T)$ belongs to the class $RN_\rho(D)$.

Thus we have the following analogs of the Hille–Yosida theorem.

Theorem 6.15   *Let D be a bounded convex domain in a complex Banach space X, and let ρ be its hyperbolic metric. Suppose that $f : D \to X$ is* *bounded and uniformly continuous on each ρ-ball in D. Then $f \in GN_\rho(D)$ if and only if $f \in RN_\rho(D)$.*

Theorem 6.16 *Let D be a bounded convex domain in a complex Banach space X, and let $f : D \to X$ be a holomorphic mapping. Then f generates a one-parameter semigroup on $\mathbb{R}^+$ of holomorphic self-mappings of D if and only if for some $T > 0$ and for each $r \in (0, T]$, the mapping $J_r = (I + rf)^{-1}$ is a well-defined holomorphic self-mapping of D. Moreover, in this case, $J_r = (I + rf)^{-1}$ is a well-defined holomorphic self-mapping of D for all $r > 0$.*

Chapter 7

Flow–Invariance Conditions

7.1 Boundary Flow Invariance Conditions

Let $\mathcal{D}$ be a convex subset of a Banach space X and let $f : \overline{\mathcal{D}} \to X$ be a continuous mapping on $\overline{\mathcal{D}}$, the closure of $\mathcal{D}$. Then the following tangency condition of flow invariance

$$\lim_{h \to 0^+} \operatorname{dist}(x - hf(x), \overline{\mathcal{D}})/h = 0, \quad x \in \overline{\mathcal{D}}, \tag{7.1}$$

is a necessary condition for the solvability of the Cauchy problem

$$\begin{cases} \dfrac{du}{dt} + f(u) = 0 \\ u(0) = x \in \overline{\mathcal{D}} \end{cases} \tag{7.2}$$

on $R^+ = [0, \infty)$.

It was shown in [Reich (1975); Reich (1976)] that condition (7.1) can be rewritten in another form which sometimes is more convenient. Namely, let X be a Banach space, and let $\mathcal{D}$ be a convex domain in X with $0 \in \mathcal{D}$. We denote by $p \ (= p_{\mathcal{D}})$ the Minkowski functional of $\mathcal{D}$, that is,

$$p(x) := \inf\{\lambda > 0 : x \in \lambda\mathcal{D}\}, \quad x \in X. \tag{7.3}$$

(Note that since $\mathcal{D}$ is open, it is an absorbing set, *i.e.*, for each $x \in X$ there is $\lambda > 0$ such that $x \in \mu\mathcal{D}$ for all μ with $|\mu| > \lambda$.) It is known that $p : X \to [0, \infty)$ is continuous on X and that

$$\mathcal{D} = \big\{x \in X : p_{\mathcal{D}}(x) < 1\big\}, \ \overline{\mathcal{D}} = \big\{x \in X : p_{\mathcal{D}}(x) \leq 1\big\}. \tag{7.4}$$

Note also that p is a gauge function:

$$p(x + y) \leq p(x) + p(y), \ p(\lambda x) = \lambda p(x), \ \lambda \geq 0, \tag{7.5}$$

199

whence each one of the sets

$$\mathcal{D}_\ell = \{x \in X : p(x) < \ell\} \tag{7.6}$$

is an open convex absorbing subset of X, the closure of which is $\overline{\mathcal{D}}_\ell = \{x \in X : p(x) \le \ell\}$. For $0 \le \ell \le 1$ we will call $\{\mathcal{D}_\ell\}$ the level sets of p.

Now it follows by the Hahn–Banach theorem that for each $x \in X$, there is a linear functional $x^* \in X$ such that

$$\begin{cases} \operatorname{Re}\langle x, x^* \rangle = p_{\mathcal{D}}(x) \\ \text{and} \\ \operatorname{Re}\langle y, x^* \rangle \le p_{\mathcal{D}}(y), \ y \in X. \end{cases} \tag{7.7}$$

We say that such a **functional** is **subordinate to** p at the point x and denote it by j_x.

In particular, if $x \in \partial\mathcal{D}$, then

$$\operatorname{Re}\langle y, j_x \rangle \le \operatorname{Re}\langle x, j_x \rangle = p(x) = 1 \quad \text{for all} \ \ y \in \overline{\mathcal{D}}, \tag{7.8}$$

and this means that j_x is a **support functional** of $\mathcal{D}$ at x.

If $\mathcal{D}$ is the open unit ball in X, then $x^* = j_x$ for $x \in \partial\mathcal{D}$ is a selection of the duality map $J : X \to 2^{X^*}$ defined by

$$J(x) = \big\{ x^* \in X^* : \langle x, x^* \rangle = \|x\|^2 = \|x^*\|^2 \big\}. \tag{7.9}$$

Note that we can assume that $j_{\lambda x} = j_x$ for all $\lambda > 0$.

If now $f : \overline{\mathcal{D}} \to X$ generates a strongly continuous semigroup $S = \{F_t\}_{t \ge 0}$, $F_t : \overline{\mathcal{D}} \to \overline{\mathcal{D}}$, $t \ge 0$, i.e.,

$$\lim_{t \to 0^+} f_t(x) = f(x), \quad x \in \overline{\mathcal{D}}, \tag{7.10}$$

where $f_t(x) = \frac{x - F_t(x)}{t}$ and the limit is taken with respect to the norm of X, then we have $\operatorname{Re}\langle f_t(x), j_x \rangle = \frac{1}{t} \operatorname{Re}\langle x - F_t(x), j_x \rangle = \frac{1}{t}\left(1 - \operatorname{Re}\langle F_t(x), j_x \rangle\right) \ge 0$, $x \in \partial\mathcal{D}$. Thus, it is evident that f satisfies the following boundary condition:

$$\operatorname{Re}\langle f(x), j_x \rangle \ge 0, \ \ x \in \partial\mathcal{D}, \ \ j_x \text{ is a support functional of } \mathcal{D} \text{ at } x. \tag{7.11}$$

As a matter of fact, it follows by a result in [Reich (1975)] that (7.1) is equivalent to condition (7.11) if we require it to hold for all support functionals j_x. To see this, we define the set

$$I(\mathcal{D}, x) = \big\{ z \in X : z = x + a(y - x) \text{ for some } y \in \mathcal{D} \text{ and } a \geq 0 \big\}. \tag{7.12}$$

The closure $\overline{I(\mathcal{D}, x)}$ of this subset is sometimes called the **support cone** to $\mathcal{D}$ at x.

Lemma 7.1 *Let $\mathcal{D}$ be a convex subset of X. For z in X and x in $\mathcal{D}$, the following are equivalent:*

(1)

$$z \in \overline{I(\mathcal{D}, x)}. \tag{7.13}$$

(2)

$$\lim_{h \to 0^+} \operatorname{dist}\big((1 - h)x + hz, \mathcal{D}\big)/h = 0. \tag{7.14}$$

(3) If $x^* \in X^*$ supports $\mathcal{D}$ at x, then $\operatorname{Re}\langle z, x^* \rangle \leq \operatorname{Re}\langle x, x^* \rangle$.

Proof. $(1) \Longrightarrow (3)$ is immediate. If z does not belong to $\overline{I(\mathcal{D}, x)}$, then there exists x^* in X^* such that $\operatorname{Re}\langle z, x^* \rangle > \sup\{\operatorname{Re}\langle y, x^* \rangle : y \in \overline{I(\mathcal{D}, x)}\}$. The functional x^* must support $\overline{I(\mathcal{D}, x)}$ at x. Consequently, $(3) \Longrightarrow (1)$. The implication $(2) \Longrightarrow (1)$ is also immediate. To prove $(1) \Longrightarrow (2)$, let $\varepsilon > 0$ be given. There are $a > 0$ and $y \in \mathcal{D}$ such that $\big|z - (x + a(y - x))\big| < \varepsilon$. Let $0 < h < 1/a$. Then $\operatorname{dist}\big((1 - h)x + hz, \mathcal{D}\big)/h \leq \big|(1 - h)x + hz - ((1 - ah)x + ahy)\big|/h < \varepsilon$. $\qquad\square$

Setting in this lemma $z = x - f(x)$, we get that (7.1) is equivalent to (7.11). In particular, if $\mathcal{D}$ is the open unit ball in X and f maps $\overline{\mathcal{D}}$ into X, then (7.11) can be written as

$$\inf_{x^* \in J(x)} \operatorname{Re}\langle f(x), x^* \rangle \geq 0, \quad x \in \partial\mathcal{D}. \tag{7.15}$$

For the classes of monotone and accretive mappings, the flow invariance condition (7.1) (or equivalently, (7.11)) was systematically used to study the null point set of f and the Cauchy problem (7.2) (see, for example, [Brézis (1973)], [Reich (1975); Reich (1976)], [Goebel and Kirk (1990)], [Brézis (1970)], [Martin (1973)] and [Webb (1996)].

A result of Martin [Martin (1973)] shows that if $\mathcal{D}$ is a convex subset of X and $f : \overline{\mathcal{D}} \to X$ is a continuous accretive mapping on $\overline{\mathcal{D}}$, then (7.1) (respectively, (7.11) or (7.15)) is also sufficient for the existence of solutions to the Cauchy problem (7.2). This yields a continuous semigroup of nonexpansive mappings on $\overline{\mathcal{D}}$.

Theorem 7.1 *Let $\mathcal{D}$ be a convex subset of a Banach space X and let $f : \overline{\mathcal{D}} \to X$ be a continuous mapping on $\overline{\mathcal{D}}$. Assume that f is an accretive mapping on $\mathcal{D}$. Then f is a generator of a continuous semigroup of norm nonexpansive mappings on $\overline{\mathcal{D}}$ if and only if it satisfies the boundary flow invariance* (7.1):

$$\lim_{h \to 0^+} \operatorname{dist}(x - hf(x), \overline{\mathcal{D}})/h = 0, \quad x \in \overline{\mathcal{D}}. \tag{7.16}$$

For holomorphic mappings an analog of Martin's theorem can be formulated as follows.

Theorem 7.2 ([Aizenberg *et al.* (1996)]) *Let $\mathcal{D}$ be a bounded convex domain in a complex Banach space X and let $f : \overline{\mathcal{D}} \to X$ be a uniformly continuous mapping on $\overline{\mathcal{D}}$, which is holomorphic in $\mathcal{D}$. Then f is a generator of a continuous semigroup of holomorphic self-mappings of $\mathcal{D}$ if and only if it satisfies the flow invariance* (7.1).

Note that in contrast with Martin's theorem, we require that f should be uniformly continuous on $\overline{\mathcal{D}}$.

The first question one can now raise is: *What happens when f is not necessarily uniformly continuous on the closure $\overline{\mathcal{D}}$ of $\mathcal{D}$?*

We will answer this question as well as prove the above theorem (in a somewhat more general setting) by using the notion of the so-called numerical range.

7.2 Numerical Range of Holomorphic Mappings

Let $\mathcal{D}$ be a convex domain in a complex Banach space X, and suppose that $\mathcal{D}$ contains the origin. Let $h : \mathcal{D} \to X$ be a holomorphic mapping. Define

$$h_s(x) = h(sx), \quad 0 < s < 1, \tag{7.17}$$

and note that h_s is holomorphic in the enveloping domain $\frac{1}{s}\mathcal{D}$. Clearly, this domain contains $\overline{\mathcal{D}}$, the closure of $\mathcal{D}$.

For $x \in \partial \mathcal{D}$, let $J(x)$ be the set of all continuous linear functionals on X which are tangent to (support) $\mathcal{D}$ at x, *i.e.*,

$$J(x) = \left\{ l \in X^* : l(x) = 1, \ \operatorname{Re} l(y) \le 1 \ \text{ for all } \ y \in \mathcal{D} \right\}. \tag{7.18}$$

Let $Q(x)$ be a non-empty subset of $J(x)$.

Definition 7.1 If h has a continuous extension to $\overline{\mathcal{D}}$, the **numerical range** of h (taken with respect to Q) is the set

$$W(h) = \{l(h(x)) : l \in Q(x), \quad x \in \partial\mathcal{D}\}. \tag{7.19}$$

We write V in place of W when the numerical range is taken with respect to J. Note that in the case where $\mathcal{D}$ is the open unit ball

$$B = \{x \in X : \|x\| < 1\}. \tag{7.20}$$

This definition agrees with that given by L. A. Harris in [Harris (1971b)]. We now define the upper and lower bounds $L^+(h)$ and $L^-(h)$ for the numerical range of holomorphic mappings as follows:

$$L^+(h) := \lim_{s \to 1^-} \sup \operatorname{Re} V(h_s), \tag{7.21}$$

respectively,

$$L^-(h) := \lim_{s \to 1^-} \inf \operatorname{Re} V(h_s). \tag{7.22}$$

Using the maximum principle one can show that the limits in the above definition exist when $\mathcal{D}$ is balance.

Definition 7.2 A holomorphic mapping $h : \mathcal{D} \to X$ is said to be **holomorphically dissipative** (respectively, **holomorphically accretive**) if

$$L^+(h) \leq 0, \tag{7.23}$$

respectively,

$$L^-(h) \geq 0. \tag{7.24}$$

It is clear that h is holomorphically dissipative if and only if the mapping $-h$ is holomorphically accretive.

Definition 7.3 A holomorphic mapping $h : \mathcal{D} \to X$ is said to be **holomorphically conservative** if both h and $-h$ are holomorphically dissipative (respectively, holomorphically accretive) *i.e.*,

$$L^-(h) = L^+(h) = 0. \tag{7.25}$$

Let $p\,(= p_\mathcal{D})$ be the Minkowski functional of $\mathcal{D}$ and $h : \mathcal{D} \to X$ be a holomorphic mapping. Put

$$\|h\|_\mathcal{D} = \sup\{\|h(x)\| : x \in \mathcal{D}\}, \tag{7.26}$$

when this is finite. Suppose h has a uniformly continuous extension to $\overline{D}$. Then h is holomorphicaly dissipative (respectively, holomorphically accretive) if and only if $\sup \operatorname{Re} V(h) \leq 0$ (respectively, $\inf \operatorname{Re} V(h) \geq 0$), since $\|h - h_s\| \to 0$ as $s \to 1^-$. Moreover, exactly as it was shown in [Harris (1971b), Theorem 2], one can prove that

$$\lim_{t \to 0^+} \frac{\|I + th\|_{\mathcal{D}} - 1}{t} = \sup \operatorname{Re} W(h), \qquad (7.27)$$

respectively,

$$\lim_{t \to 0^+} \frac{\|I - th\|_{\mathcal{D}} - 1}{t} = \inf \operatorname{Re} W(h). \qquad (7.28)$$

Thus, even though the numerical range may differ when taken with respect to different choices of Q, the number $\sup \operatorname{Re} W(h)$ (respectively, $\inf \operatorname{Re} W(h)$) remains fixed. Hence we can replace it by $\sup \operatorname{Re} V(h)$ (respectively, $\inf \operatorname{Re} V(h)$). In fact, this remains true for mappings which do not necessarily have a continuous extension to $\overline{D}$. (See the end of this section.)

For the case of the open unit ball, the only properties of $J(x)$ our arguments use are that $J(x) \neq \oslash$ and

$$\bar{\lambda} J(\lambda x) \subseteq J(x) \qquad (7.29)$$

whenever $|\lambda| = 1$ and $\|x\| = 1$.

For the case of bounded linear operators the notion of dissipativeness coincides with the classical one (see, for example, [Yosida (1974)]). In the theory of linear operators and its applications (to evolution equations, probability and ergodic theory), the classical Lumer–Phillips theorem (see, for example, [Lumer and Philips (1961)] and Theorem 6.6 in Chapter 6) plays an important role.

For the case of bounded linear operators the Lumer–Phillips theorem can be reformulated as follows:

A bounded linear operator $A : X \to X$ (where X is a Banach space) is dissipative if and only if its resolvent $(I - \lambda A)^{-1}$ is well defined on X for all $\lambda > 0$ and satisfies the condition

$$\|(I - \lambda A)^{-1}\| \leq 1, \quad \lambda > 0. \qquad (7.30)$$

In other words, for all $t \geq 0$, the operator $(I - tA)^{-1} : X \to X$ is a contraction on X.

It turns out that for holomorphic mappings the notion of dissipativeness is equivalent to the flow invariance condition.

First, we formulate the following extension of the Lumer–Phillips theorem.

Theorem 7.3 *Let $\mathbb{B}$ be the open unit ball of a complex Banach space X and let $h : B \to X$ be holomorphic. Then h is holomorphically dissipative (respectively, holomorphically accretive) if and only if $(I - th)(\mathbb{B}) \supseteq \mathbb{B}$ and $(I - th)^{-1}$ is a well-defined holomorphic mapping of B into itself for each $t \geq 0$ (respectively, $(I + th)(\mathbb{B}) \supseteq \mathbb{B}$ and $(I + th)^{-1}$ is a well-defined holomorphic mapping of B into itself for each $t \geq 0$).*

In other words, $h \in \mathrm{Hol}(\mathbb{B}, X)$ is holomorphically accretive if and only if it satisfies the range condition on $\mathbb{B}$.

We will, in fact, prove (the necessity part of) this theorem in a more general setting.

Theorem 7.4 *Let $\mathcal{D}$ be a bounded convex domain in X and let $h : \mathcal{D} \to X$ be a holomorphically dissipative mapping bounded on each subset strictly inside $\mathcal{D}$. Then for each $t \geq 0$, we have $(I - th)(\mathcal{D}) \supseteq \mathcal{D}$ and $(I - th)^{-1}$ is a well-defined holomorphic mapping of $\mathcal{D}$ into itself.*

Proof. Fix $y \in \mathcal{D}$ and $t > 0$ and define (suspending our subscript convention)

$$g_s(x) = \frac{1}{s}\left(y + th(sx)\right) \tag{7.31}$$

for $x \in \mathcal{D}$ and $0 < s < 1$. Given $x \in \partial\mathcal{D}$ and $l \in J(x)$, we have

$$\mathrm{Re}\, l(g_s(x)) \leq \frac{1}{s}\left(\mathrm{Re}\, l(y) + t\alpha_s\right), \tag{7.32}$$

where

$$\alpha_s = \sup \mathrm{Re}\, V(h_s). \tag{7.33}$$

Assume without loss of generality that $\mathcal{D}$ contains the origin and let p be the Minkowski functional of $\mathcal{D}$. Since $p(y) < 1$, it follows from our assumption that there exist a constant $k < 1$ and a number $\delta > 0$ such that

$$\frac{1}{s}\left(p(y) + t\alpha_s\right) \leq k \tag{7.34}$$

whenever $1 - \delta < s < 1$. Then $f := I - rg_s$ satisfies

$$p(f(x)) \geq \mathrm{Re}\, l(x - rg_s(x)) \geq 1 - rk \tag{7.35}$$

whenever $r \geq 0$.

The space X is an F-space with respect to p since $\mathcal{D}$ is open and bounded. Moreover, by hypothesis and the mean value theorem, there exists a number M_s satisfying

$$p(g_s(x) - g_s(y)) \leq M_s p(x - y), \quad x, y \in \mathcal{D} \tag{7.36}$$

for each s with $0 < s < 1$. Fix s with $1 - \delta < s < 1$. Then by the inverse function theorem, for all small $r > 0$ the function f maps $\overline{\mathcal{D}}$ homeomorphically onto a subset of X containing 0. Hence

$$f(\mathcal{D}) \supseteq (1 - rk)\mathcal{D}, \tag{7.37}$$

by (7.35). Therefore the function

$$F := f^{-1}((1 - r)I) \tag{7.38}$$

is a holomorphic mapping of $c\mathcal{D}$ into $\mathcal{D}$, where

$$c = (1 - rk)/(1 - r). \tag{7.39}$$

Clearly, $c > 1$. Since $\mathcal{D}$ lies strictly inside $c\mathcal{D}$, the Earle–Hamilton theorem implies that F has a unique fixed point x^* in $\mathcal{D}$. This is also a fixed point for g_s and hence $z = sx^*$ satisfies

$$(I - th)(z) = y. \tag{7.40}$$

Suppose now that there is another solution z_1 of (7.40). Choose $s_1 < 1$ so that $s_1 > s$ and $s_1 > p(z_1)$. Clearly, $x_1 = z_1/s_1$ is in $\mathcal{D}$ and satisfies $g_{s_1}(x_1) = x_1$. One can show as above that x_1 is the unique fixed point of g_{s_1} in $\mathcal{D}$. On the other hand, if we set $x_2 = z/s_1$, then $x_2 \in \mathcal{D}$ and $g_{s_1}(x_2) = x_2$ by (7.40). Hence $x_2 = x_1$, so $z_1 = z$. It can be shown (as, for example, in [Harris (1977), Theorem 5]) that the solution z of (7.40) depends holomorphically on $y \in \mathcal{D}$. This completes the proof of our assertion. $\qquad\square$

Thus we have proved the necessity part of our generalization of the Lumer–Phillips theorem. The sufficiency part is proved in [Harris *et al.* (2000)].

Another immediate consequence of this theorem and the holomorphic analog of the Hille–Yosida theorem is the following boundary flow invariance condition.

Corollary 7.1 *Let $\mathcal{D}$ be a bounded convex domain in a complex Banach space X, and let $f : \overline{\mathcal{D}} \to X$ be a uniformly continuous mapping on $\overline{\mathcal{D}}$ which is holomorphic in $\mathcal{D}$. If f satisfies the boundary condition*

$$\inf_{x^* \in J(x)} \operatorname{Re}\langle f(x), x^* \rangle \geq 0, \quad x \in \partial\mathcal{D}, \tag{7.41}$$

then it generates a semigroup of holomorphic self-mappings of $\mathcal{D}$.

This corollary proves, in fact, the last theorem of the previous section.

On the other hand, regarding the sufficiency part of the generalized Lummer–Phillips theorem, it does not seem natural to consider a boundary condition to characterize generators. This is because a bounded convex domain itself is a complete metric space with respect to its hyperbolic metric. In addition, there are many examples of holomorphic generators which have no continuous extension to $\overline{\mathcal{D}}$ (see the examples below). In particular, if $F \in \operatorname{Hol}(\mathcal{D})$, then $f = I - F$ is a generator because it satisfies the range condition.

Thus, the following question arises: *Is there an interior flow invariance condition which characterizes the class of generators?*

7.3 Interior Flow Invariance Conditions

It would be desirable to find such an interior flow invariance condition from which (7.41) could be derived in the case where $f \in \operatorname{Hol}(\mathcal{D}, X)$ has a continuous extension to $\overline{\mathcal{D}}$.

For the Euclidean ball $\mathcal{D}$ in $X = \mathbb{C}^n$, a certain condition in this direction was established by M. Abate [Abate (1992)]. Namely, he proved that $f \in \operatorname{Hol}(\mathcal{D}, \mathbb{C}^n)$ is a generator if and only if it satisfies the estimate

$$2\big[\|g(x)\|^2 - |\langle g(x), x \rangle|^2\big]\operatorname{Re}\langle g(x), x \rangle$$
$$+\big(1 - \|x\|^2\big)^2 \operatorname{Re}\langle f'(x)f(x), g(x) \rangle \geq 0, \tag{7.42}$$

where

$$g(x) = \big(1 - \|x\|^2\big)f(x) + \langle f(x), x \rangle x. \tag{7.43}$$

For $n = 1$ this condition becomes

$$\operatorname{Re} f(z)\bar{z} \geq -\frac{1}{2}\operatorname{Re} f'(z)\big(1 - |z|^2\big), \tag{7.44}$$

where $z \in \Delta$, the open unit disk in the complex plane $\mathbb{C}$ and $f \in \operatorname{Hol}(\Delta, \mathbb{C})$.

Despite the simplicity of condition (7.44) it is not clear how (7.41) can be derived from (7.44) when f has a continuous extension to $\bar{\Delta}$.

On the other hand, this condition may be useful in studying the behavior of the derivative of a semi-complete vector field in Δ. Therefore it is natural to ask the following question: *Can this condition be extended to a general Banach space in a form similar to (7.44) (instead of condition (7.42))?*

Note also that in the one-dimensional case it follows from the maximum principle for harmonic functions that if $f \in \mathrm{Hol}(\Delta, \mathbb{C})$ has a continuous extension on $\bar{\Delta}$, then the boundary flow invariance condition

$$\mathrm{Re} f(z)\bar{z} \geq 0, \quad z \in \Delta \tag{7.45}$$

implies the following interior condition:

$$\mathrm{Re} f(z)\bar{z} \geq \mathrm{Re} f(0)\bar{z}(1 - |z|^2), \quad z \in \Delta. \tag{7.46}$$

Conversely, it is clear that (7.45) does result from (7.46) if f has a continuous extension to all of $\bar{\Delta}$. Thus another question arises: *Are (7.46) and its Banach space analog necessary and sufficient for f to be a generator?*

The following result provides affirmative answers, in any Banach space, to all the questions raised above.

Theorem 7.5   *Let $\mathcal{D}$ be the open unit ball in a complex Banach space X. Then $f \in \mathrm{Hol}(\mathcal{D}, X)$ is a generator on $\mathcal{D}$ if and only if it is bounded on each subset strictly inside $\mathcal{D}$ and one of the following conditions holds:*

(a) For each $x \in \mathcal{D}$ there exists $x^* \in J(x)$ such that

$$\mathrm{Re}\langle f(x) - (1 - \|x\|^2)f(0), x^* \rangle \geq 0; \tag{7.47}$$

(b)

$$\inf_{x^* \in J(x)} \mathrm{Re}\langle 2\|x\|^2 f(x) + (1 - \|x\|^2)f'(x)x, x^* \rangle \geq 0 \quad x \in \mathcal{D}; \tag{7.48}$$

(c) for each $x \in \mathcal{D}$ and for each $x^* \in J(x)$,

$$\mathrm{Re}\left\langle \frac{1 - \|x\|}{1 + \|x\|} f'(0)x + (1 - \|x\|^2)f(0), x^* \right\rangle \leq \mathrm{Re}\langle f(x), x^* \rangle$$

$$\leq \mathrm{Re}\left\langle \frac{1 + \|x\|}{1 - \|x\|} f'(0)x + (1 - \|x\|^2)f(0), x^* \right\rangle. \tag{7.49}$$

Furthermore, equality in one of the conditions (a), (b) or (c) holds if and only if it holds in the other conditions and f is a generator of a group of automorphisms of $\mathcal{D}$.

Note that in the one-dimensional case condition (a) reduces to (7.46) and condition (b) becomes Abate's condition (7.44).

Proof. First we observe that condition (a) (as well as condition (c)) implies that f must be a holomorphically accretive mapping, hence satisfy the range condition on $\mathcal{D}$. Thus, to prove our theorem we just have to show the equivalence of conditions (a), (b) and (c), and that the property of f to be a generator implies condition (a).

So let f be a generator of the one-parameter semigroup $\{F_t\}_{t\geq 0}$ of holomorphic self-mappings of $\mathcal{D}$. We intend to prove that condition (a) of the theorem is satisfied. Indeed, fix any $x \in \mathcal{D}$ and $x^* \in J(x)$, and set $u = x/\|x\|$, $u^* = x^*/\|x\|$. Consider the holomorphic function $\hat{f}$ on the unit disk $\Delta \subset \mathbb{C}$ defined as follows:

$$\hat{f}(\lambda) = \langle f(\lambda u), u^* \rangle, \quad \lambda \in \Delta. \tag{7.50}$$

Similarly we define a family $\{\widehat{F}_t\}_{t\geq 0}$ of holomorphic self-mappings of Δ:

$$\widehat{F}_t(\lambda) = \langle F_t(\lambda u), u^* \rangle, \quad \lambda \in \Delta, \ t \geq 0. \tag{7.51}$$

It is clear that $\widehat{F}_0(\lambda) = \lambda$ and that there exists the limit

$$\lim_{t \to 0^+} \frac{1}{t} \left(\lambda - \widehat{F}_t(\lambda) \right) = \hat{f}(\lambda). \tag{7.52}$$

Now let $\mathcal{M}_b$ denote the Möbius transformation on Δ defined by

$$\mathcal{M}_b(\lambda) = \frac{\lambda - b}{1 - \lambda \bar{b}}, \quad b \in \Delta. \tag{7.53}$$

Consider the family $\{\mathcal{H}_t\}_{t\geq 0}$ of holomorphic self-mappings of Δ defined by

$$\mathcal{H}_t(\lambda) = \mathcal{M}_{\widehat{F}_t(0)}(\widehat{F}_t(\lambda)), \quad \lambda \in \Delta, \ t \geq 0. \tag{7.54}$$

Note that since $\mathcal{H}_t(0) = 0$ for all t, it follows by the Schwarz Lemma that

$$|\mathcal{H}_t(\lambda)| \leq |\lambda| \text{ for all } \lambda \in \Delta \text{ and } t \geq 0. \tag{7.55}$$

Now, by simple calculations, one can conclude that for each $\lambda \in \Delta$ the curve $\mathcal{H}_t(\lambda) : \mathbb{R}^+ \to \Delta$ is right-differentiable at zero and

$$\lim_{t \to 0^+} \frac{\lambda - \mathcal{H}_t(\lambda)}{t} = \hat{f}(\lambda) - \hat{f}(0) + \overline{\hat{f}(0)}\lambda^2 := h(\lambda), \quad \lambda \in \Delta. \tag{7.56}$$

It follows now by (7.55) and (7.56) that for all $\lambda \in \Delta$,

$$\operatorname{Re} h(\lambda)\bar{\lambda} \geq 0 \tag{7.57}$$

and

$$h(0) = 0. \tag{7.58}$$

But (7.57) means that

$$\operatorname{Re}\hat{f}(\lambda)\bar{\lambda} \geq \operatorname{Re}\hat{f}(0)\bar{\lambda}(1 - |\lambda|^2). \tag{7.59}$$

Setting in this inequality $\lambda = \|x\|$ we have by (7.50),

$$\operatorname{Re}\langle f(x), x^* \rangle \geq \operatorname{Re}\langle f(0), x^* \rangle(1 - \|x\|^2). \tag{7.60}$$

Since x and $x^* \in J(x)$ are arbitrary, this proves the following implication: If f is a generator, then f satisfies (a).

Now we will show that (a) is equivalent to (b). To do this, we return again to the function

$$\hat{f}(\lambda) = \hat{f}(0) - \overline{\hat{f}(0)}\lambda^2 + h(\lambda) \tag{7.61}$$

defined by (7.50), where $h(\lambda)$ satisfies (7.56) and (7.57). But these conditions are equivalent to the conditions

$$h(\lambda) = \lambda \cdot p(\lambda), \quad \lambda \in \Delta \tag{7.62}$$

with

$$\operatorname{Re}p(\lambda) \geq 0. \tag{7.63}$$

So, (7.59) is equivalent to (7.62) and (7.63) for $\hat{f}$ of the form (7.61). Now, in the same terms, we translate condition (b). If we define $\hat{f}$ as above by (7.50), then (b) implies

$$\operatorname{Re}\left[2\hat{f}(\lambda)\bar{\lambda} + \hat{f}'(\lambda)(1 - |\lambda|^2)\right] \geq 0. \tag{7.64}$$

If we substitute here $\hat{f}$ in the form (7.61) with $h(\lambda) = \lambda p)\lambda)$ we see that (7.64) is equivalent to the condition

$$\operatorname{Re}\left[\lambda p'(\lambda) + \frac{1 + |\lambda|^2}{1 - |\lambda|^2}p(\lambda)\right] \geq 0, \quad \lambda \in \Delta. \tag{7.65}$$

We intend to show that (7.65) is equivalent to (7.63). Then setting again $\lambda = \|x\|$ in (7.50) and (7.64) and noting that $\hat{f}'(\lambda) = \langle f'(\lambda u)u, u' \rangle = \langle f'(x)x, x^* \rangle\frac{1}{\|x\|^2}$ for $x \neq 0$, we will get by continuity the equivalence of conditions (a) and (b) of the theorem.

So, let $p \in \mathrm{Hol}(\Delta, \mathbb{C})$ satisfy (7.63). Define $F = (p-1)(p+1)^{-1}$. Since the mapping $w = \frac{z-1}{z+1}$ maps the right half-plane into Δ, F is a self-mapping of Δ. Applying the Schwarz–Pick Lemma to F, we obtain

$$\left| \left(\frac{p-1}{p+1} \right)' \right| = \frac{2|p'|}{|1+p|^2} \leq \frac{|p+1|^2 - |p-1|^2}{|p+1|^2(1-|\lambda|^2)} \tag{7.66}$$

or

$$|p'(\lambda| \leq \frac{2\,\mathrm{Re}\,p(\lambda)}{1-|\lambda|^2} \, . \tag{7.67}$$

This implies

$$\mathrm{Re}(-\lambda p'(\lambda)) \leq |\lambda p'(\lambda)| \leq \frac{2|\lambda|\,\mathrm{Re}\,p(\lambda)}{1-|\lambda|^2} \leq \frac{1+|\lambda|^2}{1-|\lambda|^2}\,\mathrm{Re}\,p(\lambda), \tag{7.68}$$

which is equivalent to (7.65).

In the opposite direction ((7.65) implies (7.63)) we prove the following somewhat more general fact:

Let $p \in \mathrm{Hol}(\Delta, \mathbb{C})$ and suppose that there is a positive function $\psi : [0,1) \to R^+$ such that the following condition holds:

$$\mathrm{Re}\big(\lambda p'(\lambda) + \psi(|\lambda|)p(\lambda)\big) \geq 0, \quad \lambda \in \Delta. \tag{7.69}$$

Then $\mathrm{Re}\,p(\lambda) \geq 0$ everywhere on Δ. Indeed, setting $\lambda = re^{i\theta}$ we have

$$\lambda p'(\lambda) = r\frac{\partial p}{\partial r} \tag{7.70}$$

and (7.69) becomes

$$\mathrm{Re}\left(r\frac{\partial p}{\partial r} \right) + \psi(r)\mathrm{Re}\,p(\lambda) \geq 0, \quad \lambda = re^{i\theta} \in \Delta. \tag{7.71}$$

Assume now that there exists $\lambda_0 = r_0 e^{i\theta_0}$ in Δ such that

$$\mathrm{Re}\,p(\lambda_0) < 0. \tag{7.72}$$

Since (7.69) implies that $\mathrm{Re}\,p(0) \geq 0$, there exist $0 \leq r_1 < r_0$ such that $\mathrm{Re}\,p(r_1 e^{i\theta_0}) = 0$ and $\mathrm{Re}\,p(r_0 e^{i\theta_0}) < 0$. Thus one can find $r_2 \in (r_1, r_0)$ such that

$$\mathrm{Re}\,p(r_2 e^{i\theta_0}) < 0 \tag{7.73}$$

and

$$\mathrm{Re}\frac{\partial p}{\partial r}(r_2 e^{i\theta_0}) < 0. \tag{7.74}$$

This implies

$$\mathrm{Re}\left(r_2\frac{\partial p}{\partial r}\right) + \psi(r_2)\mathrm{Re}\,p(r_2 e^{i\theta_0}) < 0, \qquad (7.75)$$

a contradiction. Thus $\mathrm{Re}\,p(\lambda) \geq 0$ for all $\lambda \in \Delta$ and we are done. In other words, conditions (a) and (b) are equivalent.

To obtain (c) we now represent $\hat{f}(\lambda) = \langle f(\lambda u), u^*\rangle$ in the form

$$\hat{f}(\lambda) = a + ib\lambda - \bar{a}\lambda^2 + \lambda \int_{\partial\Delta} \frac{1 + \bar{\zeta}\lambda}{1 - \bar{\zeta}\lambda}\, d\mu(\zeta), \qquad (7.76)$$

and we calculate:

$$\mathrm{Re}\,\frac{1 + \bar{\zeta}\lambda}{1 - \bar{\zeta}\lambda} = \frac{1 - |\lambda|^2}{|1 - \bar{\zeta}\lambda|^2} = \frac{1 - |\lambda|}{1 + |\lambda|}\frac{(1 + |\lambda|)^2}{|1 - \bar{\zeta}\lambda|^2} = \frac{1 + |\lambda|}{1 - |\lambda|}\cdot\frac{(1 - |\lambda|)^2}{|1 - \bar{\zeta}\lambda|^2}. \qquad (7.77)$$

Since $|\zeta| = 1$, this equality shows that

$$\frac{1 - |\lambda|}{1 + |\lambda|} \leq \mathrm{Re}\,\frac{1 + \bar{\zeta}\lambda}{1 - \bar{\zeta}\lambda} \leq \frac{1 + |\lambda|}{1 - |\lambda|}. \qquad (7.78)$$

Noting that $p(0) = \hat{f}'(0)$ (see (7.61) and (7.62)), we obtain from (7.76) and (7.78)

$$\mathrm{Re}\hat{f}(0)\bar{\lambda}(1 - |\lambda|^2) + |\lambda|^2\frac{1 + |\lambda|}{1 - |\lambda|}\,\hat{f}'(0) \geq \mathrm{Re}\hat{f}(\lambda)\bar{\lambda}$$

$$\geq \mathrm{Re}\hat{f}(0)\bar{\lambda}(1 - |\lambda|^2) + |\lambda|^2\frac{1 - |\lambda|}{1 + |\lambda|}\,\hat{f}'(0). \qquad (7.79)$$

Once again, setting $\lambda = \|x\|$ we get from the last inequality condition (c).

Since by (b), $\mathrm{Re}\langle f'(0)x, x^*\rangle \geq 0$, it is clear that (c) is stronger than (a). $\qquad\square$

The following fact is a direct consequence of Theorem 7.5.

Corollary 7.2 *Let $f \in \mathrm{Hol}(\mathcal{D}, X)$ be a holomorphic generator on $\mathcal{D}$. Then the linear operator $A = f'(0)$ is (totally) accretive, i.e.,*

$$\inf_{y^* \in J(y)} \mathrm{Re}\langle Ay, y^*\rangle \geq 0, \quad y \in X. \qquad (7.80)$$

Proof. Substitute $x = ty$, $x^* = ty^*$ in (b), where $\|y\| = \|y^*\| = 1$, $y^* \in J(y)$ and $t \in (0, 1)$. Letting t tend to zero we get (7.80).

In its turn, (7.80) implies that the left-hand inequality in (c) is sharper than (a). Moreover, it implies that (a) holds for all $x^* \in J(x)$. Thus we have that if f has a continuous extension to $\overline{\mathcal{D}}$, then (a) yields the flow invariance boundary condition (7.41) (equivalently, (7.1)). $\qquad\square$

7.4 Semi-Complete and Complete Vector Fields

In connection with flow invariance conditions we give the following definition.

Definition 7.4 A holomorphic vector field

$$T_f = f(x)\frac{\partial}{\partial x} \tag{7.81}$$

on a domain $\mathcal{D}$ is determined by a holomorphic mapping $f \in \mathrm{Hol}(\mathcal{D}, X)$ and can be regarded as a linear operator mapping $\mathrm{Hol}(\mathcal{D}, X)$ into itself, where $T_f g \in \mathrm{Hol}(\mathcal{D}, X)$ is defined by

$$(T_f g)(x) = g'(x)f(x), \quad x \in \mathcal{D}. \tag{7.82}$$

The set of all holomorphic vector fields on $\mathcal{D}$ is a Lie algebra under the commutator bracket

$$[T_g, T_h] = \left[g(x)\frac{\partial}{\partial x}, \ h(x)\frac{\partial}{\partial x} \right] := \left(g'(x)h(x) - h'(x)g(x) \right)\frac{\partial}{\partial x} \tag{7.83}$$

(see, for example, [Dineen (1989)]).

Furthermore, each vector field (7.81) is locally integrable in the following sense: for each $x \in \mathcal{D}$ there exist a neighborhood $\Omega \subset \mathcal{D}$ of x and $\delta > 0$ such that the Cauchy problem (7.2) has a unique solution $\{u(t,x)\} \subset \mathcal{D}$ defined on the set $\{|t| < \delta\} \times \Omega \subset \mathbb{R} \times \mathcal{D}$.

Definition 7.5 A holomorphic vector field T_f defined by (7.81) and (7.82) is said to be (right) **semi-complete** (respectively, **complete**) on $\mathcal{D}$ if the solution of the Cauchy problem (7.2) is well-defined on all of $\mathbb{R}^+ \times \mathcal{D}$ (respectively, $\mathbb{R} \times \mathcal{D}$), where $\mathbb{R}^+ = [0, \infty)$ (respectively, $\mathbb{R} = (-\infty, \infty)$).

Thus, if $\mathcal{D}$ is hyperbolic, then T_g is semi-complete (respectively, complete) if and only if g is the generator of a one-parameter continuous semigroup (respectively, group).

On the other hand, if $\mathcal{D}$ is bounded and $\widetilde{\mathrm{Hol}}(\mathcal{D}, X)$ is the subspace of $\mathrm{Hol}(\mathcal{D}, X)$ consisting of all those $g \in \mathrm{Hol}(\mathcal{D}, X)$ which are bounded on each ball strictly inside $\mathcal{D}$, then a semigroup (group) $\{F_t\}$, $t \in \mathbb{R}^+$

(respectively, $t \in \mathbb{R}$), induces a linear semigroup (group) $\{L(t)\}$ of linear mappings $L(t) : \widetilde{\mathrm{Hol}}(\mathcal{D}, X) \mapsto \widetilde{\mathrm{Hol}}(\mathcal{D}, X)$, defined by

$$(L(t)g)(x) := g(F_t(x)), \qquad (7.84)$$

where $t \in \mathbb{R}^+ (t \in \mathbb{R})$ and $x \in \mathcal{D}$.

This semigroup is called the semigroup of composition operators on $\widetilde{\mathrm{Hol}}(\mathcal{D}, X)$. If $\{F_t\}$, $t \in \mathbb{R}^+ (t \in \mathbb{R})$, is T-continuous (that is, differentiable), then $\{L(t)\}$, $t \in \mathbb{R}^+ (t \in \mathbb{R})$, is also differentiable and

$$\begin{cases} \dfrac{\partial L(t)g}{\partial t} + T_f(L(t)g) = 0 \\ L(0)g = g \end{cases} \qquad (7.85)$$

for all $g \in \widetilde{\mathrm{Hol}}(\mathcal{D}, X)$, where $f = \dfrac{dF_t}{dt}\Big|_{t=0}$.

In other words, a holomorphic vector field T_f, defined by (7.81) and (7.82) and considered a linear operator on $\widetilde{\mathrm{Hol}}(\mathcal{D}, X)$, is the infinitesimal generator of the semigroup $\{L(t)\}$. It is sometimes called the Lie generator. Thus a holomorphic vector field T_f is semi-complete (respectively, complete) if and only if it is the Lie generator of a linear semigroup (respectively, group) of composition operators on $\widetilde{\mathrm{Hol}}(\mathcal{D}, X)$. This follows from the observation that

$$L(t)I_{\mathcal{D}} = F_t \qquad (7.86)$$

and

$$T_f I_{\mathcal{D}} = f, \qquad (7.87)$$

where $I_{\mathcal{D}}$ is the restriction of the identity operator to $\mathcal{D}$.

Moreover, using the exponential formula representation for the linear semigroup:

$$L(t)g = \sum_{k=0}^{\infty} \frac{(-1)^k t^k}{k!} \, T_f^k g = \exp[-tT_f]g \qquad (7.88)$$

(see Chapter 6), we also have

$$S(t) = \sum_{k=0}^{\infty} \frac{(-1)^k t^k}{k!} \, T_f^k I_{\mathcal{D}} = \exp[-tT_f]I_{\mathcal{D}}. \qquad (7.89)$$

So, a locally uniformly continuous semigroup of holomorphic self-mappings can be represented in exponential form by the holomorphic vector field induced by its generator.

In particular, Theorem 7.5 (see the previous section) asserts that if $f \in \mathrm{Hol}(\mathcal{D}, X)$ then $T_f \left(= f(x)\frac{\partial}{\partial x}\right)$ is a semi-complete vector field if and only if f satisfies the interior flow invariance condition (a) (respectively, (b) or (c)).

Furthermore, condition (c) also leads to a characterization of a semi-complete vector field to be complete.

Following S. G. Krein [Krein (1971)] (see also [Vesentini (1996a]), we say that a linear operator $A : X \to X$ is conservative if for all $x \in X$ and $x^* \in J(x)$ there holds

$$\mathrm{Re}\langle Ax, x^* \rangle = 0. \tag{7.90}$$

In order to simplify our terminology we will sometimes identify a semi-complete (complete) vector field T_f with its determining holomorphic mapping f, which is, in fact, the infinitesimal generator of a semigroup (group) of holomorphic self-mappings on $\mathcal{D}$.

The following corollaries are consequences of Theorem 7.5 in the previous section.

Corollary 7.3 *Let $f \in \mathrm{Hol}(\mathcal{D}, X)$ be a semi-complete vector field. Then f is actually complete if and only if its derivative at zero, $f'(0)$, is a conservative linear operator.*

It is well known that a complete vector field g on the open unit ball $\mathcal{D}$ in a Banach space X is a polynomial of degree at most 2 (see, for example, [Arazy (1987)], [Dineen (1989)] and [Upmeier (1986)]).

More precisely, g has the form

$$g(x) = a + Ax + P_a(x), \tag{7.91}$$

where a is an element of X, A is a conservative operator on X, and P_a is a homogeneous form of the second degree such that $P_{ia} = iP_a$.

One of the consequences of this representation is an infinitesimal analog of Cartan's uniqueness theorem ([Arazy (1987)], [Dineen (1989)] and [Isidro and Stacho (1984)]): If $g \in \mathrm{Hol}(\mathcal{D}, X)$ is a complete vector field such that $g(0) = 0$ and $g'(0) = 0$, then $g \equiv 0$.

Applying Corollary 7.3, we obtain at once the following extension of this theorem.

Corollary 7.4 *If $f \in \mathrm{Hol}(\mathcal{D}, X)$ is a semi-complete vector field on $\mathcal{D}$ such that $f(0) = 0$ and $f'(0) = 0$, then $f \equiv 0$.*

We know already that if $f \in \mathrm{Hol}(\mathcal{D}, X)$ has the form

$$f = I - F, \tag{7.92}$$

where F is a self-mapping of $\mathcal{D}$, then f is semi-complete. Thus Corollary 7.4 is also a generalization of Cartan's uniqueness theorem (sometimes this theorem is also called the Generalized Schwarz Lemma (see Chapter 3)).

Suppose now that a complex Banach space X is a so-called JB^* triple system. This is equivalent to saying that its open unit ball $\mathcal{D}$ is a homogeneous domain, *i.e.*, for each pair $x, y \in \mathcal{D}$ there exists a holomorphic automorphism of $\mathcal{D}$ such that $F(x) = y$ (see, for example, [Upmeier (1986); Dineen (1989); Isidro and Stacho (1984)]). Then it is well-known that for each $a \in X$ there exists a homogeneous polynomial $P_a(x)$ such that $P_{ia} = iP_a$ and the mapping $g : \mathcal{D} \to X$ defined by

$$g(x) = a - P_a(x) \tag{7.93}$$

is a complete vector field on $\mathcal{D}$, which is called a transvection of $\mathcal{D}$. Using this fact and Corollary 4 in [Aharonov *et al.* (1999b)] we get the following representation theorem.

Corollary 7.5 *Let X be a JB^* triple system and let $\mathcal{D}$ be its open unit ball. Then the cone $\mathcal{G}$ of semi-complete vector fields on $\mathcal{D}$ admits the decomposition*

$$\mathcal{G} = \mathcal{G}_0 \bigoplus \mathcal{G}_+, \tag{7.94}$$

where $\mathcal{G}_0$ is the real Banach subspace of $H_{1,\infty}(\mathcal{D}, X)$ consisting of transvections and $\mathcal{G}_+$ is the subcone of $\mathcal{G}$ such that for each $h \in \mathcal{G}_+$,

$$\inf_{x' \in J(x)} \mathrm{Re}\langle h(x), x^* \rangle \geq 0 \quad \text{for all} \ \ x \in \mathcal{D}. \tag{7.95}$$

In other words, $f \in \mathcal{G}$ admits a unique representation

$$f = g + h \tag{7.96}$$

where $g = f(0) - P_{f(0)}(x)$ is complete, $h \in \mathcal{G}_+$ and $h(0) = 0$.

The natural examples of JB^* triple systems are a complex Hilbert space H, the space of bounded linear operators $L(H)$ on H, and its subspaces J such that $A \in J$ if and only if $AA^*A \in J$ (such subspaces are called

J^*-algebras); see [Harris (1974a); Dineen (1989)]. In the latter case the general form of transvections on $\mathcal{D}$ is

$$g(x) = a - xa^*x, \tag{7.97}$$

where $a \in J$ and a^* is its conjugate. Thus each semi-complete vector field on the open unit ball of a J^*-algebra has the form

$$f(x) = f(0) - xf^*(0)x + h(x), \tag{7.98}$$

where $h \in \mathcal{G}_+$ and $h(0) = 0$.

In particular, when $X = \mathbb{C}$ is the complex plane and $\mathcal{D} = \Delta$ the open unit disk in $\mathbb{C}$, (7.98) becomes

$$f(z) = f(0) - \overline{f(0)}z^2 + zp(z), \tag{7.99}$$

where $p(z) \in \text{Hol}(\Delta, \mathbb{C})$ and

$$\text{Re}\, p(z) \geq 0, \quad z \in \Delta. \tag{7.100}$$

That is, $p(z)$ is a function in the class of Carathéodory. Using the Riesz–Herglotz integral characterization of this class (see, for example, [Duren (1983)] and [Aleksandrov (1994)]) we deduce the following conclusion:

$f \in \text{Hol}(\Delta, \mathbb{C})$ is a semi-complete vector field if and only if it admits the representation

$$f(z) = a + ibz - \bar{a}z^2 + z \int_{\partial\Delta} \frac{1 + \bar{\zeta}z}{1 - \bar{\zeta}z}\, d\mu(\zeta), \tag{7.101}$$

where $a \in \mathbb{C}$, $b \in \mathbb{R}$ and μ is a positive measure on $\partial\Delta$.

As a matter of fact, this representation is the key to proving the results in the previous section because we have partially used a reduction to the one-dimensional case. We have also obtained (7.99) with (7.100) by an independent method.

Chapter 8

Stationary Points of Continuous Semigroups

8.1 Generalities

Let $\mathcal{D}$ be a topological space, and let a family $S = \{F_t \in (0, T)\}$, $T > 0$, of self-mappings F_t of $\mathcal{D}$ form a (one-parameter) continuous semigroup, *i.e.*,

(i)

$$F_{s+t} = F_t \cdot F_s, \quad 0 < s + t < T, \tag{8.1}$$

(ii)

$$\lim_{t \to 0+} F_t(x) = x, \tag{8.2}$$

where the limit is taken with respect to the topology of $\mathcal{D}$.

Definition 8.1 A subset $\mathcal{W}$ of $\mathcal{D}$ is said to be the **stationary point** set of S if it consists of all the points $a \in \mathcal{D}$ such that

$$F_t(a) = a \quad \text{for all} \quad t \in (0, T). \tag{8.3}$$

In other words,

$$\mathcal{W} = \bigcap_{0 < t < T} \text{Fix} F_t. \tag{8.4}$$

The first problem in this context is the existence problem of a stationary point of S, in other words, to determine when $\mathcal{W} \neq \emptyset$. The second problem is the uniqueness problem, that is, to find out when $\mathcal{W}$ consists of a unique point $a \in \mathcal{D}$. This problem is connected with a more general, rather important in applications, problem – to determine the structure of the set $\mathcal{W}$ in relation to the topological structure of $\mathcal{D}$. Another important problem is to find constructive methods for the approximation of $\mathcal{W}$. In

219

the case when $T = \infty$ this problem, in particular, is closely related to the study of the asymptotic behavior of semigroups.

Definition 8.2   A point $a \in \mathcal{D}$ is said to be a **periodic point** of a semigroup $S = \{F_t : t \in (0, T)\}$ if for some $t_0 > 0$ this point is a fixed point of F_{t_0}, i.e., $a = F_{t_0}(a)$, hence $a = F_{t_0}^n(a)$ for all $n = 0, 1, 2, \ldots$.

The simplest situation is, of course, when for some $t_0 > 0$ the mapping $F_{t_0} : \mathcal{D} \to \mathcal{D}$ has a **unique** fixed point $a = F_{t_0}(a)$ in $\mathcal{D}$. Then directly from the semigroup property (i) it follows that $\mathcal{W} = \{a\} \neq \emptyset$. But even in this case the approximation problem is still open.

Sometimes one can study the existence and the structure of $\mathcal{W}$ as the common fixed point set of the commuting family $\{F_t\}$. Indeed, if, for example, $\mathcal{D} = \mathbb{B}^n$, where $\mathbb{B}$ is the open Hilbert ball, then the following assertion, due to I. Shafrir ([Shafrir (1992a)]) is useful for the study of the common fixed point sets of one-parameter semigroups, which consist of nonexpansive mappings with respect to the hyperbolic metric ρ on $\mathcal{D}\,(= \mathbb{B}^n)$.

Lemma 8.1   *Let $\{F_\alpha : \alpha \in J\}$ be a **commuting family** of ρ-nonexpansive (holomorphic mappings) self-mappings of $\mathbb{B}^n$ and assume that there is a ρ-bounded subset $M \subset \mathbb{B}^n$ which is F_α-invariant for all $\alpha \in J$. Then there exists a **common fixed point** $a \in \mathbb{B}^n$ of this family.*

Remark 8.1   *Observe that if in Lemma 8.1 we omit the assumption that there exists a ρ-bounded subset $M \subset \mathbb{B}^n$ which is F_α-invariant for all $a \in J$, then one can construct a counterexample even for $n = 1$.*

Example 8.1 ([Kuczumow *et al.* (2001a)]) Let $\mathbb{B}$ be the open unit ball in a separable Hilbert space $(H, \langle \cdot, \cdot \rangle)$ with the orthonormal basis $\{e_n\}_{n \in \mathbb{N}}$. With each affine subset $Y = (x + \widetilde{Y}) \cap \mathbb{B}$ of $\mathbb{B}$, where $\widetilde{Y}$ is a closed subspace of X and $x \in \mathbb{B}$ is the unique point of least norm in Y, we associate the nearest point holomorphic retraction of $\mathbb{B}$ onto Y given by the formula

$$R = M_x \circ \mathrm{Proj}_{\widetilde{Y}} \circ M_{-x}, \tag{8.5}$$

where M_x is a Möbius transformation. We construct the family $\{F_n\}_{n \in \mathbb{N}}$ in the following way: we choose a sequence of positive real numbers $\{\alpha_n\}_{n \in \mathbb{N}}$ such that this sequence is strictly decreasing, $0 < \alpha_n < \frac{1}{2}\pi$ and $\lim\limits_{n \to \infty} \alpha_n = 0$. Let us set

$$X_0 = X, \; C_0 = \mathbb{B}, \; a_0 = 0, \; r_0 = \|e_1 - a_0\|. \tag{8.6}$$

For $n = 1$, the mapping R_1 is the nearest point holomorphic retraction of $\mathbb{B}$ onto the affine set C_1 in $\mathbb{B}$, where

$$f_1 = \frac{1}{\|e_1 - a_0\|}\,(e_1 - a_0), \tag{8.7}$$

$$g_1 = e_2 - a_0 - \frac{1}{\|e_1 - a_0\|^2}\,\langle e_2 - a_0,\, e_1 - a_0 \rangle (e_1 - a_0), \tag{8.8}$$

$$h_1 = \frac{1}{\|g_1\|}\,g_1, \tag{8.9}$$

$$a_1 = a_0 + r_0 \left[\left(\frac{1}{2} + \frac{1}{2}\cos\alpha_1 \right) f_1 + \left(\frac{1}{2}\sin\alpha_1 \right) h_1 \right], \tag{8.10}$$

$$r_1 = \|e_1 - a_1\|, \tag{8.11}$$

$$X_1 = \{ x \in X_0 : x = a_1 + y \ \text{ and } \ \langle a_1, y \rangle = 0 \} \tag{8.12}$$

and

$$C_1 = X_1 \cap \mathbb{B}. \tag{8.13}$$

If we already have $r_0, r_1, \ldots, r_n \in \mathbb{R}$, $a_1, a_2, \ldots, a_n \in X$, $f_1, f_2, \ldots, f_n \in X$, $g_1, g_2, \ldots, g_n \in X$, $h_1, h_2, \ldots, h_n \in X$, $X_0, X_1, \ldots, X_n$, $C_0, C_1, \ldots, C_n$ and $R_1, R_2, \ldots, R_n$, where each X_i is an affine set in X, $e_1, a_i \in C_i \subset X_i, e_k \in X_i$ for $k \geq i + 2\,(i = 0, \ldots, n), X_{i+1} \subset X_i$ for $i = 0, \ldots, n-1$, and each R_i is the nearest point holomorphic retraction of $\mathbb{B}$ onto C_i for $i = 1, \ldots, n$, then the next holomorphic retraction R_{n+1} is the nearest point holomorphic retraction onto

$$C_{n+1} = X_{n+1} \cap \mathbb{B}, \tag{8.14}$$

where

$$f_{n+1} = \frac{1}{\|e_1 - a_n\|}\,(e_1 - a_n), \tag{8.15}$$

$$g_{n+1} = e_{n+2} - a_n - \frac{1}{\|e_1 - a_n\|^2}\,\langle e_{n+2} - a_n\, e_1 - a_n \rangle (e_1 - a_n), \tag{8.16}$$

$$h_{n+1} = \frac{1}{\|g_{n+1}\|}\,g_{n+1}, \tag{8.17}$$

$$a_{n+1} = a_n + r_n \left[\left(\frac{1}{2} + \frac{1}{2}\cos\alpha_{n+1} \right) f_{n+1} + \left(\frac{1}{2}\sin\alpha_{n+1} \right) h_{n+1} \right], \tag{8.18}$$

$$r_{n+1} = \|e_1 - a_{n+1}\|, \tag{8.19}$$

and

$$X_{n+1} = \{x \in X_n : x = a_{n+1} + y \ \text{ and } \ \langle a_{n+1}, y \rangle = 0\}. \qquad (8.20)$$

Now it is sufficient to take

$$F_n = R_n \circ R_{n-1} \circ \cdots \circ R_1 \qquad (8.21)$$

for $n = 1, 2, \ldots$ to get

$$F_n \circ F_m = F_m \circ F_n \qquad (8.22)$$

for every $m, n \in \mathbb{N}$,

$$\text{Fix} F_n = C_n \qquad (8.23)$$

for each $n \in \mathbb{N}$,

$$\lim_{n \to \infty} \text{diam} C_n = 0, \qquad (8.24)$$

and

$$\bigcap_{n=1}^{\infty} \text{Fix} F_n = \emptyset. \qquad (8.25)$$

Nevertheless, for a one-parameter continuous semigroup of ρ-nonexpansive mappings the following assertion holds.

Theorem 8.1   *Let $\mathbb{B}$ be the open unit ball in a complex Hilbert space H and let ρ be the hyperbolic metric on $\mathbb{B}^n$, $n \geq 1$. If $S = \{F_t : t \geq 0\}$ is a continuous semigroup of ρ-nonexpansive self-mapping of $\mathbb{B}^n$, then the following statements are equivalent:*

(i) S has a periodic point in $\mathbb{B}^n$, i.e., there is $t_0 > 0$ such that F_{t_0} has a fixed point in $\mathbb{B}^n$;
(ii) F_t has a fixed point in $\mathbb{B}^n$ for each $t \geq 0$;
(iii) there is a stationary point $a \in \mathbb{B}^n$ of the semigroup S.

We prove this theorem in a somewhat more general setting. In this connection we give the following definition.

Definition 8.3   Let $\mathcal{D}$ be a topological space. We say that a class G of self-mappings of $\mathcal{D}$ has the **compact commuting property (CCP)** if for each commuting family $\{F_a : a \in J\}$ the set $\bigcap_{a \in J} \text{Fix} F_a$ is not empty whenever there is a compact subset M of $\mathcal{D}$ which is F_a-invariant for all $a \in J$.

Proposition 8.1   *Let $\mathcal{D}$ be a topological space, and let G be a class of self-mappings of $\mathcal{D}$ which has the compact commuting property (CCP). If $S = \{F_t : t \geq 0\} \subset G$ is a continuous semigroup of self-mappings of $\mathcal{D}$ which has a periodic point in $\mathcal{D}$, i.e., there is $s_0 > 0$ such that F_{s_0} has a fixed point in $\mathcal{D}$, then there is a stationary point of the semigroup S, i.e., $\bigcap\limits_{t \geq 0} \mathrm{Fix} F_t \neq \emptyset$.*

Proof.   Let $a \in \mathcal{D}$ be a periodic point of S, that is, a is a fixed point of F_{s_0} for some $s_0 > 0$. Consider the compact subset M of $\mathcal{D}$ defined by $M = \{F_t(a) : 0 \leq t \leq s_0\}$. We claim that $F_s(M) \subset M$ for all $s \geq 0$. Indeed, if $x \in M$, then there is $0 \leq p \leq s_0$ such that $x = F_p(a)$, and we have, by the semigroup property (i), that

$$F_s(x) = F_s(F_p(a)) = F_{s+p}(a). \tag{8.26}$$

On the other hand, the number $s + p$ can be represented as $s + p = n s_0 + t$ for some $0 \leq t \leq s_0$. Once again, by the semigroup property (i), we see that

$$F_s(x) = F_{s+p}(a) = F_{t+ns_0}(a) = F_t(F_{ns_0}(a)) =$$
$$F_t(F_{s_0}^n(a)) = F_t(a) \in M. \tag{8.27}$$

This proves our claim which, in turn, implies the existence of a common fixed point of S. $\qquad\qquad\square$

Now Theorem 8.1 is seen to be a consequence of Proposition 8.1 and Lemma 8.1.

Corollary 8.1   *Let S be a continuous semigroup of holomorphic mappings of $\mathbb{B}$ such that for some $t_0 \in (0, \infty)$, $F_{t_0} = I$. Then S is a group of automorphisms of $\mathbb{B}$ of elliptic type, i.e., $F_t = M^{-1} \circ e^{tA} \circ M$, where A is a linear conservative operator $(\mathrm{Re}\langle Ax, x \rangle = 0$ for all x in $H)$ and M is a Möbius transformation of $\mathbb{B}$.*

Proof.   Since $F_{t_0} = I$, Theorem 8.1 shows that S has a common fixed point $a \in \mathbb{B}$. Let $M = M_{-a}$ denote a Möbius transformation of $\mathbb{B}$ such that $M(a) = 0$, and consider the semigroup $\Phi(t) = M \circ F_t \circ M^{-1}$ defined on $\mathbb{B}$. It is clear that $\Phi(t)(0) = 0$ for all $t \geq 0$, and that $\Phi(t_0) = I$. Since

$$\Phi(s) \circ \Phi(t_0 - s) = \Phi(t_0) = I \tag{8.28}$$

for all $0 \leq s \leq t_0$, we see that for all $0 \leq s \leq t_0$, $\Phi^{-1}(s) = \Phi(t_0 - s)$ and $\Phi(s)$ is an automorphism of $\mathbb{B}$. Hence $\Phi(s)$ is the restriction to $\mathbb{B}$ of a linear isometry of H onto itself for each $s \geq 0$. This implies our assertion. $\square$

By using the notion of the duality mapping one can easily reformulate this result for $\mathbb{B}^n$ too. Occasionally, the mappings F_t, $t \in \mathbb{R}^+$, will also be denoted by $F(t)$.

The above approach implies also the following observation. If $\mathcal{D}$ is a strongly convex bounded C^2 domain in $\mathbb{C}^n$ and $\{F(t) : t \geq 0\}$ is a semigroup of holomorphic self-mappings of $\mathcal{D}$ such that for each $t > 0$, $F(t)$ has a fixed point in $\mathcal{D}$, then it follows by a result in [Abate (1989a)] (see also [Abate and Vigué (1991)]) that this semigroup has a common fixed point (stationary point) in $\mathcal{D}$ as a commutative family of holomorphic mappings. Moreover, as in the case of the Hilbert ball, it is enough to require the existence of an interior fixed point only for one $t_0 > 0$ to provide the existence of such a point for the whole semigroup.

Theorem 8.2 (*cf.* [Abate (1988b)]) *Let $\mathcal{D}$ be a strongly convex bounded C^2 domain in $\mathbb{C}^n$ and let $\{F_t : t \geq 0\}$ be a semigroup of holomorphic self-mappings of $\mathcal{D}$. Then the following assertions are equivalent:*

(a) *The semigroup $\{F_t\}$ has a stationary point in $\mathcal{D}$.*
(b) *The semigroup $\{F_t\}$ has a periodic point in $\mathcal{D}$, i.e., there exists $t_0 > 0$ such that F_{t_0} has a fixed point in $\mathcal{D}$.*
(c) *There exists $x \in \mathcal{D}$ and a sequence $t_n \to \infty$ such that $\{F_{t_n}(x)\}$ is strictly inside $\mathcal{D}$.*
(d) *For each $x \in \mathcal{D}$ there is a sequence $t_n \to \infty$ such that $\{F_{t_n}(x)\}$ is strictly inside $\mathcal{D}$.*

We will give a proof of this theorem which is different from the one in [Abate (1988b)] by using the infinitesimal generator of a semigroup in a somewhat more general situation.

Unfortunately, we cannot assert the full analog of this theorem for the infinite dimensional case (even for the Hilbert ball), but for some classes of ρ-nonexpansive or holomorphic mappings one can formulate a similar statement. We need the following notion.

Definition 8.4   Let $\mathcal{D}$ be a domain in a Banach space X. We will say that a class G of self-mappings of $\mathcal{D}$ has the **Denjoy–Wolff iteration property** **(DWIP)** if whenever $F \in G$ has no fixed point in $\mathcal{D}$, the sequence of iterates $\{F^n\}$ strongly converges to a point on the boundary of $\mathcal{D}$, uniformly on each compact subset of $\mathcal{D}$.

Proposition 8.2 *Let G be a class of ρ-nonexpansive mappings on $\mathbb{B}$ which satisfies the Denjoy–Wolff iteration property. Let $S = \{F_t\}_{t \geq 0}$ be a one-parameter continuous semigroup of ρ-nonexpansive mappings on $\mathbb{B}$ which has no common fixed point in $\mathbb{B}$. If $F_{t_0} \in G$ for at least one $t_0 > 0$, then there is a point $e \in \partial\mathbb{B}$ such that S converges to e as t tends to infinity, uniformly on each compact subset of $\mathbb{B}$.*

Proof. First we note that by Theorem 8.1, F_{t_0} has no fixed point in $\mathbb{B}$. Therefore there is a point $e \in \partial\mathbb{B}$ such that F_{nt_0} converges to e uniformly on each compact subset of $\mathbb{B}$. Let $\widetilde{C}$ be a compact subset of $\mathbb{B}$. Since the semigroup $S = \{F_t\}_{t \geq 0}$ is continuous, the set

$$C := \left\{ z_s = F_s(\tilde{z}) : \tilde{z} \in \widetilde{C},\ 0 \leq s \leq t_0 \right\} \tag{8.29}$$

is also a compact subset of $\mathbb{B}^n$. Therefore, for each $\varepsilon > 0$ one can find $n_0 \in \mathbb{N} = \{1, 2, \dots\}$ such that

$$\sup_{z \in C} \left\| F_{t_0}^n(z) - e \right\| = \sup_{\tilde{z} \in \widetilde{C}} \sup_{0 \leq s \leq t_0} \left\| F_{nt_0}(z_s) - e \right\|$$

$$= \sup_{\tilde{z} \in \widetilde{C}} \sup_{0 \leq s \leq t_0} \left\| F_{nt_0 + s}(\tilde{z}) - e \right\| < \varepsilon \tag{8.30}$$

for all $n \geq n_0$. If follows now that

$$\left\| F_t(\tilde{z}) - e \right\| < \varepsilon \tag{8.31}$$

for each $t \geq n_0 t_0$ and $\tilde{z} \in \widetilde{C}$. $\qquad\square$

Remark 8.2 *For the Hilbert ball $\mathbb{B}$ the following classes of self-mappings of $\mathbb{B}$ are known (see [Kuczumow et al. (2001a)]) to satisfy the Denjoy–Wolff iteration property:*

(1) the class $\mathcal{G}_1$ consisting of the condensing holomorphic mappings;
(2) the class $\mathcal{G}_2$ consisting of the firmly ρ-nonexpansive mappings of the first kind;
(3) the class $\mathcal{G}_3$ consisting of the firmly ρ-nonexpansive mappings of the second kind;
(4) the class $\mathcal{G}_4$ consisting of the averaged mappings of the first kind, i.e., $F = (1 - c)I \oplus cT$, where T is ρ-nonexpansive and $c \in (0, 1)$;
(5) the class $\mathcal{G}_5$ consisting of the averaged mappings of the second kind, i.e., $F = (1 - c)I + cT$, where T is ρ-nonexpansive and $c \in (0, 1)$.

Thus if a semigroup $S = \{F_t\}_{t \geq 0}$ contains at least one element F_{t_0}, $t_0 > 0$, of the class $\mathcal{G}_i$, $1 \leq i \leq 5$, we have the conclusion of Proposition 8.2.

In a somewhat more general situation one can say the following.

Proposition 8.3 *Let $\mathcal{D}$ be a domain in a Banach space X, and let G be a class of self-mappings of $\mathcal{D}$ which has both the (CCP) and the (DWIP) properties. Suppose that $\{F_t : t \geq 0\} \subset G$ is a semigroup of self-mappings of $\mathcal{D}$. Then the following assertions are equivalent:*

(a) The semigroup $\{F_t\}$ has a stationary point in $\mathcal{D}$.
(b) The semigroup $\{F_t\}$ has a periodic point in $\mathcal{D}$, i.e., there exists $t_0 > 0$ such that F_{t_0} has a fixed point in $\mathcal{D}$.
(c) There exists $x \in \mathcal{D}$ and a sequence $t_n \to \infty$ such that $\{F_{t_n}(x)\}$ is strictly inside $\mathcal{D}$.
(d) For each $x \in \mathcal{D}$ there is a sequence $t_n \to \infty$ such that $\{F_{t_n}(x)\}$ is strictly inside $\mathcal{D}$.

8.2 Generated Semigroups

As a matter of fact, we already know that if $\mathcal{D}$ is a bounded domain in $\mathbb{C}^n$, then each continuous semigroup of holomorphic self-mappings of $\mathcal{D}$ is locally uniformly continuous and hence differentiable at $t = 0$.

We consider now such a situation in general. That is, we assume that $\mathcal{D}$ is a domain in a Banach space X and a semigroup $S = \{F_t : t \geq 0\}$ is generated on $\mathcal{D}$, *i.e.*, for each $x \in \mathcal{D}$ there exists the strong limit

$$f(x) = \lim_{t \to 0^+} \frac{1}{t}\left(x - F_t(x)\right). \tag{8.32}$$

If in this case the mapping $f : \mathcal{D} \to X$ (the infinitesimal generator of S) is locally Lipshitzian on $\mathcal{D}$, then, by using the uniqueness of the solution to the Cauchy problem

$$\begin{cases} \dfrac{\partial u(t, x)}{\partial t} + f(u(t, x)) = 0 \\ \lim_{t \to 0^+} u(t, x) = x, \end{cases} \tag{8.33}$$

where $u(t, x) = F_t(x)$ and f is defined by (8.32), it follows that the stationary point set $\mathcal{W}$ is the null point set $(\mathrm{Null}_{\mathcal{D}} f)$ of f in $\mathcal{D}$, *i.e.*,

$$\mathcal{W} = \bigcap_{t \geq 0} \mathrm{Fix}_{\mathcal{D}} F_t = \mathrm{Null}_{\mathcal{D}} f. \tag{8.34}$$

Once again, we continue with a somewhat more general situation.

Definition 8.5 We say that a class $\mathcal{M}$ of self-mappings of $\mathcal{D}$ has the common fixed point property (CFPP) if the following condition holds:

If $\{F_s\}_{s\in\mathcal{A}}$ is a net of commuting mappings in $\mathcal{M}$ such that for each $s \in A$, F_s has a fixed point in $\mathcal{D}$, then $\bigcap\limits_{s\in\mathcal{A}} \mathrm{Fix}_{\mathcal{D}} F_s \neq \emptyset$.

Proposition 8.4 *Let X be a real or complex Banach space, and let $\mathcal{D}$ be a domain in X. Suppose that $\mathcal{M}$ is a class of self-mappings of $\mathcal{D}$ with both the (CFPP) and the (DWIP). Let $f : \mathcal{D} \to X$ be the infinitesimal generator of a one-parameter continuous semigroup $\{F(t) : t \geq 0\} \subset \mathcal{M}$ such that $\{F(t)\}$ is locally uniformly Lipschitzian on $\mathcal{D}$. If f has no null point in $\mathcal{D}$, then for each $x \in \mathcal{D}$ the net $\{F(t)(x)\}$ strongly converges to a point b on the boundary of $\mathcal{D}$.*

Proof. First we note that there is an interval $(0, \mu)$ such that for each $t \in (0, \mu)$, $F(t)$ has no null point in $\mathcal{D}$. Indeed, if we suppose that there is a sequence $t_n \to 0$ such that for each n the mapping $F(t_n)$ has a fixed point in $\mathcal{D}$, then by the (CFPP) there is a point $x \in \mathcal{D}$ such that $F(t_n)(x) = x$ for all n and hence $f(x) = 0$. This is a contradiction.

So, for each m large enough $(m > 1/\mu)$ all the mappings $F\left(\frac{1}{m}\right)$ have no fixed point in $\mathcal{D}$. This means that for each $x \in \mathcal{D}$ the sequence $F\left(\frac{n}{m}\right)(x) \left(= F^n\left(\frac{1}{m}\right)\right)$ converges strongly as $n \to \infty$ to a point b_m on the boundary of $\mathcal{D}$. But it follows from the semigrcup property that $F^{nm}\left(\frac{1}{m}\right)(x) = F^n(1)(x) \to b_m$, as $n \to \infty$, and hence $b_m = b$ does not depend on m.

We are now able to show that the semigroup $\{F(t)\}$ strongly converges to b as t tends to infinity. In fact, for any given $\varepsilon > 0$ and $x \in \mathcal{D}$, we can choose $\delta > 0$ such that $\|F(t)(x) - F(t)(y)\| < \varepsilon/2$ for all $t > 0$ whenever $y \in \mathcal{D}$ and $\|y - x\| < \delta$. For such δ we take $m \in \mathbb{N}$ so large that $m^{-1} \in (0, \mu)$ and $\|F(h)(x) - x\| < \delta$ for all $h \in [0, 1/m)$.

Finally, for such m and $t > 0$, setting $n = [tm]$, we have

$$\left\|F\left(\frac{n}{m}\right)(x) - b\right\| = \left\|F^n\left(\frac{1}{m}\right)(x) - b\right\| < \varepsilon/2 \tag{8.35}$$

for $t > 0$ big enough. Since $h = t - \frac{n}{m} \in \left[0, \frac{1}{m}\right)$, we get for such $t > 0$,

$$\|F(t)(x) - b\| = \left\|F\left(\frac{n}{m}\right)(F(h)(x)) - b\right\| \leq$$
$$\left\|F\left(\frac{n}{m}\right)(F(h)(x)) - F\left(\frac{n}{m}\right)(x)\right\| + \left\|F\left(\frac{n}{m}\right)(x) - b\right\|$$
$$\leq \frac{\varepsilon}{2} + \frac{\varepsilon}{2} = \varepsilon, \tag{8.36}$$

and we are done.   □

Corollary 8.2   *Under the conditions of Proposition 8.4, assume in addition that f is also locally Lipschitzian. Then the semigroup $\{F(t)\}$ has a stationary point in $\mathcal{D}$ if and only if for some $t_0 > 0$ the mapping $F(t_0)$ has a fixed point in $\mathcal{D}$.*

Note now that by results in [Abate (1988a)] (see also [Kuczumow and Stachura (1990)]) for each strongly convex bounded C^2 domain $\mathcal{D}$ in $\mathbb{C}^n$ the class of holomorphic self-mappings of $\mathcal{D}$ has the Denjoy–Wolff iteration and the common fixed point properties. In addition, it is known [Abate (1992)] that each one-parameter semigroup of holomorphic self-mappings on $\mathcal{D}$ is differentiable with respect to the parameter. In this way we are led to a proof of Theorem 8.2.

Remark 8.3   *The following example shows that formula (8.34) is no longer true for the closure of $\mathcal{D}$ even in the case when f is continuous on $\overline{\mathcal{D}}$.*

Example 8.2   Let $\mathcal{D}$ be the open unit disk in the complex plane $\mathbb{C}$, *i.e.*, $\mathcal{D} = \{x \in \mathbb{C} : |x| < 1\}$. Consider $f(x) = x - 1 + \sqrt{1 - x}$. It is clear that $f \in \mathrm{Hol}(\mathcal{D}, \mathbb{C})$ and that it is continuous on $\overline{\mathcal{D}}$.

In addition, $\mathrm{Null}_{\overline{\mathcal{D}}} f = \{0, 1\}$. Ho"wever, the Cauchy problem (8.33) has the solution

$$F_t : \overline{\mathcal{D}} \mapsto \overline{\mathcal{D}}, \quad t \geq 0 \tag{8.37}$$

defined by the formula

$$F_t(x) = 1 - \left[1 - e^{-\frac{1}{2}t} + e^{-\frac{1}{2}t}\sqrt{1 - x}\right]^2, \tag{8.38}$$

and for all $t > 0$ we have

$$F_t(1) = 1 - \left[1 - e^{-\frac{1}{2}t}\right]^2 < 1. \tag{8.39}$$

Thus $F_{\overline{\mathcal{D}}} \neq \mathrm{Null}_{\overline{\mathcal{D}}} f$.

We will study boundary null points of generators later.

8.3   The Resolvent Method

Since in most situations the generator of a one-parameter semigroup satisfies the range condition, one can study the common fixed point set of the semigroup as the fixed point set of a single self-mapping defined by the resolvent of the generator.

Lemma 8.2 *Let $\mathcal{D}$ be a domain in X, and let $f : \mathcal{D} \to X$ satisfy the range condition on $\mathcal{D}$, i.e., the resolvent $J_r = (I + rf)^{-1}$ is well-defined for all $r > 0$. Then for each $r > 0$,*

$$\mathrm{Fix}_{\mathcal{D}} J_r = \mathrm{Null}_{\mathcal{D}} f. \tag{8.40}$$

Proof. Indeed, by definition, for each $x \in \mathcal{D}$ and each $r > 0$ we have the identity

$$J_r(x) + rf(J_r(x)) = x. \tag{8.41}$$

If $a \in \mathrm{Fix}_{\mathcal{D}} J_r$, then we have by (8.41) $f(a) = f(J_r(a)) = 0$, i.e., $a \in \mathrm{Null}_{\mathcal{D}} f$. Conversely, if $f(a) = 0$, then $(I + rf)(a) = a$ for all $r > 0$. Thus $J_r(a) = (I + rf)^{-1}(a) = a$. $\qquad\square$

We now describe the resolvent method for the study of the null point set of generators. This method can be used to determine its structure, to solve the existence problem, as well as to find methods for approximating null points of generators.

First we will trace a parallel with an implicit method for the study of the fixed point set of a single self-mapping of a domain.

Let $\mathcal{D}$ be a bounded convex metric domain in X endowed with a metric ρ compatible with the convex structure of $\mathcal{D}$ (in particular, ρ may be the hyperbolic metric on $\mathcal{D}$). Suppose that $F : \mathcal{D} \to \mathcal{D}$ is a ρ-nonexpansive self-mapping of $\mathcal{D}$. For $t \in [0, 1)$ and a fixed $y \in \mathcal{D}$, consider the mapping $\Phi(x) = tF(x) + (1 - t)y$, $x \in \mathcal{D}$. Since Φ is a strict contraction on $(\mathcal{D}, \rho)$ (in particular, Φ maps $\mathcal{D}$ strictly inside itself), the Banach principle implies that there exists a unique fixed point $z = z_t(y) \in \mathcal{D}$ of the mapping Φ. Moreover,

$$z_t(y) = \lim_{n \to \infty} \Phi^n(y), \tag{8.42}$$

where the limit is uniform on each ρ-ball in $(\mathcal{D}, \rho)$. (For holomorphic mappings the same conclusion follows from the Earle–Hamilton theorem.)

The X-valued function $z_t(y)$, $0 \le t < 1$, is called an **approximating curve**.

At the same time, changing our point of view, formula (8.42) implies that for each $t \in [0, 1)$, $z_t = z_t(y)$ can be considered a ρ-nonexpansive self-mapping of $\mathcal{D}$ depending on $y \in \mathcal{D}$ (in particular, if F is holomorphic, then

z_t depends holomorphically on $y \in \mathcal{D}$). We denote this mapping by $\mathcal{T}_t$. In other words, $\mathcal{T}_t$ is the unique solution of the nonlinear operator equation

$$\mathcal{T}_t = tF \circ \mathcal{T}_t + (1-t)I. \tag{8.43}$$

The mapping $\mathcal{T}_t$ belongs to $N_\rho(\mathcal{D})$ (in particular, to $\mathrm{Hol}(\mathcal{D})$), and as we already know

$$\mathrm{Fix}(\mathcal{T}_t) = \mathrm{Fix}(F), \quad t \in (0,1). \tag{8.44}$$

On the other hand, we know that the mapping $g = I - F$ is a generator of a continuous flow of ρ-nonexpansive self-mappings of D. In addition, $g : \mathcal{D} \to X$ satisfies the range condition, *i.e.*, for each $r > 0$ the resolvent $G_r = (I + rg)^{-1}$ is a well defined ρ-nonexpansive (holomorphic) self-mapping of $\mathcal{D}$.

Claim 8.1　$G_r = \mathcal{T}_t$ with $t = \frac{r}{1+r}$.

Indeed, since $G_r = (I + r(I - F))^{-1}$ is the solution of the equation $(I + r(I - F)) \circ G_r = I$, we have $(1 + r)G_r - rF \circ G_r = I$ or

$$G_r = \frac{r}{1+r} F \circ G_r + \frac{1}{1+r} I. \tag{8.45}$$

Setting $t = \frac{r}{1+r}$, we get our claim by the uniqueness of the solution of this equation.

Returning to the general case, let $f : \mathcal{D} \to X$ be the generator of a continuous flow of ρ-nonexpansive self-mappings of $\mathcal{D}$, and let $J_s = (I + sf)^{-1}$, $s > 0$, be its resolvent. Setting $F = J_s$ in our previous considerations, we see that for each $r > 0$ the mapping $G_r = (I + r(I - J_s))^{-1}$ is a well-defined ρ-nonexpansive self-mapping of $\mathcal{D}$.

The following relation is the key to our approach in the sequel.

Lemma 8.3　*Let $f : \mathcal{D} \to X$ be the generator of a continuous flow of ρ-nonexpansive self-mappings of $\mathcal{D}$, and let $J_s = (I + sf)^{-1}$, $s > 0$, be its resolvent. Define $G_r = (I + r(I - J_s))^{-1} : \mathcal{D} \to \mathcal{D}$. Then*

$$J_{(r+1)s} = J_s(G_r) = J_s(I + r(I - J_s))^{-1}. \tag{8.46}$$

Proof.　Fix $s > 0$ and set $F = J_s$. By the previous claim we know that for each $r > 0$, the mapping $G_r = (I + r(I - J_s))^{-1}$ is a well-defined ρ-nonexpansive self-mapping of $\mathcal{D}$ and for each $x \in \mathcal{D}$ this mapping can be defined as the unique solution of the equation

$$G_r(x) = \frac{r}{r+1} J_s(G_r(x)) + \frac{1}{r+1} x. \tag{8.47}$$

On the other hand, since $[I + r(I - J_s)]G_r = I$ and $I - J_s = sf(J_s)$ we have

$$[I + rsf(J_s)]G_r(x) = G_r(x) + rsf(J_s(G_r(x))) = x, \quad x \in \mathcal{D}. \quad (8.48)$$

Hence it follows from (8.47) that

$$J_s(G_r(x)) + (1 + r)sf(J_s(G_r(x))) = x, \quad x \in \mathcal{D}. \quad (8.49)$$

Since the equation

$$z + (1 + r)sf(z) = x, \quad x \in \mathcal{D}, \quad (8.50)$$

has the unique solution $z = J_{(r+1)s}(x)$ this means that

$$J_{(r+1)s}(x) = J_s(G_r(x)). \quad (8.51)$$
$$\square$$

Definition 8.6 Let $\mathcal{D}$ be a bounded convex metric domain in X endowed with a metric ρ compatible with the convex structure of $\mathcal{D}$. We say that $\mathcal{D}$ satisfies the **implicit approximation property** if for each $x \in \mathcal{D}$ and for each ρ-nonexpansive self-mapping F of $\mathcal{D}$ its approximating curve $\{z_t(x) : t \in [0,1)\}$ defined by the equation

$$z_t(x) = tF(z_t(x)) + (1 - t)x \quad (8.52)$$

is convergent to a point of $\overline{\mathcal{D}}$, the closure of $\mathcal{D}$.

Proposition 8.5 *Let $\mathcal{D}$ be as above and let $f : \mathcal{D} \to X$ satisfy the range condition, i.e., for each $r > 0$ the resolvent $J_r = (I + rf)^{-1}$ is a well-defined ρ-nonexpansive self-mapping of $\mathcal{D}$. Then*

(1) For each $x \in \mathcal{D}$, there exists a $(= a(x)) \in \overline{\mathcal{D}}$ such that $J_r(x) \to a$ and $f(J_r(x)) \to 0$ in the topology of X as $r \to \infty$.
(2) The following hypotheses are equivalent:

(a) $\text{Null}_{\mathcal{D}} f \neq \emptyset$;
(b) for some $x \in \mathcal{D}$, the net $\{J_r(x)\}$, where $J_r = (I + rf)^{-1}$ is strictly inside $\mathcal{D}$;
(c) for each $x \in \mathcal{D}$, the net $\{J_r(x)\}$ is strictly inside $\mathcal{D}$.

Proof. Fix $s > 0$ and $x \in \mathcal{D}$. For $r > 0$, consider $G_r(x) = (I + r(I - J_s))^{-1}(x)$. By our assumptions and the above claim, $G_r(x) \to a$ for some $a (= a(x)) \in \overline{\mathcal{D}}$. In turn, by (8.47), we have that

$$J_s(G_r(x)) - G_r(x) \to 0, \quad r \to \infty. \quad (8.53)$$

Together with (8.46) we get that for a given $x \in \mathcal{D}$,

$$\lim_{r \to \infty} J_r(x) = a \ (= a(x)) \in \overline{\mathcal{D}}. \tag{8.54}$$

On the other hand, it follows by the definition of the resolvent that

$$\frac{x - J_r(x)}{r} = f(J_r(x)), \quad r > 0. \tag{8.55}$$

Thus $f(J_r(x)) \to 0$ as $r \to \infty$, and assertion 1 is proved.

To prove assertion 2, we just have to show that (b) implies (a) and (a) implies (c) since the implication (c) $\Rightarrow$ (b) is trivial. Indeed, if we assume that for some $x \in \mathcal{D}$ the net $\{J_r(x)\}$ is strictly inside $\mathcal{D}$, then $a = \lim_{r \to \infty} J_r(x) \in \mathcal{D}$, and it follows from assertion 1 that a is a null point of f.

Conversely, if f has a null point $a \in \mathcal{D}$, then for each $x \in \mathcal{D}$ we have by (8.40), $\rho(J_r(x), a) = \rho(J_r(x), J_r(a)) \leq \rho(x, a)$. Hence for each $x \in \mathcal{D}$, the net $\{J_r(x)\}$ is ρ-bounded in $(\mathcal{D}, \rho)$, or, which is the same, this net is strictly inside $\mathcal{D}$. $\qquad\square$

Remark 8.4 *Two standard examples of metric domains which satisfy the **implicit approximating property** are once again the open Hilbert ball and bounded convex domains in $\mathbb{C}^n$ endowed with the hyperbolic metric (see Chapters 4 and 5). In addition, it follows that $\mathrm{Null}_{\mathcal{D}} f$ is a ρ-nonexpansive retract of $\mathcal{D}$, and therefore, by Theorem 5 in [Kuczumow and Stachura (1990)] it is a metrically convex subset of $(\mathcal{D}, \rho)$. We consider these two cases independently in the next section in the situation where a generator f has no interior null points in $\mathcal{D}$.*

8.4   Null Point Free Generators

Let $\mathcal{D}$ be a bounded convex metric domain in X endowed with a metric ρ compatible with the convex structure of $\mathcal{D}$. Assume also that $\mathcal{D}$ satisfies the implicit approximating property. Let f be the generator of a one parameter semigroup $S = \{F_t : t \in (0, \infty)\}$ of ρ-nonexpansive self-mappings of $\mathcal{D}$. Suppose that f is null point free, *i.e.*, the stationary point set

$$\mathcal{W} = \bigcap_{0 < t < \infty} \mathrm{Fix}_{\mathcal{D}} F_t = \mathrm{Null}_{\mathcal{D}} f = \emptyset. \tag{8.56}$$

In this case, we already know by Proposition 8.5 that for each $x \in \mathcal{D}$ the net of resolvents $\{J_r(x) \, (= (I + rf)^{-1}(x)) : t \geq 0\}$ strongly converges to a boundary point a $(= a(x))$. We will now show that for the two cases when $\mathcal{D}$ is the open Hilbert ball or a bounded strongly convex domain in $\mathbb{C}^n$ this point $a \in \partial \mathcal{D}$ does not depend on $x \in \mathcal{D}$ and is the **sink point** (Wolff point) for the semigroup $\{F_t : t \in (0, \infty)\}$ generated by f.

Now let $X = H$ be a complex Hilbert space with the inner product $\langle \cdot, \cdot \rangle$, and let $\mathcal{D} = \mathbb{B}$ be the open unit ball in H with the hyperbolic metric ρ.

We recall that a boundary point $a \in \partial \mathbb{B}$ is called the sink (or Wolff) point for a self-mapping $F \in N_\rho(\mathbb{B})$ if all the ellipsoids

$$E(a, K) = \left\{ x \in \mathbb{B} : \left| 1 - \langle x, a \rangle \right|^2 / (1 - \|x\|^2) < K \right\}, \quad K > 0, \quad (8.57)$$

internally tangent to $\partial \mathbb{B}$ at the point a are invariant under F.

Theorem 8.3   *Let f be the generator of a one parameter semigroup $S = \{F_t : t \in (0, \infty)\}$ of ρ-nonexpansive self-mappings of $\mathcal{D}$ and let f have no null point in $\mathbb{B}$. Then there is a unique boundary point $a \in \partial \mathbb{B}$ such that*

(i) for each $x \in \mathcal{D}$ the net of resolvents $\{J_r(x) \, (= (I + rf)^{-1}(x)) : r > 0\}$ strongly converges to a;

(ii) for each $r > 0$ the point $a \in \partial \mathbb{B}$ is the sink point for the resolvent mapping $J_r : \mathbb{B} \to \mathbb{B}$;

(iii) for each $t > 0$ the point $a \in \partial \mathbb{B}$ is the sink point for the mapping $F_t : \mathbb{B} \to \mathbb{B}$, an element of the semigroup S.

Proof.   If $\mathrm{Null}_{\mathbb{B}} f = \emptyset$, then we claim that there exists a point $a \in \partial \mathbb{B}$ such that for each $x \in \mathbb{B}$, $J_r(x)$ strongly converges to a as $r \to \infty$. Indeed, for each $x \in \mathbb{B}$, the net $\{J_r(x)\}$ is ρ-unbounded, *i.e.*, $\|J_r(x)\| \to 1$ as $r \to \infty$. Now fix $r_0 > 0$ and consider the mapping $J_{r_0} : \mathbb{B} \to \mathbb{B}$. Since J_{r_0} has no fixed points in $\mathbb{B}$, it is known (see Chapter 5) that there exists a point $a \in \partial \mathbb{B}$ (the sink point) such that all the ellipsoids

$$E(a, K) = \left\{ x \in \mathbb{B} : \left| 1 - \langle x, a \rangle \right|^2 / (1 - \|x\|^2) < K \right\}, \quad K > 0, \quad (8.58)$$

are invariant under J_{r_0}. But it follows from the resolvent identity

$$J_r x = J_{r_0} \left(\frac{r_0}{r} x + \left(1 - \frac{r_0}{r} \right) J_r(x) \right), \quad (8.59)$$

$r > r_0$, that $E(a, k)$ is also invariant under J_r for all $r > r_0$. Now if for a fixed $x \in \mathbb{B}$ we choose $K > 0$ such that $x \in E(a, K)$, then we get

$$\left| 1 - \langle J_r(x), a \rangle \right|^2 < K \left(1 - \|J_r(x)\|^2 \right) \quad (8.60)$$

and consequently, $\left| 1 - \langle J_r(x), a \rangle \right| \to 0$ as $r \to \infty$. Hence $J_r(x)$ converges strongly to a as $r \to \infty$. If we now assume that for some $0 < r < r_0$ there exists $K > 0$ such that the ellipsoid $E(a, K)$ is not invariant under J_r we get that there is another boundary sink point $b \in \partial \mathbb{B}$ of J_r for such $r \in (0, r_0)$. That is, all the ellipsoids

$$E(b, K) = \left\{ x \in \mathbb{B} : \left| 1 - \langle x, b \rangle \right|^2 / (1 - \|x\|^2) < K \right\}, \quad K > 0, \quad (8.61)$$

are invariant under J_r. Once again, using the resolvent identity

$$J_{r_0}(x) = J_r \left(\frac{r}{r_0} x + \left(1 - \frac{r}{r_0} \right) J_{r_0}(x) \right), \quad (8.62)$$

we get that all the ellipsoids $E(b, K), K > 0$, are invariant under J_{r_0}. Clearly, we can find $K > 0$ such that the set $M = E(a, K) \cap E(b, K)$ is not empty and is contained in $\mathbb{B}$. But this set is a ρ-bounded subset of $\mathbb{B}$ which is also invariant under J_{r_0}. Thus J_{r_0} must have an interior fixed point in M. This is a contradiction. So, for all $r > 0$, all the ellipsoids $E(a, K)$ are invariant under J_r.

Finally, by using the exponential formula

$$F_t = \lim_{n \to \infty} \left(J_{\frac{t}{n}} \right)^n \quad (8.63)$$

we see that these ellipsoids $E(a, K)$ are also invariant under F_t for each $t \geq 0$. The theorem is proved. $\qquad \square$

Corollary 8.3 *Let $\mathbb{B}$ be the open unit ball in a complex Hilbert space H, and let $f : \mathbb{B} \to H$ satisfy the range condition on $\mathbb{B}$. If f has a continuous extension to $\overline{\mathbb{B}}$, then it has a null point in $\overline{\mathbb{B}}$.*

Proof. If f has no null point in $\mathbb{B}$, then as we saw above there exists a sink point $a \in \partial \mathbb{B}$ such that for each $x \in \mathbb{B}$, $J_r(x)$ strongly converges to a, as $r \to \infty$. But since $f(J_r(x))$ converges to zero as $r \to \infty$, it follows by continuity that $f(a) = 0$ and we are done. $\qquad \square$

Remark 8.5 *We will see below that, in fact, for holomorphic generators the following assertion is true: if f has no null point in $\mathbb{B}$, then there is a point $a \in \partial \mathbb{B}$ such that*

$$\lim_{r \to 1^-} f(ra) = 0. \quad (8.64)$$

Moreover, we will see that for the finite dimensional case (when $\mathbb{B}$ is the open unit ball in $\mathbb{C}^n$), if $a \in \partial \mathbb{B}$ is the sink point of the semigroup $S = \{ F_t :$

$t \in (0, \infty)\}$, *then there exists the so-called angular derivative*

$$\angle f'(a) = \beta, \tag{8.65}$$

*which is a real nonnegative number. It is clear that (8.65) implies (8.64).
In addition, in this case, the semigroup* $S = \{F_t : t \in (0, \infty)\}$ *strongly
converges to* a *as* $t \to \infty$.

Unfortunately, we cannot claim the latter fact in general. More details
on the asymptotic behavior of semigroups in general Hilbert spaces can be
found in Chapter 9.

Here we show that in the particular case when the generator f has the
special form $f = I - T$, where T is a self-mapping of $\mathbb{B}$, then the semigroup
$S = \{F_t : t \in (0, \infty)\}$ strongly converges to the sink point $a \in \partial\mathbb{B}$ of the
mapping T as $t \to \infty$.

Theorem 8.4 *Let* T *be a* ρ-*nonexpansive mapping of* $\mathbb{B}$.

(1) The Cauchy problem

$$\begin{cases} \dfrac{\partial u(t, x)}{\partial t} + u(t, x) - T(u(t, x)) = 0 \\ u(0, x) = x \end{cases} \tag{8.66}$$

has a unique solution $\{u(t, x) : t \in \mathbb{R}^+\} \subset \mathbb{B}$ *for each* $x \in \mathbb{B}$.
(2) The semigroup $\{F(t)\}_{t \geq 0}$ *defined by*

$$F(t)x := u(t, x) \tag{8.67}$$

consists of ρ-*nonexpansive mappings* $F(t)$, $t \geq 0$.
(3) If T *has no fixed point in* $\mathbb{B}$, *then* $F(t)$ *is also fixed-point-free for all*
$t > 0$, *and* $F(t)x$ *converges as* t *tends to infinity to a point* $a \in \partial\mathbb{B}$, *the
boundary of* $\mathbb{B}$, *for all* $x \in \mathbb{B}$.

Proof. First we note that the curve $G_t = (1 - t)I + tT$ converges to I as
$t \to 0^+$ and

$$\frac{1}{t}(I - G_t) = I - T. \tag{8.68}$$

Since

$$\rho(G_t(x), G_t(y)) \leq \max\left[\rho(x, y), \rho(T(x), T(y))\right], \tag{8.69}$$

each G_t is ρ-nonexpansive. Hence it follows again by the product formula that for each $t > 0$ the mapping $F(t)$ given by

$$F(t) = \lim_{n \to \infty} G_{\frac{t}{n}}^n \tag{8.70}$$

is the solution operator for the Cauchy problem (8.66) and is also a ρ-nonexpansive mapping on $\mathbb{B}$. If now T has no fixed point in $\mathbb{B}$, then it follows that

$$\bigcap_{t>0} \mathrm{Fix} F(t) = \emptyset. \tag{8.71}$$

By Theorem 8.1 we now see that each F_t has no fixed point in $\mathbb{B}$.

At the same time direct calculations show that the following formula holds:

$$F(t)(x) = e^{-t}x + \int_0^t e^{-(t-\xi)} T(F(\xi)(x)) d\xi. \tag{8.72}$$

Since

$$\|T(F(\xi)(x))\| < 1 \tag{8.73}$$

for all $\xi \in [0, t]$ and $x \in \mathbb{B}$, we can write

$$F(t) = e^{-t}I + (1 - e^{-t})A, \tag{8.74}$$

where

$$A = \frac{1}{1 - e^{-t}} \int_0^t e^{-(t-\xi)} (T \circ F(\xi)) d\xi \tag{8.75}$$

is a ρ-nonexpansive mapping on $\mathbb{B}$. Hence F_t is an averaged mapping of the second kind and the result follows by Proposition 8.2 and Remark 8.2. $\square$

In the setting of finite dimensional Banach spaces it is known that each convex domain in $\mathbb{C}^n$ satisfies the implicit approximation property. If, in particular, $\mathcal{D}$ is a strongly convex C^2 domain, and $f \in \mathrm{Hol}(D, X)$ has no null point in $\mathcal{D}$, then there exists a unique point $a \in \partial \mathcal{D}$ such that for each $x \in \mathcal{D}$, the net $\{J_r(x)\}$ converges to a, as $r \to \infty$. Moreover, in this case the class of holomorphic self-mappings of $\mathcal{D}$ satisfies the Denjoy–Wolff iteration property, so the semigroup generated by f strongly converges to a boundary point $b \in \partial \mathcal{D}$. The question is, of course, whether these two points a and b coincide.

Theorem 8.5 *Let $\mathcal{D}$ be a strongly convex bounded C^2 domain in $\mathbb{C}^n$ and let $\{F(t) : t \geq 0\}$ be a semigroup of holomorphic self-mappings of $\mathcal{D}$. If $\{F(t)\}$ has no stationary point in $\mathcal{D}$, then there exists a unique point $a \in \partial\mathcal{D}$ such that*

(a) For all $x \in \mathcal{D}$ the net $\{F(t)(x)\}$ strongly converges, as $t \to \infty$, to a.

(b) For all $x \in \mathcal{D}$ the net $J_r(x) = I - rF'(0)^{-1}(x)$ strongly converges as $r \to \infty$ to the same point a.

Proof. To establish our assertion it is enough to identify in our situation the points a and b obtained in Propositions 8.4 and 8.5. Setting $f = -F'(0)$, we consider again the resolvents $J_s = (I + sf)^{-1}$, $s > 0$, which are holomorphic self-mappings of $\mathcal{D}$. Since f has no null points in $\mathcal{D}$, J_s has no fixed point in $\mathcal{D}$ and as we saw above, the net $\{G_r(0)\}$ defined by formula (8.47) converges to $a \in \partial\mathcal{D}$ as $r \to \infty$ $\left(G_r(0) = \frac{r}{r+1} J_s(G_r(0))\right)$. It follows from the proof of Theorem 2.3 in [Abate (1988a)] that for each $n \in \mathbb{N}$ and for every $x \in \mathcal{D}$ and $R > 0$, the following inclusion holds:

$$J_s^n(E_x(a, R)) \subset F_x(a, R), \tag{8.76}$$

where $E_x(a, R)$ and $F_x(a, R)$ are the small and the big horospheres of center a, pole x, and radius R defined by

$$E_x(a, R) = \left\{ z \in \mathcal{D} : \limsup_{w \to a} \left[\rho(z, w) - \rho(x, w)\right] < \frac{1}{2} \log R \right\}, \tag{8.77}$$

$$F_x(a, R) = \left\{ z \in \mathcal{D} : \liminf_{w \to a} \left[\rho(z, w) - \rho(x, w)\right] < \frac{1}{2} \log R \right\}, \tag{8.78}$$

where ρ is the hyperbolic metric on $\mathcal{D}$.

Therefore it follows again by the exponential formula (8.63) that for all $t > 0$,

$$F(t)(E_x(a, R)) \subset \overline{F_x(a, R)}. \tag{8.79}$$

Hence, if $b = \lim_{t \to \infty} F(t)x$, then $b = a$ and we are done. $\qquad\square$

8.5 The Structure of Null Point Sets of Holomorphic Generators. Retractions

By $\mathrm{Null}_{\mathcal{D}} f$ we denote the analytic set defined as the null point set of $f \in \mathrm{Hol}(\mathcal{D}, X)$. Even in the finite dimensional case it is a complicated problem to recognize when an analytic set N consists only of irreducible components (see, for example [Aizenberg and Yuzhakov (1983)] and [Chirka (1985)]). It is known that this is the case when N is locally a complex analytic manifold.

By using the above approach one can study the null point set of a generator f as the common fixed point set of the commuting family $\{F_t\}$, the semigroup generated by f.

Indeed, if $\mathcal{D}$ is a bounded convex domain in $\mathbb{C}^n$ and $\{F_t\}$ is a semigroup of holomorphic mappings, then it is well known ([Abate and Vigué (1991)]) that the common fixed point set of this commutative family is a holomorphic retract of $\mathcal{D}$. This approach becomes less transparent if X is an infinite dimensional space.

Equality (8.40) and the Mazet–Vigué theorem [Mazet and Vigué (1991)] (for example) immediately imply that the null point set of a holomorphic generator on a convex bounded domain in a complex reflexive Banach space is also a holomorphic retract of $\mathcal{D}$. (This, in turn, implies that $\text{Null}_{\mathcal{D}} f$ is a connected submanifold of $\mathcal{D}$ [Cartan (1986)].)

Theorem 8.6 *Let $\mathcal{D}$ be a convex bounded domain in a complex Banach space X and let $f \in \mathcal{G}\text{Hol}(\mathcal{D})$. Suppose that $a \in \text{Null}_{\mathcal{D}} f$ and that one of hte following hypotheses holds:*

(1) X is reflexive,
(2) $\text{Ker}A \oplus \text{Im}\, A = X$, where $A = f'(a)$.

Then $\text{Null}_{\mathcal{D}} f$ is a connected complex analytic submanifold in $\mathcal{D}$, which is tangent to $\text{Ker}A$.

Proof.   It is sufficient to note the $a \in \text{Null}_{\mathcal{D}} f = \text{Fix}_{\mathcal{D}} J_r(f)$ for all $r > 0$, where $\text{Fix}_{\mathcal{D}} J_r(f)$ is the fixed point set of the resolvent $J_r(f)$ of f in $\mathcal{D}$. In addition, it follows by the chain rule that $I - J_r(A) = r J_r(A)$, where $J_r(A) = [J_r(f)]'(a)$ is the resolvent of the linear operator A for $r > 0$. Thus $\text{Ker}\big(I - [J_r(f)]'(a)\big) = \text{Ker}A$, and the theorem follows from the Mazet–Vigué theorem. $\square$

Corollary 8.4 *Let $\mathcal{D}$ be a bounded convex domain in a reflexive complex Banach space X, and let $\{F_t : t \in (0, \infty)\}$ be a locally uniformly continuous semigroup of holomorphic self-mappings of $\mathcal{D}$. Then*

$$\mathcal{W} = \bigcap_{0 < t < \infty} \text{Fix}_{\mathcal{D}} F_t \tag{8.80}$$

is a holomorphic retract of $\mathcal{D}$.

Thus $\mathcal{W}$ is a connected analytic submanifold of $\mathcal{D}$. Moreover, if $\mathcal{D} = \mathbb{B}$ is the open unit ball in a Hilbert space H, then $\mathcal{W}$ is an affine submanifold of $\mathbb{B}$.

In the context of the above theorem we will also need the following notion.

Definition 8.7 Let f be a holomorphic mapping from $\mathcal{D}$ into X and let $\mathcal{W} = \mathrm{Null}_{\mathcal{D}} f \neq \emptyset$. A point $a \in \mathcal{W}$ is said to be **quasi-regular** if the condition

$$\mathrm{Ker} f'(a) \oplus \mathrm{Im}\, f'(a) = X \tag{8.81}$$

holds. If, in addition, $\mathrm{Ker} f'(a) = \{0\}$, then we say that a is a **regular null point** of f.

Corollary 8.5 *Let $\mathcal{D}, X$ and f be as in Theorem 8.6. If $a \in \mathcal{D}$ is an isolated point of $\mathrm{Null}_{\mathcal{D}} f$, then it is unique. In particular, if $a \in \mathrm{Null}_{\mathcal{D}} f$ is regular, i.e., $f'(a)$ is invertible, then a is unique.*

Thus from Theorem 8.6 we can obtain the global description of the (interior) stationary point set of a semigroup $\{F_t\}$, $t \geq 0$, generated by a holomorphic mapping. But even in this case we only know that a retraction exists, but we have no constructive approximation process for finding the points in $\mathcal{W}$.

So, the question is how to construct a retraction onto this set.

A possible way is to investigate the asymptotic behavior of the semigroup as $t \to \infty$. It will become clear that this may be done only if we know a priori at least one point $a \in \mathcal{W}$ and the spectrum of the linear operator $f'(a)$ satisfies certain conditions (*i.e.*, it does not intersect the imaginary axis with, perhaps, the exception of zero).

Another way would be to apply equality (8.40), and for a fixed $r > 0$ to construct the sequence of the discrete Cesàro averages

$$G_n = \frac{1}{n} \sum_{j=0}^{n-1} \mathcal{J}_r^j, \tag{8.82}$$

so that a subsequence $\{G_{n_k}\}$ weakly converges to a mapping $G : \mathcal{D} \to \mathcal{W}$ which is a holomorphic retraction of $\mathcal{D}$ onto $\mathcal{W}$.

As a matter of fact, this method turns out to be supefluous because as we will see below, the iterates of the resolvents $\mathcal{J}_r$ strongly converge to a holomorphic retraction of $\mathcal{D}$ onto $\mathcal{W}$.

Theorem 8.7   *Let $\mathcal{D}$ be a bounded convex domain in X, and let $f \in$ $\mathrm{Hol}(\mathcal{D}, X)$ be the generator of a one-parameter semigroup of holomorphic self-mappings of $\mathcal{D}$. Suppose that $\mathcal{W} = \mathrm{Null}_{\mathcal{D}} f \neq \emptyset$. Then*

(i) *If $\mathcal{W}$ contains a quasi-regular point $a \in \mathcal{D}$, then for each $s > 0$ the sequence $\{\mathcal{J}_s^n = (I + sf)^{-n}\}_1^\infty$ converges to a holomorphic retraction of $\mathcal{D}$ onto $\mathcal{W}$, as $n \to \infty$, uniformly on each ball strictly inside $\mathcal{D}$.*

(ii) *If $\mathcal{W}$ contains a regular point $a \in \mathcal{D}$, then $\mathcal{W} = \{a\}$ and the net $\{\mathcal{J}_s = (I + sf)^{-1}\}_{s>0}$ converges to a as $s \to \infty$, uniformly on each ball strictly inside $\mathcal{D}$.*

To prove our theorem we need the following lemma.

Lemma 8.4 (*cf.* [Lyubich and Zemanek (1994)]) *Let A be a bounded linear operator on a Banach space X such that*

$$\|(I - A)^n\| \leq M, \quad n = 1, 2, \ldots \tag{8.83}$$

for some $M < \infty$. Then the following conditions are equivalent:

$$\mathrm{Ker}\, A \oplus \mathrm{Im}\, A = X \tag{8.84}$$

$$\overline{\mathrm{Im}\, A} = \mathrm{Im}\, A. \tag{8.85}$$

Proof.   The implication $(8.84) \Rightarrow (8.85)$ is obvious. Now let (8.85) hold, and let a functional $x^* \in X^*$ vanish on the sum $\mathrm{Ker}\, A \oplus \mathrm{Im}\, A$. Then $x^* \in \mathrm{Ker}\, A^*$. Furthermore, it follows from (8.85) and the Banach–Hausdorff theorem that the condition $\langle u, x^* \rangle = 0$ for all $u \in \mathrm{Ker}\, A$ implies that $x^* \in \mathrm{Im}\, A^*$. Thus $x^* \in \mathrm{Ker}\, A^* \cap \mathrm{Im}\, A^*$. But because of (8.83), $\mathrm{Ker}\, A^* \cap \mathrm{Im}\, A^* = \{0\}$ by the Yosida mean ergodic theorem ([Yosida (1974)]). So $x^* = 0$, and this implies (8.84).

We recall that a linear operator $A : X \to X$ is said to be **m-accretive** if for each $r > 0$, the operator $\mathcal{I}_r = (I + rA)^{-1}$ is well defined on X and $\|(I + rA)^{-1}\| \leq 1$. $\square$

Lemma 8.5   *Let A be a bounded linear operator on X which is m-accretive with respect to some norm equivalent to the norm of X. If A satisfies condition (8.84), then for each $r > 0$, the linear operator $\mathcal{I}_r = (I + rA)^{-1}$ satisfies the condition*

$$\mathrm{Ker}(I - \mathcal{I}_r) \oplus \mathrm{Im}(I - \mathcal{I}_r) = A. \tag{8.86}$$

Proof. Returning to the original norm of X, we have by the definition

$$\|\mathcal{I}_r^n\| \le M < \infty, \quad n = 1, 2, \ldots, \quad r > 0. \tag{8.87}$$

By Lemma 8.4 and (8.87), it is sufficient to show that $\mathrm{Im}(I - \mathcal{I}_r)$ is closed in X. Indeed, if $y_n \in \mathrm{Im}(I - \mathcal{I}_r)$ converge to $y \in X$, we get a sequence $\{x_n\} \subset X$ such that

$$(I - \mathcal{I}_r)x_n = rA\mathcal{I}_r x_n \to y \in \mathrm{Im}A. \tag{8.88}$$

Note that it follows by the definition of $\mathcal{I}_r$ that

$$\mathrm{Ker}(I - \mathcal{I}_r) = \mathrm{Ker}A. \tag{8.89}$$

Therefore, if we represent $x_n \in X$ in the form $x_n = u_n + v_n$, where $u_n \in \mathrm{Im}A$ and $v_n \in \mathrm{Ker}A$ (see (8.84)), we get from (8.88) and (8.89),

$$\begin{aligned}
(I - \mathcal{I}_r)u_n &= r(A\mathcal{I}_r u_n + A\mathcal{I}_r v_n) = r(A\mathcal{I}_r u_n + Av_n) \\
&= rA\mathcal{I}_r u_n \to y \in \mathrm{Im}A.
\end{aligned} \tag{8.90}$$

Denote $z_n = \mathcal{I}_r u_n \in X$. We have by (8.90),

$$z_n = u_n - rAz_n \tag{8.91}$$

and hence $z_n \in \mathrm{Im}A$. Since $\mathrm{Im}A$ is closed and invariant under A, and $Az_n \to \frac{1}{r}y \in \mathrm{Im}A$, the sequence $\{z_n\}$ converges to some element $z \in X$ and hence $u_n = z_n + rAz_n \to z + y = x$. Once again, it follows by (8.90) that $(I - \mathcal{I}_r)x = y$. $\qquad\square$

Lemma 8.6 *Let the conditions of Lemma 8.5 hold. Then $\sigma(\mathcal{I}_r)$, the spectrum of the operator $\mathcal{I}_r$, is contained in the open unit disk Δ, except perhaps, for 1, i.e.,*

$$\sigma(\mathcal{I}_r) \subset \Delta \cup \{1\}. \tag{8.92}$$

Proof. Fix $r > 0$. It follows by (8.87) that $\sigma(\mathcal{I}_r) \subseteq \bar{\Delta}$. It is known that $\sigma(A)$, the spectrum of the accretive operator A, lies in the right half-plane. Therefore there is an open domain $\Omega \subset \mathbb{C}$ such that $\sigma(A) \subset\subset \Omega$, and $\Gamma = \partial\Omega$ separates $\sigma(A)$ and the real number $\lambda = -r^{-1}$. Thus the function $f(\lambda) = (1 + r\lambda)^{-1}$ is holomorphic in Ω and $\mathcal{I}_r$ can be represented in the form

$$(I + rA)^{-1} = \frac{1}{2\pi i} \int_\Gamma (1 + r\lambda)^{-1}(\lambda I - A)^{-1} d\lambda = \tilde{f}(A). \tag{8.93}$$

It follows by the spectral mapping theorem (see, for example, [Dunford and Schwarz (1958)]) that $\sigma(\tilde{f}(A)) = f(\sigma(A))$. Thus, if we assume that $e^{i\varphi} \in \sigma(\mathcal{I}_r)$ we get $\lambda = r^{-1}(e^{-i\varphi} - 1) \in \sigma(A)$ and hence $\mathrm{Re}\,\lambda \geq 0$. This implies that $\varphi = 0$ and the lemma is proved. $\qquad\qquad\square$

Proof of Theorem 8.7   (i) Let $f \in \mathrm{Hol}(\mathcal{D}, X)$ be a generator and let $\{F_t\}_{t \geq 0}$ be the semigroup of holomorphic self-mappings of $\mathcal{D}$ generated by f. Suppose that $\mathcal{W} = \mathrm{Null}_{\mathcal{D}} f = \bigcap_{r>0} \mathrm{Fix}_{\mathcal{D}} F_t$ contains a quasi-regular point $a \in \mathcal{D}$ (see Definition 8.7). Observe that the linear operator $A = f'(a)$ is the infinitesimal generator of the semigroup $\{U_t\}_{t \geq 0}$, where $U_t = (F_t)'(a)$. Since $\mathcal{D}$ is bounded, it follows by the Cauchy inequalities that $\{U_t = e^{-At}\}_{t \geq 0}$ is uniformly bounded and therefore A is an m-accretive operator with respect to some norm equivalent to the norm of X. In addition, for each $r > 0$, $(I + rA)^{-1} = \left[(I + rf)^{-1}\right]'(a)$ by the chain rule. Thus by applying the Vesentini theorem [Vesentini (1985)] and Lemmata 8.4–8.6, we see that for each $r > 0$, the sequence $\{\mathcal{T}_r^n\}_{n=1}^{\infty}$, where $\mathcal{T}_r = (I + rf)^{-1} : \mathcal{D} \to \mathcal{D}$ converges locally uniformly to some holomorphic mapping $\varphi : \mathcal{D} \to \mathcal{D}$ which is a retraction onto the fixed point set of $\mathcal{T}_r$. But this set coincides with $\mathcal{W}$, and we are done.

    (ii) Now let $a \in \mathcal{D}$ be a regular null point of f. Once again, by Lemmata 8.5 and 8.6, this means that for each $r > 0$, the spectral radius of the operator $(\mathcal{T}_r)'(a)$ is less than 1. Thus there is an equivalent norm $\|\cdot\|_1$ on X such that $\left\|(\mathcal{T}_r'(a)\right\|_1 < 1$. It follows by continuity that in this norm there is a ball $B_R(a)$ centered at a with radius $R(= R(r))$ such that $B_R(a) \subset\subset \mathcal{D}$ and $\left\|(\mathcal{T}_r)'(x)\right\|_1 \leq q_r < 1$ for each $x \in B_r(a)$. Fix $r > 0$ and take any $t \geq r$. Using the resolvent identity and the equality $\mathcal{T}_r(a) = a$, $t > 0$, we have for all $x \in B_r(a)$,

$$\left\|\mathcal{T}_t(x) - a\right\|_1 = \left\|\mathcal{T}_t(x) - \mathcal{T}(a)\right\|_1$$

$$= \left\|\mathcal{T}_r\left(\frac{r}{t}\,x + \left(1 - \frac{r}{t}\right)\mathcal{T}_t(x)\right) - \mathcal{T}_r(a)\right\|_1$$

$$\leq q_r\left\|\frac{r}{t}\,x + \left(1 - \frac{r}{t}\right)\mathcal{T}_t(x) - a\right\|_1$$

$$\leq q_r\frac{r}{t}\|x - a\|_t + q_r\left(1 - \frac{r}{t}\right)\left\|\mathcal{T}_t(x) - a\right\|_1. \qquad (8.94)$$

Thus we obtain the inequality

$$\left\|\mathcal{T}_t(x) - a\right\|_1 \leq \frac{q_r\frac{r}{t}}{1 - q_r\left(1 - \frac{r}{t}\right)}\,\|x - a\|_1, \qquad (8.95)$$

which implies that $\mathcal{T}_t(x)$ converges to a as $t \to \infty$, uniformly on $B_R(a)$.

Now it follows by the Vitali property that $\{T_t\}_{t>0}$ converges to a, locally uniformly on all of $\mathcal{D}$. The theorem is proved. $\qquad\square$

Another retraction method is based on the following observation.

Let $S = \{F_t\}_{t>0}$ be a uniformly continuous semigroup with $\mathcal{F} = \text{Null}_{\mathcal{D}} f = \bigcap_{0<t<\infty} \text{Fix} F_t \neq \emptyset$.

Pick $a \in \mathcal{F}$ and denote $A_t = (F_t)'(a)$, $t > 0$. It follows from the Cauchy inequalities that $\{A_t\}_{t>0}$ is a uniformly continuous semigroup of bounded linear operators, *i.e.*, A_t converges to the identity in the operator norm as $t \to 0^+$. Therefore it has an infinitesimal generator $B : X \to X$, *i.e.*,

$$B = \lim_{t\to 0^+} \frac{I - A_t}{t}. \tag{8.96}$$

Moreover, since the infinitesimal generator $f : \mathcal{D} \to X$ of $S = \{F_t\}_{t>0}$ is defined by the formula

$$f = \lim_{t\to 0^+} \frac{I - F_t}{t}, \tag{8.97}$$

we have $f'(a) = B$. Now we can state our assertion.

Theorem 8.8   *Let $\mathcal{D}$ and $S = \{F_t\}_{t>0}$ be as above. Suppose that for some $a \in \mathcal{F} = \cap_{t>0}\text{Fix}(F_t)$, the following condition holds:*

$$\text{Ker} B \oplus \text{Im} B = X, \tag{8.98}$$

where B is defined by (8.96) with $A_t = (F_t)'(a)$. Then, for each $t > 0$, the continuous Cesàro average

$$\Phi_t = \frac{1}{t} \int_0^t F_s ds \tag{8.99}$$

is power convergent to a retraction onto $\mathcal{F}$. Thus $\mathcal{F}$ is a connected submanifold of $\mathcal{D}$.

Proof.   It follows from (8.99) that $\Phi_t(a) = a$ for each $t > 0$, and hence $\Phi_t \in \text{Hol}(\mathcal{D})$. In addition, $(\Phi_t)'(a) = C_t$, where C_t is defined by

$$C_t = \frac{1}{t} \int_0^t A_s ds = \frac{1}{t} \int_0^t e^{-Bs} ds. \tag{8.100}$$

Therefore, if we set

$$\Omega = \{\lambda \in \mathbb{C} : \text{Re}\lambda > 0\}, \quad \lambda_0 = 0, \tag{8.101}$$

and

$$f_t(\lambda) = \frac{1 - e^{-t\lambda}}{t\lambda}, \quad t > 0, \tag{8.102}$$

we get that

$$C_t = \frac{1}{2\pi i} \int_\Gamma f_t(\lambda)(\lambda I - B)^{-1} d\lambda, \tag{8.103}$$

where Γ is a contour that surrounds $\sigma(B)$ in $\Omega_\varepsilon = \{\lambda \in : \mathrm{Re}\lambda > -\varepsilon\}$ for some $\varepsilon > 0$. Furthermore, it follows by the maximum modulus principle that $f(\overline{\Omega}\backslash\{0\}) \subset \Delta$ and hence $\lambda = 0$ is the unique and simple root of the equation $f_t(\lambda) = 1$ in $\overline{\Omega}$. So, C_t is power convergent to a projection onto $\mathrm{Ker}B$. But $\mathcal{F} = \bigcap_{t>0} \mathrm{Fix}(F_t)$ coincides with the null point set of f in $\mathcal{D}$, where f is the generator of $S = \{F_t\}_{t\geq 0}$ and $f'(a) = B$. Since $(\Phi_t)'(a) = C_t$, Φ_t is power convergent to a retraction onto $\mathrm{Fix}(\Phi_t)$, which is tangent to $\mathrm{Ker}B$. Together with the inclusion $\mathcal{F} \subseteq \mathrm{Fix}(\Phi_t)$ this implies the equality $\mathcal{F} = \mathrm{Fix}(\Phi_t)$, which proves our assertion. $\square$

8.6   A Stabilization Phenomenon

Here we establish another interesting feature of the null point set of a holomorphic generator and then use it to construct holomorphic retractions [Khatskevich *et al.* (1998)].

Let us consider a semigroup $S = \{F_t\}$, $t \in \mathbb{R}^+$, generated by $f \in \mathrm{Hol}(\mathcal{D}, X)$. Let f_t be the difference approximations of f, *i.e.*,

$$f_t = \frac{1}{t}(I - F_t), \quad t > 0. \tag{8.104}$$

If $\mathrm{Null}_{\mathcal{D}} f$ is not empty, then

$$\mathrm{Null}_{\mathcal{D}} f \subseteq \mathrm{Null}_{\mathcal{D}} f_t, \quad t > 0. \tag{8.105}$$

Moreover, it is natural to expect that for sufficiently small t, the set $\mathrm{Null}_{\mathcal{D}} f_t$ will approximate $\mathrm{Null}_{\mathcal{D}} f$ in some sense. As a matter of fact, in the linear case, as well as in the holomorphic case, there is a stabilization phenomenon of $\mathrm{Null}_{\mathcal{D}} f_t$ for sufficient small t.

Theorem 8.9   *Let $S = \{F_t\}_{t\geq 0}$ be a uniformly continuous semigroup generated by $f \in \mathrm{Hol}(\mathcal{D}, X)$ and let $f_t = t^{-1}(I - F_t)$. Suppose that $\mathrm{Null}_{\mathcal{D}} f \neq \emptyset$ and that one of the following conditions holds:*

(1) X is reflexive;

(2) $\operatorname{Ker} f'(a) \oplus \operatorname{Im} f'(a) = X$ for some $a \in \operatorname{Null}_{\mathcal{D}} f$.

Then there exists $\delta > 0$ such that for all $t \in (0, \delta)$,

$$\operatorname{Fix}_{\mathcal{D}} F_t \, (= \operatorname{Null}_{\mathcal{D}} f_t) = \operatorname{Null}_{\mathcal{D}} f. \tag{8.106}$$

Proof.   Since both $\operatorname{Null}_{\mathcal{D}} f_t$ and $\operatorname{Null}_{\mathcal{D}} f$ are connected complex submanifolds of $\mathcal{D}$ and

$$\operatorname{Null}_{\mathcal{D}} f \subseteq \operatorname{Null}_{\mathcal{D}} f_t, \tag{8.107}$$

it suffices to show that their tangent spaces coincide. A simple calculation shows that for $a \in \operatorname{Null}_{\mathcal{D}} f$, $(f_t)'(a) = \frac{1}{t}(I - e^{-tA})$, where $A = f'(a)$. Thus our claim is that there exists a positive δ such that for all $t \in (0, \delta)$

$$\operatorname{Fix}(e^{-tA}) = \operatorname{Ker} A. \tag{8.108}$$

In order to prove (8.108) when X is reflexive, we first note that the semigroup $e^{-tA} = (F_t)'(a)$ is uniformly bounded by the Cauchy inequalities. We then let P denote the projection of X onto $\operatorname{Ker} A$ obtained from the mean ergodic theorem.

Now let $g_t = \frac{1}{t} \int_0^t e^{-sA} ds$. There is a positive δ such that g_t is invertible for all $0 < t < \delta$. For such t, let P_t be the mean ergodic projection onto $\operatorname{Fix}(e^{-tA})$. A computation shows that for all natural numbers m

$$g_{mt} = \left(\frac{1}{m} \sum_{j=1}^{m} e^{-(j-1)tA} \right) g_t. \tag{8.109}$$

Letting $m \to \infty$, we see that $P = P_t g_t = g_t P_t$. Hence $P_t = g_t^{-1} P$ and $\operatorname{Ker} P \subset \operatorname{Ker} P_t$. Since

$$X = \operatorname{Fix} P \oplus \operatorname{Ker} P = \operatorname{Fix} P_t \oplus \operatorname{Ker} P_t \tag{8.110}$$

and $\operatorname{Fix} P \subset \operatorname{Fix} P_t$, it follows that $\operatorname{Ker} A = \operatorname{Fix} P = \operatorname{Fix} P_t = \operatorname{Fix} e^{-tA}$. When hypothesis (2) holds, the following simple direct argument is due to V. Khatskevich. In this case there is a positive ε such that

$$\|Az\| \geq \varepsilon \|z\| \tag{8.111}$$

for all $z \in \operatorname{Im} A$.

Let $x = y + z$, where $y \in \mathrm{Ker}A$ and $z \in \mathrm{Im}A$, belong to $\mathrm{Fix}(e^{-tA})$. Then $e^{-tA}z = z$ and

$$Az = t\left(\frac{A^2}{2!} - \frac{tA^3}{3!} + \ldots \right)z. \tag{8.112}$$

If $0 < t < \min\{1, \varepsilon/(e^{\|A\|} - 1 - \|A\|)\}$ and $z \neq 0$, it follows that $\|Az\| < \varepsilon\|z\|$, which contradicts (8.111). Hence $z = 0$ and $x = y$ belongs to $\mathrm{Ker}A$.

$\square$

Theorem 8.10   *Let $\mathcal{D}$ be a bounded convex domain in a complex Banach space X, and let $f : \mathcal{D} \to X$ be a holomorphic generator of a flow $\{F_t\}_{t\geq 0}$ on $\mathcal{D}$.*

Suppose that $\mathcal{F} = \bigcap\limits_{t\geq 0} \mathrm{Fix}F_t$ is not empty and that $a \in \mathcal{F}$. Then $\{F_t\}$ converges, as $t \to \infty$, in the topology of locally uniform convergence over $\mathcal{D}$ to a retraction $\rho : \mathcal{D} \to \mathcal{F}$ if and only if one of the following conditions holds: either

(i) $\mathrm{Re}\lambda \geq \varepsilon > 0$ *for all* $\lambda \in \sigma(f'(a))$,

or

(ii) $\{\lambda \in \mathbb{C} : \mathrm{Re}\lambda = 0\} \cap \sigma(f'(a)) = \{0\}$, 0 *is an isolated point of* $\sigma(f'(a))$,

and

$$\mathrm{Ker}(f'(a)) \oplus \mathrm{Im}f'(a) = X. \tag{8.113}$$

In particular, when X is finite–dimensional, $\{F_t\}$ converges uniformly on compact subsets of $\mathcal{D}$ to a retraction $\rho : \mathcal{D} \to X$ if and only if the spectrum of $f'(a)$ does not intersect the imaginary axis except perhaps at the origin.

Note also that in the presence of (i), $\mathcal{F} = \{a\}$, and a is an asymptotically stable stationary point of $\{F_t\}$.

Proof.   Let $\mathcal{D}$ and f be as in the statement of the theorem, let $S = \{F_t\}$, $t \geq 0$, be the semigroup generated by f, and choose

$$a \in \mathcal{F} = \bigcap\limits_{t\geq 0} \mathrm{Fix}F_t \neq \emptyset. \tag{8.114}$$

Denote $A_t = (F_t)'_x(a)$, $t \geq 0$, and $B = f'(a)$. It is easy to see that $A_t = \exp(-tB)$, $t \geq 0$. Suppose now that one of the hypotheses, either (i) or (ii), holds. Then using the Cauchy inequalities and direct computations, we see that A_t satisfies either condition (i) or (ii). In addition, for $t > 0$ small enough, we have

$$\operatorname{Ker}(I - A_t) \oplus \operatorname{Im}(I - A_t) = X. \tag{8.115}$$

Indeed, it follows from Theorem 8.9 that $\operatorname{Ker}(I - A_t) = \operatorname{Ker}B\left(= \bigcap_{t \geq 0} \operatorname{Fix}A_t\right)$ for each $t > 0$ which is small enough. Once again, the Cauchy inequalities imply that $B = \lim_{t \to 0+} \frac{1}{t}(I - A_t)$ in the uniform operator topology. Finally, each A_t commutes with B, i.e., $A_t B = B A_t$. Thus $\operatorname{Im}(I - A_t) = \operatorname{Im}B$.

Now for such $t > 0$ the mapping F_t satisfies all the conditions of Vesentini's theorem (see also Lemma 5.2 in [Mazet and Vigué (1991)]). Thus we can conclude that for a small enough fixed $\tau > 0$, say $\tau \in (0, \delta)$, the sequence $\{F_{n\tau}\}$ converges in the topology of locally uniform convergence over $\mathcal{D}$ to a mapping $\rho_\tau : \mathcal{D} \to \mathcal{D}$, which is a retraction onto the set $\operatorname{Fix}F_\tau = \mathcal{F}$.

Taking $\tau = \frac{1}{q}$ with q a big enough integer, we deduce that $\{F_n\} = \{F_{nq\tau}\}$ converges to ρ_τ as $n \to \infty$ in the topology of locally uniform convergence. Hence $\rho_{\frac{1}{q}}$ is independent of such q. We denote this retraction by ρ.

Let $B_1 \subset\subset B_2 \subset\subset \mathcal{D}$ be two arbitrary balls strictly inside $\mathcal{D}$. Choose q_1 large enough, so that $F_s(x) \in B_2$ for all $x \in B_1$ and $0 \leq s \leq \frac{1}{q_1} < \delta$, and choose M such that $\{F_t\}$ is Lipschitzian on B_2 with constant M.

We claim that $\{F_t\}$ converges to ρ uniformly on B_1. Indeed, given $\varepsilon > 0$ we first choose $q_2 \geq q_1$ such that $\|F_s(x) - x\| < \varepsilon/2M$ for all $0 \leq s \leq \frac{1}{q_2}$ and $x \in B_1$. Now for $k = [tq_2]$ and for $0 \leq s = t - \frac{k}{q_2} \leq \frac{1}{q_2}$ we have

$$\begin{aligned}
\|F_t(x) - \rho(x)\| &= \left\|F_{\frac{k}{q_2}}(F_s(x)) - \rho(x)\right\| \\
&\leq \left\|F_{\frac{k}{q_2}}(F_s(x)) - F_{\frac{k}{q_2}}(x)\right\| + \left\|F_{\frac{k}{q_2}}(x) - \rho(x)\right\| \\
&< \frac{\varepsilon}{2} + \frac{\varepsilon}{2} = \varepsilon,
\end{aligned} \tag{8.116}$$

wherever t (hence k) is large enough.

Conversely, let the net $\{F_t\}$ converge in the topology of locally uniform convergence to a mapping $\rho : \mathcal{D} \to \mathcal{D}$. Then, by Vesentini's theorem, we have that F_t satisfies the conditions analogous to either condition (i) or condition (ii) for each fixed $t > 0$. Repeating the above considerations and using the spectral mapping theorem we obtain either (i) or (ii) (with (8.113)), respectively. This concludes the proof of the theorem. $\qquad\square$

Theorem 8.11 *Let $\mathcal{D}$ be a bounded convex domain in a complex Banach*

space X, and let $F : \mathcal{D} \to \mathcal{D}$ be a holomorphic self-mapping such that
$\mathrm{Fix}F \neq \emptyset$. Then

(1) the mapping $f = I - F$ is the generator of a flow on $\mathcal{D}$;
(2) if for some $a \in \mathrm{Fix}F$,

$$\mathrm{Ker}(I - F'(a)) \oplus \mathrm{Im}(I - F'(a)) = X, \qquad (8.117)$$

then $\{F_t\}$, the semigroup generated by f, converges, as $t \to \infty$, to a
retraction $\rho : \mathcal{D} \to \mathrm{Fix}F$ in the topology of locally uniform convergence
over $\mathcal{D}$.

Proof. 1. Let $\mathcal{D}$ and $F : \mathcal{D} \to \mathcal{D}$ satisfy the hypotheses. Then $f = I - F$ is bounded and has a null point in $\mathcal{D}$. for arbitrary $r > 0$, setting $t = r(r + 1)^{-1}$ we have that the equation $x + r(x - F(x)) = z$ is exactly the equation $x = (1 - t)z + tF(x)$, which has a unique solution $x = x_r(z)$ for each $z \in \mathcal{D}$ and $t \in (0, 1)$ by the Earle–Hamilton theorem because $\mathcal{D}$ is convex. Since this solution may be obtained by iteration, it is holomorphic in $z \in \mathcal{D}$. Thus by [Reich and Shoikhet (1996)], we see that $I - F$ is a generator of a flow $\{F_t\}_{t \geq 0}$.

2. We already know that $\sigma(F'(a)) \subset \bar{\Delta}$. Therefore $\sigma(f'(a)) = \sigma(I - F'(a))$ satisfies either (i) or (ii). The result now follows from Theorem 8.10.

$\square$

8.7 Local and Spectral Characteristics of Stationary Points

8.7.1 *Cartan's uniqueness theorem*

The following simple consequences of the above results indicate that some local characteristics of a null point of a generator can influence the global structure of the whole null point set and the global behavior of the semigroup.

Theorem 8.12 *Let $\mathcal{D}$ be a convex bounded domain in X, and let f be the holomorphic generator of a uniformly continuous one–parameter semigroup and have a null point $a \in \mathcal{D}$. If $f'(a) = 0$, then $f \equiv 0$.*

Proof. Indeed, it is clear that $a \in \mathrm{Null}_{\mathcal{D}}f$ is a fixed point of the resolvent $J_r = (I + rf)^{-1} \in \mathrm{Hol}(\mathcal{D})$, $r > 0$. In addition $(J_r)'(a) = I$ is the identity mapping on X. Thus, by Cartan's theorem (see, for example, [Franzoni and Vesentini (1980)], [Khatskevich and Shoikhet (1994a)]), $J_r \equiv I_\mathcal{D}$ is the identity mapping on $\mathcal{D}$. This implies that $f \equiv 0$ in $\mathcal{D}$.

Moreover, we can establish a continuous form of this assertion. it is a generalization of the Harris–Schwarz Lemma [Harris (1971a)]. $\qquad\square$

Theorem 8.13 *Let $\mathcal{D}$ be a convex bounded domain in X, and let $\{f_n\} \subset \mathrm{Hol}(\mathcal{D}, X)$ be a sequence of holomorphic generators which is uniformly bounded on each subset strictly inside $\mathcal{D}$. Assume that for some $a \in \mathcal{D}$, the following conditions hold:*

(a) $\{f_n(a)\}$ strongly converges to zero;
(b) $\{f_n'(a)\}$ converges to 0 in the operator topology.

Then $\{f_n\}$ T–converges to 0, i.e., $T-\lim\limits_{n\to\infty} f_n \equiv 0$.

8.7.2 *Harris' spectrum of a semi-complete vector field*

Following L. A. Harris [Harris (1971b)] we give the following definition.

Definition 8.8 Let $\mathcal{D}$ be an open subset of X, $a \in \mathcal{D}$, and let $h \in \mathrm{Hol}(D, X)$. The spectrum of h with respect to a, denoted by $\sigma_a(h)$, is the set of all $\lambda \in \mathbb{C}$ such that it is not possible to find open sets $U \subset D$, with $a \in U$ and $V \subseteq X$ with the property that $\lambda I - h$ is a biholomorphism of U onto V.

Proposition 8.6 ([Harris (1971b)]) *$\sigma_a(h) = \sigma(h'(a))$ is the spectrum of the linear operator $h'(a)$.*

Theorem 8.14 *Let $f \in \mathrm{Hol}(\mathcal{D}, X)$ be the holomorphic generator of a one-parameter semigroup on $\mathcal{D}$ and let $a \in \mathrm{Null}_\mathcal{D} f$. Then*

(1) $\sigma_a(f)$ lies in the right half-plane;
(2) if $0 \notin \sigma_a(f)$, then a is the unique null point of f in $\mathcal{D}$;
(3) $\sigma_a(f)$ lies strictly inside the right hall-plane iff a is a globally asymptotically stable (in the Lyapunov sense) stationary point of the semigroup $S_f = \{F_t\}$, $t \geq 0$, i.e., $\{F_t\}$ converges to a locally uniformly in $\mathcal{D}$ as $t \to \infty$.

Proof. Set $A = f'(a)$. It is easy to see that A is the infinitesimal generator of a uniformly continuous semigroup $U_t = e^{-tA}$ and that $U_t = (F_t(x))'_{x=a}$. Thus it follows by the Cauchy inequalities that U_t is a uniformly bounded semigroup of linear operators. it is well known that the resolvent $R(\lambda, A) = (\lambda I - A)^{-1}$ is defined on the open left half-plane, *i.e.*, $\mathrm{Re}\lambda \geq 0$ for all $\lambda \in \sigma(A)$. Thus assertion (1) follows from Proposition 8.6.

(2). If $0 \notin \sigma_a(f)$, then the operator $A = f'(a)$ is invertible. Hence a is an isolated null point of f in $\mathcal{D}$, and our assertion follows from Corollary 8.5.

(3). Suppose now that $\sigma_a(f) = \sigma(A)$ lies strictly inside the right half-plane of $\mathbb{C}$. As it is well known, this fact implies the estimate

$$\|e^{-tA}\| \leq Ne^{-\nu t} \tag{8.118}$$

for some $N > 0$ and $\nu > 0$ (see, for example, [Daletskii and Krein (1970)], [Yosida (1974)]).

Rewrite now the Cauchy problem in the form of a perturbed equation:

$$\begin{aligned} x'(t) &= -Ax(t) + g(x(t)), \\ x(0) &= x \in \mathcal{D}, \end{aligned} \tag{8.119}$$

where $g = A - f$.

Since $f(a) = 0$, there is some ball $B_r(a) \subset\subset \mathcal{D}$ centered at a with radius r, such that g admits the representation

$$g(s) = \sum_{k=2}^{\infty} P_f^{(k)}(a) \circ (x - a), \tag{8.120}$$

where $P_f^{(k)}$, $k \geq 2$, are homogeneous forms of order k. Setting $M = \sup_{x \in \mathcal{D}} \|g(x)\|$, we have, by the generalized Schwarz Lemma,

$$\|g(x)\| \leq Mr^{-2}\|x - a\|^2 \tag{8.121}$$

for all $x \in B_r(a)$. Choosing now $\rho < r\sqrt{\nu(MN)^{-1}}$, where ν and N are as in (8.118), we obtain the inequality

$$\|g(x)\| < \frac{\nu}{N}\|x - a\| \tag{8.122}$$

for all $x \in \overline{\mathbb{B}_\rho(a)} = \{x \in \mathcal{D} : \|x - a\| \leq \rho\}$. Thus Theorem VII.2.1 from [Daletskii and Krein (1970), p. 403] implies that problem (8.119) has a uniformly asymptotically stable solution on $\overline{B_\rho(a)} \times \mathbb{R}^+$. In other words, the net $F_{t|B_\rho(a)} = x(t)$ converges uniformly to the point a, uniformly on $\overline{B_\rho(a)}$. An appeal to Vitali's property concludes now the proof of our assertion in one direction.

Conversely, let $\{F_t\}_{t \geq 0}$ T–converge to $a \in \mathrm{Null}_\mathcal{D} f$. Then it follows from the Cauchy inequalities that the linear semigroup $U_t = e^{-tA} = (F_t)'_{x=a}$

uniformly converges to zero as $t \to \infty$. This is equivalent to that fact that for all $t > 0$, the spectral radius $r_\sigma(U_t) < 1$.

By Dunford's theorem on the spectrum, it follows that $\sigma(A) = \sigma_a(f)$ lies strictly inside the right half-plane and we are done. $\qquad\square$

Definition 8.9 Let $\mathcal{D}$ be a domain in X and let $f \in \mathrm{Hol}(\mathcal{D}, X)$. A point $a \in \mathrm{Null}_{\mathcal{D}} f$ is said to be regular if $0 \notin \sigma(f'(a))$, *i.e.*, $f'(a)$ is an invertible linear operator. It is said to be **strictly regular** if $\sigma(A)$ does not intersect the imaginary axis of the complex plane $\mathbb{C}$.

According to this definition we obtain the following direct consequence of Theorems 8.14 and 8.6.

Corollary 8.6 *Let $\mathcal{D}$ be a bounded convex domain in X, and let f be a bounded semi-complete vector field in $\mathcal{D}$. Suppose that f is a Fredholm mapping and that $a \in \mathrm{Null}_{\mathcal{D}} f$. Then $\mathrm{Null}_{\mathcal{D}} f = \{a\}$ if and only if the point a is regular.*

Remark 8.6 *If f is not Fredholm, but X is reflexive, we have, in general, two singular situations. Namely, if $a \in \mathrm{Null}_{\mathcal{D}} f$ and $0 \in \sigma_a(f)$, then either*

(1) a is the unique null point in $\mathcal{D}$, or
(2) there are infinitely many null points of f in $\mathcal{D}$, and they form a connected complex submanifold of $\mathcal{D}$.

The following example shows that situation (1) actually may exist in the case of an infinite dimensional space (even if it is reflexive). Despite its uniqueness, such a point has no "good" property such as regularity.

Example 8.3 Let X be the complex Hilbert space l^2 with basis $\{e_i\}_{i=1}^{\infty}$, and let $0 < a_i < 1$ satisfy $a_i \to 1$ as $i \to \infty$. Let $\mathcal{D}$ be the unit ball in X and define the linear mapping $A : \overline{\mathcal{D}} \mapsto X$ by $Ae_i = (1 - a_i)e_i$. This mapping has a unique null point $x = 0$, but it is not regular (0 is a point in the continuous spectrum of A).

It is clear that A is the generator of a semigroup of self-mappings of $\mathcal{D}$. Now we turn to the same questions concerning the approximation of fixed points.

Remark 8.7 *Let $\mathcal{D}$ be as above, and let $F : \mathcal{D} \mapsto \mathcal{D}$ be a holomorphic self-mapping of $\mathcal{D}$. Its iterates $F^n : \mathcal{D} \mapsto \mathcal{D}$, $F^n = F^{n-1} \circ F$, $n = 1, 2, \ldots$, $F^0 = I$, are well defined and holomorphic. However, even when X is finite dimensional and F has a unique fixed point, there are many* *situations when the sequence of iterates $\{F^n(x)\}_{n=0}^{\infty}$ does not converge to the fixed point a for $x \neq a$.*

For example, let $\mathcal{D}$ be a unit ball and $F = e^{i\varphi}I$, $0 < \varphi < 2\pi$. More generally, such a situation arises when the spectrum $\sigma(B)$ of the linear operator $B = f'(a)$ contains points of the unit circle other than 1 (see, for example, [Abate (1988a); Vesentini (1983)] and [Vesentini (1985)]).

There are many other approximative methods (explicit and implicit) for finding the fixed point (see the next chapter). We include here only one observation in this direction.

Let $F \in \mathrm{Hol}(\mathcal{D})$ have a fixed point $a \in \mathcal{D}$ such that $1 \notin \sigma(B)$, where $B = F'(a)$. Then this point is a regular null point for the mapping $f = I - F \in \mathrm{Hol}(\mathcal{D}, X)$, which is a semi-complete vector field. As a matter of fact, it is also strictly regular. Indeed, if $\lambda \in \sigma(A)$, where $A = f'(a) = I - B$, then $1 - \lambda \in \sigma(B)$ and $\mathrm{Re}(1 - \lambda) \leq 1$ by Theorem 8.14. But $|1 - \lambda| \leq 1$ and $1 - \lambda \neq 1$ according to our assumption. Hence $\mathrm{Re}\,\lambda > 0$ and we are done.

Thus we have proved the following assertion.

Proposition 8.7   *Let $\mathcal{D}$ be a bounded convex domain in X, and let $F \in \mathrm{Hol}(\mathcal{D})$ have a fixed point $a \in \mathcal{D}$ which is a regular null point of $f = I - F$. Then the semigroup family $\{F_t\}$ defined by the Cauchy problem*

$$\begin{cases} \dfrac{\partial F_t}{\partial t} = F(F_t) - F_t, \\ F_0 = I_{\mathcal{D}}, \end{cases} \qquad (8.123)$$

T-converges to a when t tends to infinity.

As a simple example, consider again the mapping $F = iI$ mentioned above, the iterates of which do not converge to zero for each $x \neq 0$. At the same time, the Cauchy problem (8.123) has the solution $F_t(x) = e^{it} \cdot e^{-t}x$ which evidently uniformly converges to zero as t tends to infinity.

Chapter 9

Asymptotic Behavior of Continuous Flows

9.1 Strongly Semi-Complete Vector Fields in Banach Spaces

Let $\mathcal{D}$ be a domain in a Banach space X and let f be a semi-complete vector field on a domain $\mathcal{D}$ in X with $\text{Null}_{\mathcal{D}} f \neq \emptyset$, *i.e.*, the Cauchy problem

$$\begin{cases} \dfrac{du}{dt} + f(u) = 0 \\ u(0) = x \end{cases} \tag{9.1}$$

has a solution $u(\cdot, x) : \mathbb{R}^+ \to \mathcal{D}$ which is well-defined on all of $\mathbb{R}^+$ for each initial datum $x \in \mathcal{D}$. In other words, f is the generator of a one-parameter semigroup $S = \{F(t)\}_{t \geq 0}$ of self-mappings $F(t) = u(t, \cdot) \in \text{Hol}(\mathcal{D})$.

Definition 9.1 A point $a \in \text{Null}_{\mathcal{D}} f$ is said to be locally uniformly attractive if the semigroup $S = \{F(t)\}_{t \geq 0}$ generated by f converges to a in the topology of locally uniform convergence over $\mathcal{D}$.

Definition 9.2 Let $\mathcal{D}$ be a domain in a Banach space X and let $\mathcal{G}(\mathcal{D})$ be the family of all semi-complete vector fields on $\mathcal{D}$. A mapping $f \in \mathcal{G}(\mathcal{D})$ is said to be a strongly semi-complete vector field if it has a unique null point in $\mathcal{D}$ which is a **locally uniformly attractive** fixed point for the semigroup generated by f.

Let $\sigma(A)$ denote the spectrum of a bounded linear operator $A : X \to X$. It is known that if $\mathcal{D}$ is a bounded domain, then $f \in \mathcal{G}(\mathcal{D})$ with $f(\tau) = 0$, $\tau \in \mathcal{D}$, is strongly semi-complete if and only if there is $\varepsilon > 0$ such that $\text{Re}\lambda \geq \varepsilon > 0$ for all $\lambda \in \sigma(f'(\tau))$. Such a point τ is sometimes said to be strictly regular .

In this section we will give several sufficient conditions for $f \in \mathrm{Hol}(\mathcal{D}, X)$ to be strongly semi-complete on the open unit ball $\mathcal{D}$ of X and obtain rates of convergence for the semigroups generated by such mappings.

In this context boundary flow invariance conditions are seen to be quite useful. However, it does not seem natural to consider only boundary conditions because there are many examples of semi-complete vector fields defined on a domain $\mathcal{D}$ which have no continuous extension to $\overline{\mathcal{D}}$. Since we will mainly concentrate our discussions on domains which are biholomorphically equivalent to a ball, we will assume in the sequel that $\mathcal{D}$ is the open unit ball of a complex Banach space X. If $X = H$ is a Hilbert space with the inner product $\langle \cdot, \cdot \rangle$, then we will use the letter $\mathbb{B}$ to denote its open unit ball. This will enable us to point out special features of semi-complete vector fields in this case. Our approach to the search for different (but equivalent) characterizations of the class of (strongly) semi-complete vector fields on $\mathcal{D}$ is based on the following lemma.

Lemma 9.1   *Let $\mathcal{D}$ be the open unit ball in a complex Banach space X and let $f \in \mathrm{Hol}(\mathcal{D}, X)$ satisfy the inequality*

$$\mathrm{Re}\langle f(x), x^* \rangle \geq a(\|x\|)\|x\| \tag{9.2}$$

for all $x \in \mathcal{D}$ and some $x^* \in J(x)$, where a is a real continuous function on $[0, 1)$ such that for all $\mu \in [0, 1)$ and for all $r > 0$ the equation

$$s + \lambda a(s) = \mu \tag{9.3}$$

has a unique solution $s(\mu)$ in $[0, 1)$. Then

(i) f is a semi-complete vector field on $\mathcal{D}$;
(ii) if $\beta(t, s)$ is the solution of the Cauchy problem

$$\begin{cases} \dfrac{\partial \beta(t, s)}{\partial t} + a(\beta(t, s)) = 0 \\ \beta(0, s) = s[0, 1) \end{cases} \tag{9.4}$$

and $u(t, x)$ is the solution of (9.1) then the following estimate holds:

$$\|u(t, x)\| \leq \beta(t, \|x\|), \quad x \in \mathcal{D}. \tag{9.5}$$

Proof.   Fix $r \in [0, 1)$ and $\lambda \geq 0$, and consider the equations

$$x + \lambda f(x) = y \tag{9.6}$$

and

$$s + \lambda a(s) = |y|, \tag{9.7}$$

where $y \in \mathcal{D}_r = \{x \in X : \|x\| \leq r < 1\}$ and $s \in [0, 1)$. It follows from our assumption that equation (9.7) has a unique solution $s_0 = s_0(y) \in [0, 1)$. Setting $\gamma(s) = s + \lambda a(s) - |y|$ and $\delta > 0$, we can find $\varepsilon > 0$ such that $\gamma(s + \delta) \geq \varepsilon$. Taking $x \in \mathcal{D}$ such that $\|x\| = s = s_0 + \delta$, we have, by (9.2), for such x and some $x^* \in J(x)$,

$$\mathrm{Re}\langle x + \lambda f(x) - y, x^* \rangle \geq s^2 + \lambda a(s)s - \|y\|s = s\gamma(s) \geq s \cdot \varepsilon. \tag{9.8}$$

Now it follows from Theorem 3 in [Aizenberg *et al.* (1996)] that equation (9.6) has a unique solution $x = x(y)$ such that $\|x(y)\| \leq s_0 + \delta$. Since δ is arbitrary, we must have

$$\|x(y)\| \leq s_0. \tag{9.9}$$

In terms of nonlinear resolvents the latter inequality can be rewritten as

$$\|J_\lambda(y)\| \leq (1 + \lambda a)^{-1}\|y\|. \tag{9.10}$$

Now we obtain our assertion by [Reich and Shoikhet (1996)] and the exponential formula given there. $\qquad\square$

As a consequence of Lemma 9.1 we also get the following assertion [Elin *et al.* (2004)].

Theorem 9.1 *Let $\mathcal{D}$ be the open unit ball in an arbitrary complex Banach space X, and let $f \in \mathrm{Hol}(\mathcal{D}, X)$. Then*

(1) f is semi-complete on $\mathcal{D}$ if and only if

    (i) it is bounded on each subset strictly inside $\mathcal{D}$;

    (ii) $A = f'(0)$ is an accretive linear operator on X, i.e.,

$$\mathrm{Re}\langle Ax, x^* \rangle \geq 0, \quad x \in X, \quad x^* \in J(x), \tag{9.11}$$

    and

    (iii) for each $x \in \mathcal{D}$ and $x^* \in J(x)$, the following inequality holds:

$$\mathrm{Re}\left\langle \frac{1 - \|x\|}{1 + \|x\|} f'(0)x + (1 - \|x\|^2)f(0), x^* \right\rangle$$

$$\leq \mathrm{Re}\langle f(x), x^* \rangle$$

$$\leq \mathrm{Re}\left\langle \frac{1 + \|x\|}{1 - \|x\|} f'(0)x + (1 - \|x\|^2)f(0), x^* \right\rangle. \tag{9.12}$$

(2) If $\{F(t)\}_{t\geq 0}$ is the semigroup of holomorphic self-mappings of $\mathcal{D}$ generated by f, then it satisfies the following estimate:

$$\|F(t)(x)\| \leq \beta(t, \|x\|, a, m), \tag{9.13}$$

where $m = \inf\limits_{\|x\|=1} \mathrm{Re}\langle f(0), x^*\rangle$, $a = \inf\limits_{\|x\|=1} \langle f'(0)a, a^*\rangle$, and $\beta(t, s, a, m)$ satisfies the following algebraic equations:

$$\frac{\beta}{(1-\beta)^2} = e^{-at}\frac{s}{(1-s)^2} \quad m = 0;$$

$$\frac{\beta}{(1-\beta)^2} = -2mt + \frac{1}{(1-s)^2} \quad a = -4m \neq 0;$$

$$\frac{\beta^2 + \beta\left(2 + \frac{a}{m}\right) + 1}{(1-\beta)^2} = e^{-t(4m+a)}\frac{s^2 + s\left(2 + \frac{a}{m}\right) + 1}{(1-s)^2}$$

$$a \neq -4m, \ m \neq 0. \tag{9.14}$$

Now if $f(0) = 0$ we have $m = 0$, and we get that $\{F(t)\}_{t\geq 0}$ satisfies the following explicit estimate which gives a rate of convergence of $\{F(t)\}_{t\geq 0}$ to the origin, its stationary point.

Corollary 9.1   *Let $f \in \mathcal{G}(\mathcal{D})$ be such that $f(0) = 0$ and $A = f'(0)$ is accretive with $\mathrm{Re}\langle Ax, x^*\rangle \geq k\|x\|^2$, $k \geq 0$. Suppose that $\{F(t)\}_{t\geq 0}$ is the semigroup generated by f. Then the following estimates hold:*

$$(i) \quad \|F(t)x\| \leq \|x\| a^{-k\frac{1-\|x\|}{1+\|x\|}t}, \quad x \in \mathcal{D}, \ t \geq 0; \tag{9.15}$$

$$(ii) \quad \frac{\|F(t)x\|}{(1 - \|F(t)x\|)^2} \leq e^{-kt}\frac{\|x\|}{(1 - \|x\|)^2}. \tag{9.16}$$

Proof.   Both estimates follow directly from Lemma 9.1 (or Lemma 9.2) if we set

$$a(s) = ks\frac{1-s}{1+s}. \tag{9.17}$$

In this case

$$\beta(t, s) \leq se^{-k\frac{1-s}{1+s}t}, \tag{9.18}$$

where $\{\beta(t, \cdot)\}_{t\geq 0}$ is the real-valued semigroup generated by a. $\square$

Remark 9.1   *The estimate (i) is due to Gurganus [Gurganus (1975)], while (ii) was obtained by Poreda [Poreda (1987)]. Note that the condition $f(0) = 0$ is essential in their considerations as well as in our approach above.*

Remark 9.2 *For the case of Hilbert space we will show below how more general estimates can be obtained when f has an arbitrary null point which is strictly regular .*

Remark 9.3 *If k in Corollary 9.1 is positive, then the operator A is strongly accretive, hence its spectrum lines strictly in the right-half plane. Thus in this case f is strongly semi-complete. The question is whether a one-sided estimate or a flow invariance condition could recognize f a priori to be strongly semi-complete.*

We now give some sufficient conditions for f to be strongly semi-complete on the open unit ball $\mathcal{D}$ of X, and obtain rates of convergence for the semigroups generated by such mappings in terms of a metric on $\mathcal{D}$ assigned to it by a Schwarz–Pick system.

We recall in passing that for a bounded convex domain $\mathcal{D}$ in X, all metrics assigned to it by a Schwarz–Pick system coincide. We call this unique metric the hyperbolic metric on $\mathcal{D}$.

Lemma 9.2 *Let $\mathcal{D}$ be the open unit ball in X and let $f \in \mathrm{Hol}(\mathcal{D}, X)$ satisfy the condition*

$$\mathrm{Re}\langle f(x), x^* \rangle \geq a(\|x\|) \cdot \|x\|, \quad x \in \mathcal{D}, \quad x^* \in J(x), \tag{9.19}$$

where a is a real continuous function on $[0, 1]$ such that

$$a(1) = \omega > 0. \tag{9.20}$$

Then

(i) f is strongly semi-complete;

(ii) if $\{F_t\}_{t \geq 0}$ is the semigroup generated by f, then for each pair of points x and y in $\mathcal{D}$, the estimate

$$\rho(F_t(x), F_t(y)) \leq e^{-\frac{\omega}{2}t}\rho(x, y), \tag{9.21}$$

where ρ is the hyperbolic metric on $\mathcal{D}$, holds. In particular, if $\tau \in \mathcal{D}$ is the null point of f, then

$$\rho(F_t(x), \tau) \leq e^{-\frac{\omega}{2}t}\rho(x, \tau) \tag{9.22}$$

for all $x \in \mathcal{D}$.

Remark 9.4 *Lemma 9.2 is different from Lemma 9.1 because we impose different conditions on the function a. However, the proof of Lemma 9.2 is a modification of the proof of Lemma 9.1.*

Proof of Lemma 9.2 Consider for each $n = 1, 2, \ldots$ the mappings $f_n \in$ $\mathrm{Hol}(\mathcal{D}, X)$ defined by

$$f_n(x) = x + \frac{t}{n} f(x) - y, \quad x \in \mathcal{D}, \tag{9.23}$$

where $t \geq 0$ and $y \in \mathcal{D}$. Let $\mathcal{D}_r$ be the open ball centered at the origin of radius $r \in [0, 1)$. For all $x \in \partial \mathcal{D}_r = \{x \in X : \|x\| = r\}$ and for all $x^* \in J(x)$ we have, by (9.19),

$$\mathrm{Re}\langle f_n(x), x^* \rangle = \|x\|^2 + \frac{t}{n} \mathrm{Re}\langle f(x), x^* \rangle - \mathrm{Re}\langle y, x^* \rangle$$

$$\geq r^2 + \frac{t}{n} ra(r) - r\|y\| = r\left(r + \frac{t}{n} a(r) - \|y\|\right). \tag{9.24}$$

Since $\omega = a(1) > 0$, it follows that for n big enough the equation

$$\varphi_n(r) := r + \frac{t}{n} a(r) = 1 \tag{9.25}$$

has a solution $r_n \in [0, 1)$.

Indeed, $\varphi_n(0) = \frac{t}{n} a(0) \leq 1$ for $n \geq t|a(0)|$ and $\varphi_n(1) = 1 + \frac{t}{n} \omega > 1$. The inequality (9.24) implies in turn that for such n and r_n, and for all x with $\|x\| = r_n$ and $x^* \in J(x)$, the inequality

$$\mathrm{Re}\langle f_n(x), x^* \rangle \geq r_n(1 - \|y\|). \tag{9.26}$$

holds.

Since f_n is bounded on $\overline{\mathcal{D}_{r_n}}$, it follows from [Aizenberg *et al.* (1996)] that the equation

$$f_n(x) = x + \frac{t}{n} f(x) - y = 0 \tag{9.27}$$

has a unique solution $x = J_{\frac{t}{n}}(y) := \left(I + \frac{t}{n} f\right)^{-1}(y) \in \mathcal{D}_{r_n}$ for each $y \in \mathcal{D}$. In other words, the resolvent mapping $J_{\frac{t}{n}}$ maps $\mathcal{D}$ into $\mathcal{D}_{r_n}$.

It now follows from the well-known Earle–Hamilton fixed point theorem [Earle and Hamilton (1970)] that $J_{\frac{t}{n}}$ has a unique fixed point τ in $\mathcal{D}$. This point is also a null point of f. In addition, repeating the proof of the Earle–Hamilton theorem as presented in [Goebel and Reich (1984)], we obtain the estimate

$$\rho\left(J_{\frac{t}{n}}(x), J_{\frac{t}{n}}(y)\right) \leq \frac{1}{1 + \left(\frac{t}{n}\right)\left(\frac{a(r_n)}{2}\right)} \rho(x, y) \tag{9.28}$$

for each pair of points x and y in $\mathcal{D}$.

Since $a(r)$ is continuous on the interval $[0, 1]$, it follows from (9.25) that $r_n \to 1$ and $a(r_n) \to \omega$ as $n \to \infty$. Therefore, by using the exponential formula

$$F_t(x) = \lim_{n \to \infty} J_{\frac{t}{n}}^n(x) \tag{9.29}$$

and (9.28), we get by induction the estimates (9.21) and (9.22). Lemma 9.2 is proved.

Example 9.1 Let $\mathcal{D} = \Delta$ be the open unit disk in the complex plane $\mathbb{C}$ and let $f \in \operatorname{Hol}(\Delta, \mathbb{C})$ be defined by

$$f(z) = a - \bar{a}z^2 + bz\frac{1 - cz}{1 + cz}, \tag{9.30}$$

where $a \in \mathbb{C}$, $\operatorname{Re} b > 0$ and $0 \le c < 1$. If we take

$$\alpha(s) := -|a|(1 - s^2) + (\operatorname{Re} b)s\frac{1 - cs}{1 + cs}, \tag{9.31}$$

then we get

$$\operatorname{Re} f(z)\bar{z} \ge \alpha(|z|)|z| \tag{9.32}$$

and $\alpha(1) = \operatorname{Re} b\frac{1-c}{1+c} > 0$. Hence $f(z)$ is a strongly semi-complete vector field on Δ.

Example 9.2 In the theory of autonomous systems the following system is often considered:

$$\begin{cases} \dot{x}_1 - x_2 + x_1\varphi(x_1, x_2) = 0 \\ \dot{x}_2 + x_1 + x_2\varphi(x_1, x_2) = 0. \end{cases} \tag{9.33}$$

We assume that the function φ is holomorphic in the open unit ball $|x_1|^2 + |x_2|^2 < 1$. It is clear that for any point $x = (x_1, x_2) \in \mathbb{B}$ the support functional x^* is defined by

$$\langle y, x^* \rangle = y_1\bar{x}_1 + y_2\bar{x}_2. \tag{9.34}$$

Hence, for the mapping $f(x) = \big(-x_2 + x_1\varphi(x), x_1 + x_2\varphi(x)\big)$, we have

$$\langle f(x), x^* \rangle = \big(|x_1|^2 + |x_2|^2\big)\varphi(x). \tag{9.35}$$

Thus we have to examine three cases:

(1) There exists a point $x^0 = (x_1^0, x_2^0) \in \mathbb{B}$ such that $\operatorname{Re} \varphi(x^0) < 0$; if this does not hold, then either

(2) $\varphi(x) = i\beta$, $\beta \in \mathbb{R}$;
or
(3) $\operatorname{Re}\varphi(x) > 0$.

In the first case the mapping f is not a generator. In the third case f is a strongly semi-complete vector field. In the second case f is a group generator. The third case applies to the often used function $\varphi(x) = 1 + x_1^2 + x_2^2$ (see, for example, [Jordan and Smith (1987)]). Thus the solution to the system

$$\begin{cases} \dot{x}_1 - x_2 + x_1(1 + x_1^2 + x_2^2) = 0 \\ \dot{x}_2 + x_1 + x_2(1 + x_1^2 + x_2^2) = 0 \end{cases} \tag{9.36}$$

is well defined for all $t \geq 0$ and for all initial values in $\mathbb{B}$, and converges globally on $\mathbb{B}$ to the origin.

To illustrate another application of Lemma 9.2 we consider a question regarding the solvability of an autonomous differential equation of order n.

Example 9.3   Let X be the n-dimensional complex space $\mathbb{C}^n = \{(z_1, z_2, \ldots, z_n) : z_j \in \mathbb{C}\}$ with the ℓ_p-norm

$$\|z\| = \left(\sum_{k=1}^{n} |z_k|^p \right)^{\frac{1}{p}}, \tag{9.37}$$

and let $\mathcal{D}$ be the open unit ball in X. Suppose that $g : \mathcal{D} \mapsto \mathbb{C}$ is a holomorphic function on $\mathcal{D}$ which has a continuous extension to $\overline{\mathcal{D}}$. Defining $f = (f_1, \ldots, f_n) : \overline{\mathcal{D}} \mapsto \mathbb{C}^n$ in (9.1) by the formulae

$$f_i(z_1, z_2, \ldots, z_n) = -z_{i+1}, \quad 1 \leq i \leq n - 1,$$
$$f_n(z_1, z_2, \ldots, z_n) = g(z_1, \ldots, z_n), \tag{9.38}$$

and using the standard method of rewriting an n-th order differential equation as a first order system of n equations, we deduce that the boundary condition

$$\operatorname{Re} f(z)\bar{z}_n \geq |z_n|^{2-p} \operatorname{Re} \sum_{k=1}^{n-1} z_{k+1} \frac{|z_k|^p}{z_k}, \quad z \in \partial\mathcal{D}, \tag{9.39}$$

is fulfilled if and only if the equation

$$x^{(n)} + g\big(x, x', \ldots, x^{(n-1)}\big) = 0 \tag{9.40}$$

with the initial data

$$x(0) = z_1, \ x'(0) = z_2, \ldots, x^{(n-1)}(0) = z_n \tag{9.41}$$

has a unique solution $x = x(t, z_1, \ldots, z_n)$, defined for all $t \geq 0$ and $(z_1, \ldots, z_n) \in \mathcal{D}$, which satisfies the estimate,

$$\|x(t)\|_{p,T} = \max_{0 \leq t \leq T} \left(|x(t)|^p + |x'(t)|^p + \cdots + |x^{(n-1)}(t)|^p \right)^{\frac{1}{p}} < 1 \tag{9.42}$$

for each $T > 0$.

It is clear that if we set

$$a(s) := \inf\left\{ \Phi(x, x_1, \ldots, x_{n-1}) : |x|^p + |x_1|^p + \cdots + |x_{n-1}|^p = s^p \right\}, \tag{9.43}$$

where

$$\Phi(x, x_1, x_2, \ldots, x_{n-1}) = \mathrm{Re}\left[\frac{g(x, x_1, \ldots, x_{n-1})|x_{n-1}|^p}{x_{n-1}} \right.$$
$$\left. - \frac{x_1|x|^p}{x} - \frac{x_2|x_1|^p}{x_1} - \cdots - \frac{x_{n-1}|x_{n-2}|^p}{x_{n-2}} \right], \tag{9.44}$$

then inequality (9.19) holds. Consequently, if

$$\liminf_{s \to 1^-} \left\{ \Phi(x, x_1, \ldots, x_{n-1}) : |x|^p + \cdots + |x_{n-1}|^p = s^p \right\} > 0, \tag{9.45}$$

then, by Lemma 9.2, the solution to the Cauchy problem in this example converges to the point $(x_0, 0, \ldots, 0)$ which is the unique null point of the generator f.

Remark 9.5 *Note that if $f \in \mathrm{Hol}(\mathcal{D}, X)$ is known to be a semi-complete vector field on $\mathcal{D}$, then condition (9.20) can be replaced by a slightly more general condition, namely,*

$$a(l) > 0 \quad \text{for some} \quad l \in (0, 1], \tag{9.46}$$

which will still ensure assertion (i) of Lemma 9.2. This implies the following very simple and interesting sufficient condition.

Recall that a bounded operator $A : X \mapsto X$ is said to be strongly accretive if

$$\mathrm{Re}\langle Ax, x^* \rangle \geq k\|x\|^2 \tag{9.47}$$

for some $k > 0$ and all $x \in X, \ x^* \in J(x)$.

Corollary 9.2 *Let $f \in \mathcal{G}(\mathcal{D})$, and suppose that the linear operator $A = f'(0)$ is strongly accretive, that is, it satisfies (9.47) for some $k > 0$. If*

$$a > 4\|f(0)\|, \tag{9.48}$$

then f is a strongly semi-complete vector field.

Proof. Consider the function $a(s) = -\|f(0)\|(1 - s^2) + ks\frac{1-s}{1+s}$. Using (9.48) we see that $a(1) = 0$ and $a'(1) < 0$. Hence there is $l \in (0,1)$ such that $a(l) > 0$. Since Theorem 9.1 shows that

$$\mathrm{Re}\langle f(x), x^* \rangle \geq a(\|x\|) \cdot \|x\|, \tag{9.49}$$

the result follows by Remark 9.5.

Note that if $A = f'(0)$ is strongly accretive and $f(0) = 0$, then condition (9.48) is fulfilled automatically. Hence the origin is an attractive fixed point of the semigroup generated by f. Actually, this fact also follows from Corollary 9.1 above and the exponential rate of convergence obtained there.

$\square$

9.2 Asymptotic Behavior of Flows of ρ-Nonexpansive Mappings on the Hilbert Ball

Let $\mathbb{B}$ be the open unit ball of a complex Hilbert space $\mathcal{H}$ with inner product $\langle \cdot, \cdot \rangle$, and let $\rho : \mathbb{B} \times \mathbb{B} \mapsto \mathbb{R}^+$ be the hyperbolic metric on $\mathbb{B}$, *i.e.*,

$$\rho(x, y) = \tanh^{-1} \sqrt{1 - \sigma(x, y)},$$

$$\sigma(x, y) = \frac{(1 - \|x\|^2)(1 - \|y\|^2)}{|1 - \langle x, y \rangle|^2}, \quad x, y \in \mathbb{B}. \tag{9.50}$$

As above, we denote by $\mathcal{N}_\rho$ the class of all self-mappings $F : \mathbb{B} \mapsto \mathbb{B}$ which are nonexpansive with respect to ρ (ρ-nonexpansive), *i.e.*,

$$\rho\big(F(x), F(y)\big) \leq \rho(x, y). \tag{9.51}$$

This class $\mathcal{N}_\rho$ properly contains the class $\mathrm{Hol}(\mathbb{B})$ of all holomorphic self-mappings of $\mathbb{B}$.

Definition 9.3 A mapping $f : \mathbb{B} \mapsto \mathcal{H}$ is said to be **strongly ρ-monotone** (ρ-monotone) if for each pair $x, y \in \mathbb{B}$ there is $\varepsilon = \varepsilon(x, y) > 0$ ($\varepsilon = 0$) such that

$$\rho\big(x + rf(x), \, y + rf(y)\big) \geq (1 + r\varepsilon(x, y))\rho(x, y) \tag{9.52}$$

for all $r \geq 0$ such that $x + rf(x)$ and $y + rf(y)$ belong to $\mathbb{B}$.

it was shown in [Reich and Shoikhet (1997a)] that $f : \mathbb{B} \mapsto \mathcal{H}$ is ρ-monotone if and only if it satisfies the condition

$$\mathrm{Re}\left[\frac{\langle f(x), x\rangle}{1 - \|x\|^2} + \frac{\langle y, f(y)\rangle}{1 - \|y\|^2}\right] \geq \mathrm{Re}\left[\frac{\langle f(x), y\rangle + \langle x, f(y)\rangle}{1 - \langle x, y\rangle}\right]. \tag{9.53}$$

It is also known [Reich and Shoikhet (1997a), (1998b)] that if $\mathcal{S} \subset \mathcal{N}_\rho$ is the flow generated by a generator $f \in \mathcal{GN}_\rho(\mathbb{B})$ and f is bounded and uniformly continuous on each ρ-ball in $\mathbb{B}$, then the following relations hold: for each $r \geq 0$,

$$W = \mathrm{Null}(f) = \mathrm{Fix}(\mathcal{J}_r), \tag{9.54}$$

where $\mathcal{J}_r = (I + rf)^{-1}$.

In the study of the asymptotic behavior of flows of ρ-nonexpansive (or holomorphic) self-mappings of $\mathbb{B}$, the two cases $W \neq \emptyset$ and $W = \emptyset$ are usually considered separately. In particular, if $f \in \mathcal{GN}_\rho(\mathbb{B})$, then one can look for conditions which would imply the strong ρ-monotonicity of f with $\varepsilon(x, y) = \varepsilon = \text{const.}$ (see formula (9.52)). If this is the case, then the exponential formula implies that W contains a unique point τ which is globally attractive with an exponential rate of convergence: $\rho\big(F(t)x, \tau\big) \leq \exp(-\varepsilon t)\rho(x, \tau)$.

However, such an approach cannot work when $W = \emptyset$. Therefore, our aim is to find some sufficient (and perhaps necessary) conditions for global convergence of the flow generated by f which do not depend on W being empty or not. We now recall the following definition.

Definition 9.4 Let $\mathcal{S} = \{F(t)\}_{t \geq 0}$ be a flow on $\mathbb{B}$ which is generated by f. We will say that a point $\tau \in \overline{\mathbb{B}}$, the closure of $\mathbb{B}$, is a globally attractive point for $\mathcal{S}$ if for each $x \in \mathbb{B}$ the strong limit

$$\lim_{t \to \infty} F(t)x = \tau, \tag{9.55}$$

uniformly on each ρ-ball in $\mathbb{B}$.

If $\tau \in \mathbb{B}$, then τ is the unique asymptotically stable stationary point of $\mathcal{S}$. If $\tau \in \partial\mathbb{B}$, the boundary of $\mathbb{B}$, we will call it the attractive sink point of $\mathcal{S}$.

As we have seen in the previous section, in the case of holomorphic generators the attractivity of a stationary point can be completely described in terms of their derivatives. For holomorphic generators with no null point the situation can be described by using the so-called angular derivative at the boundary sink point (see the next section).

However, in both cases ($W \neq \emptyset$ and $W = \emptyset$), the characteristics of the derivatives are not relevant in general, since f' does not exist for $f \in \mathcal{GN}_\rho(\mathbb{B})$ in the complex sense if f is not holomorphic.

In this section we study such a situation.

For a fixed $\tau \in \overline{\mathbb{B}}$, the closure of $\mathbb{B}$, and an arbitrary $x \in \mathbb{B}$, we define a non-Euclidean "distance" between x to τ by the formula

$$d_\tau(x) = \frac{|1 - \langle x, \tau \rangle|^2}{1 - \|x\|^2}\,(1 - \sigma(x, \tau)), \tag{9.56}$$

where $\sigma(x, \tau)$ is defined by formula (9.50).

We already know that geometrically, the sets

$$E(\tau, s) = \big\{ x \in \mathbb{B} : d_r(x) < s \big\}, \quad s > 0, \tag{9.57}$$

are ellipsoids. If $\tau \in \mathbb{B}$, then these sets are exactly the ρ-balls

$$E(\tau, s) = \big\{ x \in \mathbb{B} : \rho(x, \tau) < r \big\} \tag{9.58}$$

centered at $\tau \in \mathbb{B}$ and of radius $r = \tanh^{-1}\sqrt{\frac{s}{s+1-\|\tau\|^2}}$. If $\tau \in \partial\mathbb{B}$, the boundary of $\mathbb{B}$, then these sets

$$E(\tau, s) = \left\{ x \in \mathbb{B} : d_\tau(x) = \frac{|1 - \langle x, \tau \rangle|^2}{1 - \|x\|^2} < s \right\}, \quad s > 0, \tag{9.59}$$

are ellipsoids which are internally tangent to the unit sphere $\partial\mathbb{B}$ at τ.

Now for fixed $\tau \in \overline{\mathbb{B}}$ and $x \in \partial E(\tau, s)$, $x \neq \tau$, consider the non-zero vector

$$x^* = \frac{1}{1 - \sigma(x, \tau)}\left(\frac{1}{1 - \|x\|^2}\,x - \frac{1}{1 - \langle \tau, x \rangle}\,\tau \right). \tag{9.60}$$

As in [Aharonov *et al.* (1999a)], it can be shown that x^* is a support functional of the smooth convex set $E(\tau, s)$ at x, normalized by the condition

$$\lim_{x \to \tau} \langle x - \tau, x^* \rangle = 1. \tag{9.61}$$

Then for a mapping $f : \mathbb{B} \to \mathcal{H}$, the flow–invariance condition

$$\mathrm{Re}\langle f(x), x^* \rangle \geq 0 \tag{9.62}$$

is necessary for f to be the generator of a continuous flow for which the sets $E(\tau, s)$ are invariant.

In our situation, when $f \in \mathcal{GN}_\rho(\mathbb{B})$, this is exactly the case if $\tau \in \mathbb{B}$ is a null point of f, since

$$\rho(F(t)x, \tau) = \rho\big(F(t)x, F(t)\tau\big) \leq \rho(x, \tau). \tag{9.63}$$

Note also that condition (9.62) can be obtained directly from the ρ-monotonicity of f if we substitute $y = \tau$ and $f(\tau) = 0$ into (9.53).

In fact, inequality (9.63) shows that if condition (9.62) holds for some $\tau \in \mathbb{B}$ and all $x \in \mathbb{B}$, then τ must be a stationary point of $\mathcal{S} = \{F(t)\}_{t \geq 0}$, hence a null point of f.

If f has no null point, then there is a unique boundary point $\tau \in \partial\mathbb{B}$ such that (9.62) holds. This point τ is the sink point for the flow generated by f.

In order to classify the asymptotic behavior of flows we will consider a finer condition than (9.62). More precisely, for a point $\tau \in \overline{\mathbb{B}}$ and $f \in \mathcal{GN}_\rho(\mathbb{B})$ we consider the following two real nonnegative functions on $(0, \infty)$:

$$\omega_b(s) := \inf_{d_\tau(x) \leq s} 2\mathrm{Re}\langle f(x), x^* \rangle, \quad s > 0, \tag{9.64}$$

and

$$\omega^\sharp(s) := \inf_{d_\tau(x) = s} 2\mathrm{Re}\langle f(x), x^* \rangle, \quad s > 0, \tag{9.65}$$

where x^* is defined by (9.60).

It is clear that

$$\omega^\sharp(s) \geq \omega_b(s) \geq 0 \tag{9.66}$$

and that $\omega_b(s)$ is decreasing on $(0, \infty)$.

Let $\mathcal{M}(0, \infty)$ denote the class of all positive functions ω on $(0, \infty)$ such that $\frac{1}{\omega}$ is Riemann integrable on each closed interval $[a, b] \subset (0, \infty)$ and

$$\int_{0+} \frac{ds}{\omega(s)s} \quad \text{is divergent.} \tag{9.67}$$

Note that for each $\omega \in \mathcal{M}(0, \infty)$, the function Ω defined by

$$\Omega(s) := \int_s^{d_\tau(x)} \frac{d\lambda}{\omega(\lambda)\lambda} \tag{9.68}$$

is a strictly decreasing positive function on $(0, d_\tau(x)]$ which maps this interval onto $[0, \infty)$. We denote its inverse function by $V : [0, \infty) \mapsto (0, d_\tau(x)]$.

Theorem 9.2 *Let $f \in \mathcal{GN}_\rho(\mathbb{B})$ be continuous and let $\mathcal{S} = \{F(t)\}_{t \geq 0}$ be the flow generated by f. Given a point $\tau \in \overline{\mathbb{B}}$ and a function $\omega \in \mathcal{M}(0, \infty)$, the following conditions are equivalent:*

(i) For all $s \in (0, \infty)$,

$$\omega^\sharp(s) \geq \omega(s) \tag{9.69}$$

where $\omega^\sharp(s)$ is defined by (9.65);

(ii) for any differentiable function W on $[0, \infty)$ such that $V(t) \leq W(t)$, $V(0) = W(0)$ and $V'(0) = W'(0)$,

$$d_\tau(F(t)x) \leq W(t), \quad x \in \mathbb{B}, \ t \geq 0, \tag{9.70}$$

where $V = \Omega^{-1}$ and Ω is defined by (9.68).

In particular, $d_\tau(F(t)x) \leq V(t)$; hence τ is a globally attractive point for $\mathcal{S}$.

Proof. Consider the function $\Psi : \mathbb{R}^+ \times \mathbb{B} \mapsto \mathbb{R}^+$ defined by

$$\Psi(t, x) = d_\tau(F(t)x). \tag{9.71}$$

By direct calculations we have

$$\left.\frac{\partial \Psi}{\partial t}\right|_{t=0^+} = -2\Psi(0, x)\mathrm{Re}\langle f(x), x^*\rangle. \tag{9.72}$$

Let us first assume that condition (ii) holds. Since $\Phi(0, x) = d_\tau(x) = W(0)$, we get by (9.72) and (ii) that

$$2\Psi(0, x)\mathrm{Re}\langle f(x), x^*\rangle = -\left.\frac{\partial \Psi}{\partial t}\right|_{t=0^+} \geq -\frac{d}{dt}[W(t)]_{t=0^+}$$

$$= -\frac{d}{dt}[V(t)_{t=0^+} = -\frac{1}{\Omega'(d_\tau(x))} = d_\tau(x)\omega(d_\tau(x)). \tag{9.73}$$

Varying $x \in \partial E(\tau, s) = \{x \in \mathbb{B} : d_\tau(x) = s\}$, we see that this inequality immediately implies (i).

Conversely, let condition (i) hold. It follows from (9.71) and the semigroup property that for all $x \in \mathbb{B}$ and $s, t \geq 0$,

$$\Psi(s + t, x) = \Psi(s, F(t)x). \tag{9.74}$$

Hence by (9.72) and the continuity of f, Ψ is differentiable at each $t \geq 0$ and we deduce from (i) and (9.72) that

$$\frac{\partial \Psi(t, x)}{\partial t} \leq -\Psi(t, x)\omega^\sharp(\Psi(t, x)) \leq -\Psi(t, x)\omega(\Psi(t, x)). \tag{9.75}$$

Separating variables we get

$$\int_{d_\tau(F(t)x)}^{d_\tau(x)} \frac{d\Psi}{\omega(\Psi)\Psi} = \Omega\big(d_\tau(F(t)x)\big) \geq t, \tag{9.76}$$

which is equivalent to condition (ii). Theorem 9.2 is proved. $\square$

We will call a function $\omega \in \mathcal{M}(0,\infty)$ which satisfies condition (i), an appropriate lower bound for $f \in \mathcal{GN}_\rho(\mathbb{B})$.

Remark 9.6 *Of course, if the function $\omega_\flat$ defined by (9.64) belongs to $\mathcal{M}(0,\infty)$, then one can use it as an appropriate lower bound.*

However, examples show that sometimes $\omega_\flat$ may be identically zero, while $\omega^\sharp$ itself belongs to the class $\mathcal{M}(0,\infty)$. Moreover, we will see below that for a semigroup of holomorphic mappings with a boundary sink point, $\omega_\flat$ is always a constant which determines the best rate of uniform exponential convergence of the flow.

To illustrate Theorem 9.2 and to motivate our next definition, we now present several one-dimensional examples [Elin et al. (2002)].

Example 9.4 Let Δ be the open unit desk of the complex plane $\mathbb{C}$, let n be a positive integer and let $f : \Delta \mapsto \mathbb{C}$ be defined by

$$f(z) = -(1-z)^2 \frac{1+z^n}{1-z^n}. \tag{9.77}$$

If we set $\tau = 1$ and

$$z^* = \frac{z}{1-|z|^2} - \frac{1}{1-\bar{z}}, \tag{9.78}$$

then we get

$$\mathrm{Re}f(z)\bar{z}^* = \frac{|1-z|^2}{1-|z|^2} \mathrm{Re} \frac{1+z^n}{1-z^n} = d_1(z)\mathrm{Re}\frac{1+z^n}{1-z^n} > 0. \tag{9.79}$$

Since f is holomorphic on Δ, this inequality implies that f generates a flow $\mathcal{S} = \{F(t)\}_{t\geq 0}$ of holomorphic self-mappings of Δ. In addition, it can be shown (see Theorem 9.4 below) that

$$\omega(s) = \omega_\flat(s) = \inf_{d_1(z)\leq s} 2\mathrm{Re}f(z)\bar{z}^* = \mathrm{const.} = \frac{2}{n}. \tag{9.80}$$

Hence f satisfies the conditions of Theorem 9.2. In this case,

$$\Omega(s) = \frac{n}{2} \int_s^{d_1(x)} \frac{d\lambda}{\lambda} = -\frac{n}{2} \, ln\frac{s}{d_1(z)} \tag{9.81}$$

and

$$V(t) = \Omega^{-1}(t) = \exp\left\{-\frac{2}{n}t\right\} d_1(z). \tag{9.82}$$

Thus we have an exponential rate of convergence of the flow $\mathcal{S}$ to the boundary point $\tau = 1$:

$$\frac{|1 - F(t)z|^2}{1 - |F(t)z|^2} \leq \exp\left\{-\frac{2}{n}t\right\} \frac{|1 - z|^2}{1 - |z|^2}. \tag{9.83}$$

Note also that although f has $n+1$ null points $\{a_k : k = 1, 2, \ldots, n+1\}$ on the unit circle, only $a_1 = 1$ is an attractive point of $\mathcal{S} = \{F(t)\}_{t \geq 0}$. The reason is that $\mathrm{Re} f'(a_1) > 0$, while $\mathrm{Re} f'(a_k) < 1$, $k = 2, 3, \ldots, n+1$ (see Theorem 9.5 below).

Example 9.5   Let Δ be as above and let $f : \Delta \to \mathbb{C}$ be defined by

$$f(z) = -(1-z)^2 \frac{1 + cz^n}{1 - cz^n} \tag{9.84}$$

with $|c| < 1$.

Once again, if we define z^* as in Example 9.4, we have

$$\mathrm{Re} f(z)\overline{z^*} = \frac{|1-z|^2}{1-|z|^2} \mathrm{Re} \frac{1+cz^n}{1-cz^n} \geq d_1(z) \frac{1-|c|}{1+|c|} > 0. \tag{9.85}$$

In this case, $\omega_b(s) = 0$ for all $s \in (0, \infty)$ and we cannot use it as an appropriate lower bound. However, we can define $\omega(s) = as$, where $a = \frac{1-|c|}{1+|c|}$, and we find

$$\Omega(s) = \frac{1}{a} \int_s^{d_1(z)} \frac{d\lambda}{\lambda^2} = \frac{1}{a}\left(\frac{1}{s} - \frac{1}{d_1(z)}\right). \tag{9.86}$$

Thus we get, by Theorem 9.2, the following rate of convergence:

$$\frac{|1 - F(t)z|^2}{1 - |F(t)z|^2} \leq \frac{1}{1 + atd_1(z)} \frac{|1-z|^2}{1 - |z|^2}. \tag{9.87}$$

Example 9.6   Let Δ be as above and let $z = x + iy \in \Delta$. Define $f : \Delta \to \mathbb{C}$ by

$$f(z) = x^{\frac{7}{3}} + iy^{\frac{7}{3}}. \tag{9.88}$$

Since

$$\mathrm{Re} f(z)\bar{z} = x^{\frac{10}{3}} + y^{\frac{10}{3}} \geq 0, \tag{9.89}$$

f is ρ-monotone and the origin is the unique null point of f. Hence, if we set $\tau = 0$, then we have

$$d_0(z) = \frac{|z|^2}{1 - |z|^2} \tag{9.90}$$

and

$$\omega^\sharp(s) = 2 \inf_{d_0(z)=s} \frac{1}{|z|^2(1 - |z|^2)} \operatorname{Re} f(z)\bar{z}$$

$$= 2 \inf_{x^2+y^2=\frac{s}{s+1}} \frac{x^{\frac{10}{3}} + y^{\frac{10}{3}}}{(x^2 + y^2)(1 - x^2 - y^2)} = 2^{\frac{1}{3}} s^{\frac{2}{3}}(1 + s)^{\frac{1}{3}}. \tag{9.91}$$

Setting $\omega(s) = \omega^\sharp(s)$, we get

$$\Omega(s) = \int_s^{d_0(z)} \frac{d\lambda}{\omega(\lambda)\lambda} = \frac{1}{2^{\frac{1}{3}}} \int_s^{d_0(z)} \frac{d\lambda}{\lambda^{\frac{5}{3}}(\lambda + 1)^{\frac{1}{3}}}. \tag{9.92}$$

Finally, we obtain the estimate

$$d_0(F(t)z) \leq V(t) = \frac{d_0(z)}{\left[\frac{2^{\frac{4}{3}}}{3} t d_0(z)^{\frac{2}{3}} + (d_0(z) + 1)^{\frac{2}{3}}\right]^{\frac{3}{2}} - d_0(z)}. \tag{9.93}$$

The latter inequality is equivalent to the estimate

$$|F(t)z| \leq \frac{|z|}{\left[\frac{(2|z|)^{\frac{4}{3}}}{3} t + 1\right]^{\frac{3}{4}}}. \tag{9.94}$$

Note that one can calculate $F(t)$ directly by solving the Cauchy problem and get

$$|F(t)z|^2 = \frac{x^2}{\left(\frac{4}{3} x^{\frac{4}{3}} t + 1\right)^{\frac{3}{2}}} + \frac{y^2}{\left(\frac{4}{3} y^{\frac{4}{3}} t + 1\right)^{\frac{3}{2}}}. \tag{9.95}$$

Thus for $x = y$ we obtain

$$|F(t)z| = \frac{|z|}{\left[\frac{(2|z|)^{\frac{4}{3}}}{3} t + 1\right]^{\frac{3}{4}}}. \tag{9.96}$$

So, the rate of nonexponential convergence we have obtained is sharp.

Remark 9.7 *We will see below that a similar phenomenon is impossible for holomorphic mappings, namely:*

If a flow of holomorphic self-mappings converges locally uniformly to an interior stationary point, then the convergence must be of exponential type.

Definition 9.5   Let $S = \{F(t)\}_{t \geq 0}$ be a flow with a stationary (or sink) point $\tau \in \overline{B}$. We will say that the asymptotic behavior of S at τ is of order not less than $\alpha > 0$ if there is a function $\omega \in \mathcal{M}(0, \infty)$ such that

$$\liminf_{s \to 0^+} \left\{ \frac{\omega(s)}{s^{\frac{1}{\alpha}}} \right\} > 0, \tag{9.97}$$

and

$$d_\tau(F(t)x) \leq \frac{1}{\left(1 + \frac{t}{\alpha}\, \omega(d_\tau(x))\right)^\alpha}\, d_\tau(x) \tag{9.98}$$

for all $x \in \mathbb{B}$ and $t \geq 0$.

Definition 9.6   We will say that the asymptotic behavior of S at τ is of **exponential type** if there is a decreasing function $\omega \in \mathcal{M}(0, \infty)$ such that

$$d_\tau(F(t)x) \leq \exp(-t\omega(d_\tau(x)))d_\tau(x) \tag{9.99}$$

for all $x \in \mathbb{B}$ and $t \geq 0$.

In particular, if ω can be chosen to be a positive constant a, then we will say that S has a global uniform rate of convergence:

$$d_\tau(F(t)x) \leq \exp(-ta)d_\tau(x). \tag{9.100}$$

The following assertion is a consequence of Theorem 9.2 [Elin *et al.* (2002)].

Theorem 9.3   *Let $S = \{F(t)\}_{t \geq 0}$ be a flow generated by $f \in \mathcal{GN}_\rho(\mathbb{B})$ with a null (or sink) point $\tau \in \overline{\mathbb{B}}$. Then the asymptotic behavior of S at τ is of order not less than $\alpha > 0$ if and only if there exists an appropriate lower bound $\omega \in \mathcal{M}(0, \infty)$ for f such that*

$$\frac{\omega(s)}{s^{\frac{1}{\alpha}}} \quad \text{is decreasing on } (0, \infty). \tag{9.101}$$

Proof.   We first observe that condition (9.98) with some $\omega \in \mathcal{M}(0, \infty)$ satisfying (9.97) is equivalent to the same condition with a function $\omega_1 \in \mathcal{M}(0, \infty)$ which satisfies both (9.97) and (9.101). Indeed, for a given $\omega \in \mathcal{M}(0, \infty)$, define a function $\mu : (0, \infty) \mapsto (0, \infty)$ by

$$\mu(s) = \inf\left\{ \frac{\omega(l)}{l^{\frac{1}{\alpha}}} : l \in (0, s] \right\}, \quad s > 0. \tag{9.102}$$

It is clear that $\mu(s)$ is decreasing. Setting now $\omega_1(s) = s^{\frac{1}{\alpha}} \cdot \mu(s)$, we clearly see that ω_1 satisfies (9.97) and that $\omega_1(s) \leq \omega(s)$. Hence

$$\int_{0+} \frac{ds}{\omega_1(s)s} \tag{9.103}$$

is divergent and $\omega_1 \in \mathcal{M}(0, \infty)$. In addition, we have the inequality

$$\frac{1}{\left[1 + \frac{t}{\alpha}\omega(s)\right]^{\alpha}} \leq \frac{1}{\left[1 + \frac{t}{\alpha}\omega_1(s)\right]^{\alpha}} \tag{9.104}$$

which proves our claim.

Thus we can assume for the rest of the proof that ω satisfies (9.101). It remains to show that ω is an appropriate lower bound for f.

Indeed, defining $\Omega : (0, d_\tau(x)] \to [0, \infty)$ by (9.68) and using (9.101) we have

$$\begin{aligned}
\Omega(s) &= \int_s^{d_\tau(x)} \frac{d\lambda}{\omega(\lambda)\lambda} = \int_s^{d_\tau(x)} \frac{\lambda^{\frac{1}{\alpha}} d\lambda}{\omega(\lambda)\lambda^{\frac{1}{\alpha}+1}} \\
&\leq \frac{[d_\tau(x)]^{\frac{1}{\alpha}}}{\omega(d_\tau(x))} \int_s^{d_\tau(x)} \frac{d\lambda}{\lambda^{\frac{1}{\alpha}+1}} \\
&= \frac{\alpha}{\omega(d_\tau(x))} \left[s^{-\frac{1}{\alpha}} (d_\tau(x))^{\frac{1}{\alpha}} - 1\right].
\end{aligned} \tag{9.105}$$

Inverting this expression we get

$$V(t) := \Omega^{-1}(t) \leq \frac{1}{\left(1 + \frac{t}{\alpha}\omega(d_\tau(x))\right)} d_\tau(x) := W(t). \tag{9.106}$$

It is clear that the function $W(t)$ satisfies all the conditions of Theorem 9.2. This completes the proof of Theorem 9.3. $\qquad\square$

Our next assertion is a direct consequence of Theorems 9.2 and 9.3.

Corollary 9.3 *Let $S = \{F(t)\}_{t \geq 0}$ be the flow generated by a mapping f with a null (or sink) point $\tau \in \overline{\mathbb{B}}$. Then*

(i) the asymptotic behavior of S at τ is of exponential type if and only if

$$\inf\{w^{\sharp}(l) : l \in (0, s]\} > 0, \quad s > 0; \tag{9.107}$$

(ii) the flow S has a global uniform rate of exponential convergence if and only if

$$\omega^{\sharp}(s) \geq a \tag{9.108}$$

for some $a > 0$.

Indeed, in both cases (i) and (ii), there is one function $\omega \in \mathcal{M}(0, \infty)$ such that the asymptotic behavior of $\mathcal{S}$ at τ is of order not less than α for all positive α. In case (i), ω can be chosen to be

$$\omega(s) := \inf\{\omega^{\sharp}(l) : l \in (0, s]\} > 0, \quad s > 0, \tag{9.109}$$

while in case (ii) ω can be chosen to be a constant a.

Remark 9.8　*However, we will see in the next section that, for holomorphic mappings, condition* (9.108) *holds, in fact, for some $a > 0$ whenever condition* (9.107) *holds. In other words, for holomorphic flows any convergence of exponential type implies global uniform exponential convergence.*

The following example shows that for a semigroup of ρ-nonexpansive (but not holomorphic!) mappings an asymptotic behavior of exponential type does not imply, in general, a global uniform exponential rate of convergence.

Example 9.7　Define a continuous mapping $f : \Delta \mapsto \mathbb{C}$ by the following formula:

$$f(x + iy) = x(1 - x)^2 + iy(1 - y)^2. \tag{9.110}$$

Since $\mathrm{Re}f(z)\bar{z} \geq 0$ for all $z = x + iy \in \Delta$, it follows that f is the generator of a semigroup $\mathcal{S} = \{F(t)\}_{t \geq 0}$ of ρ-nonexpansive mappings such that each disk $\Delta_r = \{z \in \mathbb{C} : |z| < r < 1\}$ is $F(t)$-invariant. Setting $\tau = 0$ and $z^* = \frac{z}{|z|^2(1 - |z|^2)}$, we have

$$\omega^{\sharp}(s) = \inf_{d_0(z)=s} \mathrm{Re}f(z)\bar{z}^* = \inf_{x^2+y^2=\frac{s}{s+1}} \frac{x^2(1 - x)^2 + y^2(1 - y)^2}{(x^2 + y^2)(1 - x^2 - y^2)} \tag{9.111}$$

It is easy to see that $\lim\limits_{s \to 0^+} \omega^{\sharp}(s) = 1$ while $\omega^{\sharp}(s) \to 0$ as $s \to \infty$ (take, for example, $y = 0$ and $x = \sqrt{\frac{s}{s+1}} \to 1$).

9.3　Flows of Holomorphic Mappings on the Hilbert Ball

In this section we will study in more detail flows $\mathcal{S} = \{F(t)\}_{t \geq 0}$ of self-mappings generated by holomorphic mappings $f \in \mathcal{G}\mathrm{Hol}(\mathbb{B})$ with stationary (or sink) points $\tau \in \overline{\mathbb{B}}$. We already know that the asymptotic behavior of a flow $\mathcal{S}$ at τ is of exponential type (Definition 9.6) if and only if the function

$$\omega^{\sharp}(s) = \inf_{d_\tau(x)=s} 2\mathrm{Re}\langle f(x), x^* \rangle, \quad s > 0, \tag{9.112}$$

satisfies (9.107). It turns out (see Theorem 9.4) that in this case this function and even the function

$$\omega_b(s) = \inf_{d_\tau(x) \leq s} 2\operatorname{Re} \langle f(x), x^* \rangle \qquad (9.113)$$

are bounded from below by a positive number. Moreover, for a boundary sink point the function ω_b is just a constant.

In both cases (interior stationary point or boundary sink point) the asymptotic behavior of a flow is completely determined by the value $\omega^\sharp(0) := \liminf_{s \to 0^+} \omega^\sharp(s)$ which is related to the value of the derivative of f at its null point (for the interior case) or the so-called angular derivative (for the boundary case).

We begin with the following general assertion.

Theorem 9.4 (Theorem on universal rates of convergence) *Let $f \in \mathcal{G}\mathrm{Hol}(\mathbb{B})$ and let $\{F(t)\}_{t \geq 0}$ be the flow generated by f. If for some point $\tau \in \overline{\mathbb{B}}$ there is a decreasing function $\omega : (0, \infty) \mapsto (0, \infty)$ such that*

$$d_\tau(F(t)x) \leq e^{-t\omega(d_\tau(x))} d_\tau(x), \quad x \in \mathbb{B},\ t \geq 0, \qquad (9.114)$$

then there exists a number $\mu > 0$ such that

$$d_\tau(F(t)x) \leq e^{-t\mu} d_\tau(x), \quad x \in \mathbb{B},\ t \geq 0. \qquad (9.115)$$

Moreover,

(i) if $\tau \in \mathbb{B}$, then μ can be chosen as $\mu = \frac{\omega_b(0)}{4}$, but μ cannot be larger than $\omega_b(0)$ $(= \lim_{s \to 0^+} \omega_b(s))$;

(ii) if $\tau \in \partial\mathbb{B}$, then the maximal μ, for which (9.115) holds, is exactly $\omega_b(0)$, that is, $0 < \mu \leq \omega_b(0)$.

We will prove and discuss this theorem separately for the case where $\tau \in \mathbb{B}$ is a null point of f and for the case where $\tau \in \partial\mathbb{B}$, that is, when f is null point free.

9.3.1 *Interior stationary point*

Lemma 9.3 *Let $f \in \mathcal{G}\mathrm{Hol}(\mathbb{B})$ with $f(0) = 0$ and let $\omega^\sharp$ and ω_b be defined by (9.112) and (9.113). Then*

(i) $\omega^\sharp(0) = \omega_b(0) = 2 \inf_{\|x\|=1} \operatorname{Re} \langle f'(0)x, x \rangle$;

(ii) $\frac{\omega_b(0)}{4} \leq \omega_b(s) \leq \omega_b(0)$.

Proof. First we show that

$$\omega^\sharp(0) \le 2\nu, \tag{9.116}$$

where

$$\nu = \inf_{\|x\|=1} \mathrm{Re}\langle f'(0)x, x\rangle. \tag{9.117}$$

Since in our case $\tau = 0$, we have

$$\mathrm{Re}\langle f(x), x^*\rangle = \frac{1}{\|x\|^2(1 - \|x\|^2)} \, \mathrm{Re}\langle f(x), x\rangle. \tag{9.118}$$

Now fixing $u \in \partial\mathbb{B}$, we set $x = ru$, where $r \in (0,1)$. Then we get

$$\mathrm{Re}\langle f(x), x^*\rangle = \mathrm{Re} \, \frac{1}{1 - r^2} \left\langle \frac{1}{r} f(ru), u \right\rangle. \tag{9.119}$$

Therefore,

$$\omega^\sharp(s) \le 2\mathrm{Re} \, \frac{1}{1 - r^2} \left\langle \frac{1}{r} f(ru), u \right\rangle, \quad \text{where } r^2 = \|x\|^2 = \frac{s}{s+1}. \tag{9.120}$$

Letting s (hence, r) tend to zero we obtain

$$\omega^\sharp(0) \le 2\mathrm{Re}\langle f'(0)u, u\rangle. \tag{9.121}$$

Since u is arbitrary, (9.116) follows. On the other hand, it follows from the generalized Harnack inequality (see, for example, [Aharonov *et al.* (1999b)]) that for all $x \in \mathbb{B}$,

$$\mathrm{Re}\langle f(x), x\rangle \ge \mathrm{Re}\langle f'(0)x, x\rangle \frac{1 - \|x\|}{1 + \|x\|} \ge \nu\|x\|^2 \frac{1 - \|x\|}{1 + \|x\|}. \tag{9.122}$$

This implies that

$$
\begin{aligned}
2\mathrm{Re}\langle f(x), x^*\rangle &= \frac{2}{\|x\|^2(1 - \|x\|^2)} \, \mathrm{Re}\langle f(x), x\rangle \\
&\ge \frac{2\nu\|x\|^2}{\|x\|^2(1 - \|x\|^2)} \cdot \frac{1 - \|x\|}{1 + \|x\|} = \frac{2\nu}{(1 + \|x\|)^2}.
\end{aligned}
\tag{9.123}
$$

Hence

$$
\begin{aligned}
\omega_b(s) &= \inf_{d_\tau(x)\le s} 2\mathrm{Re}\langle f(x), x^*\rangle \ge \inf_{d_\tau(x)\le s} \frac{2\nu}{(1 + \|x\|)^2} \\
&= \inf_{\|x\|^2 \le \frac{s}{s+1}} \frac{2\nu}{(1 + \|x\|)^2} = \frac{2\nu}{\left(1 + \sqrt{\frac{s}{s+1}}\right)^2} \left(\ge \frac{\nu}{2}\right).
\end{aligned}
\tag{9.124}
$$

Letting s tend to 0^+ in (9.124), we see that $\omega_b(0) \geq 2\nu$. Since obviously $\omega^\sharp(0) \geq \omega_b(0)$, comparing the latter inequality with (9.116) we obtain (i). On the other hand, substituting now in (9.124) $\nu = \frac{\omega_b(0)}{w}$, we get assertion (ii). This completes the proof. $\qquad\square$

To proceed, we denote by M_τ the Möbius transformation of $\mathbb{B}$ defined by

$$M_\tau(x) = \frac{1}{1 - \langle x, \tau \rangle} \left(\tau - \frac{\langle x, \tau \rangle}{\|\tau\|^2} - \sqrt{1 - \|\tau\|^2} \left(x - \frac{\langle x, \tau \rangle}{\|\tau\|^2} \right) \right). \tag{9.125}$$

Note that M_τ is an automorphism of $\mathbb{B}$ (see, for example, [Rudin (1980)] and [Goebel and Reich (1984)]) which has the following properties:

(a) $M_\tau^{-1} = M_\tau$ (involution property) with $M_\tau(0) = \tau$ and $M_\tau(\tau) = 0$;
(b) $1 - \|M_\tau(x)\|^2 = \sigma(x, y)$;
(c) $1 - \langle M_\tau(x), \tau \rangle = \frac{1 - \|\tau\|^2}{1 - \langle x, \tau \rangle}$.

These properties imply the equality

$$d_\tau(M_\tau(x)) = (1 - \|\tau\|^2)\, d_0(x). \tag{9.126}$$

Now let us consider the flow $\{G(t)\}_{t \geq 0} \subset \mathrm{Hol}(\mathbb{B})$ defined by

$$G(t) = M_\tau \circ F(t) \circ M_\tau, \tag{9.127}$$

and let $g \in \mathcal{G}\mathrm{Hol}(\mathbb{B})$ be its generator, i.e.,

$$g(x) = -\frac{\partial}{\partial t} G(t)x|_{t=0^+} = [(M_\tau)'(x)]^{-1} f(M_\tau(x)). \tag{9.128}$$

Then $G_t(0) = 0$ for all $t \geq 0$ and $g(0) = 0$.

Lemma 9.4 *The following equality holds:*

$$\frac{1}{1 - \sigma(x, \tau)} \mathrm{Re} \left\langle f(x), \frac{x}{1 - \|x\|^2} - \frac{\tau}{1 - \langle \tau, x \rangle} \right\rangle$$
$$= \frac{1}{\|y\|^2} \mathrm{Re} \left\langle g(y), \frac{y}{1 - \|y\|^2} \right\rangle, \tag{9.129}$$

where $y = M_\tau(x)$. Thus the functions $\omega^\sharp(s)$ and $\omega_b(s)$ are invariant under the transformations (9.127) and (9.128).

Proof. We have already seen in (9.71) and (9.72) that

$$\frac{\partial}{\partial t}\left[d_\tau(F(t)x\right]_{t=0+} = -2d_\tau(x)\mathrm{Re}\,\langle f(x), x^*\rangle$$

$$= -2\,\frac{d_\tau(x)}{1-\sigma(x,\tau)}\,\mathrm{Re}\Big\langle f(x), \frac{x}{1-\|x\|^2} - \frac{\tau}{1-\langle\tau,x\rangle}\Big\rangle. \tag{9.130}$$

On the other hand, by (9.126) and (9.127) and property (a) of M_τ we have

$$\frac{\partial}{\partial t}\left[d_\tau(F(t)x)\right]_{t=0+} = \frac{\partial}{\partial t}\left[d_\tau(M_\tau G(t)y)\right]_{t=0+}$$

$$= \frac{\partial}{\partial t}\left[(1-\|\tau\|^2)d_0(G(t)y)\right]_{t=0+}. \tag{9.131}$$

Since

$$\frac{\partial}{\partial t}\left[d_0(G(t)y)\right]_{t=0+} = -\frac{d_0(y)}{\|y\|^2}\,2\mathrm{Re}\Big\langle g(y), \frac{y}{1-\|y\|^2}\Big\rangle$$

$$= -\frac{d_\tau(x)}{(1-\|\tau\|^2)\|y\|^2}\,2\mathrm{Re}\Big\langle g(y), \frac{y}{1-\|y\|^2}\Big\rangle, \tag{9.132}$$

we obtain (9.129) from (9.130) and (9.131). $\square$

Now we are able to complete the proof of Theorem 9.4 for the case of an interior stationary point $\tau \in \mathbb{B}$ of $\mathcal{S}$.

To this end, let us assume that condition (9.114) holds. Then it follows from Corollary 9.3 (i) and Lemma 9.3 (i) that $\omega_b(0) = \omega^\sharp(0) > 0$.

Let the flow $\{G(t)\}_{t\geq 0} \subset \mathrm{Hol}(\mathbb{B})$ and its generator $g \in \mathcal{G}\mathrm{Hol}(\mathbb{B})$ be defined by (9.127) and (9.128). By Lemmata 9.4 and 9.3 we have

$$\inf_{d_0(x)=s}\,\mathrm{Re}\,\langle g(y), y^*\rangle\,\frac{\omega_b(0)}{4}. \tag{9.133}$$

Then, by Corollary 9.3 (ii), we have

$$d_0(G(t)y) \leq d_0(y)e^{-t\frac{\omega_b(0)}{4}} \quad \text{for all } y \in \mathbb{B}. \tag{9.134}$$

Finally, setting $y = M_\tau x$ and using (9.126) we conclude that

$$d_\tau(F(t)x) \leq d_\tau(x)e^{-t\frac{\omega_b(0)}{4}}. \tag{9.135}$$

Thus, the proof of Theorem 9.4 for the case where $\tau \in \mathbb{B}$ is an interior stationary point is complete.

Corollary 9.4 *Let $f \in \mathcal{G}\mathrm{Hol}(\mathbb{B})$ with $f(\tau) = 0$, $t \in \mathbb{B}$, and let $\{F(t)\}_{t \geq 0}$ be the flow generated by f. Then $\{F(t)\}_{t \geq 0}$ has a global uniform rate of exponential convergence if and only if $\omega^\sharp(0) > 0$.*

Corollary 9.5 *Let $\{F(t)\}_{t \geq 0}$ be the flow generated by $f \in \mathcal{G}\mathrm{Hol}(\mathbb{B})$ and let $\tau \in \mathbb{B}$. Then the following estimates are equivalent:*

(i) $d_\tau(F(t)x) \leq e^{-t\mu} d_\tau(x)$, $x \in \mathbb{B}$, $t \geq 0$;

(ii) $\|M_\tau(F(t)x)\| \leq \|M_\tau(x)\| \cdot e^{-\mu \frac{1 - \|M_\tau(x)\|^2}{2} t}$, $x \in \mathbb{B}$, $t \geq 0$;

(iii) $\|M_\tau(F(t)x)\| \leq \|M_\tau(x)\| \cdot e^{-\nu \frac{1 - \|M_\tau(x)\|}{1 + \|M_\tau(x)\|} t}$, $x \in \mathbb{B}$, $t \geq 0$,

where the numbers μ in (i) and (ii) can be chosen to be one and the same such that $0 \leq \frac{\omega_b(0)}{4} \leq \mu \leq \omega_b(0)$ and ν in (iii) is defined by

$$\nu = \frac{1}{2}\omega_b(0) = \frac{1}{2}\omega^\sharp(0) = \inf_{\|x\|=1} \mathrm{Re}\langle Bf'(\tau)B^{-1}x, x\rangle. \qquad (9.136)$$

Here B is the linear operator defined by $B = P_\tau + \sqrt{1 - \|\tau\|^2}\,(I - P_\tau)$ if $\tau \neq 0$ and $B = I$ if $\tau = 0$.

Proof. First we note that inequalities (ii) and (iii) are equivalent to the following ones:

(ii*) $\|G(t)y\| \leq \|y\| \cdot e^{-\mu \frac{1 - \|y\|^2}{2} t}$, $t \geq 0$;

(iii*) $\|G(t)y\| \leq \|y\| \cdot e^{-\nu \frac{1 - \|y\|}{1 + \|y\|} t}$, $t \geq 0$,

where $y = M_\tau x \in \mathbb{B}$ and the flow $\{G(t)\}_{t \geq 0}$ is defined by (9.127). First, let us suppose that estimate (i) holds. By using (9.126) for the flow G we have

$$d_0(G(t)y) \leq e^{-t\mu} d_0(y). \qquad (9.137)$$

Rewriting the latter inequality in the form

$$\frac{\|G(t)y\|^2}{1 - \|G(t)y\|^2} \leq \frac{\|y\|^2}{1 - \|y\|^2} \cdot e^{-t\mu}, \qquad (9.138)$$

we get by direct calculations,

$$\|G(t)y\|^2 \leq \|y\|^2 \frac{1}{\|y\|^2 + (1 - \|y\|^2)e^{\mu t}} \leq \|y\|^2 \cdot e^{-t\mu(1 - \|y\|^2)}, \qquad (9.139)$$

which coincides with (ii*).

Now we will assume that inequality (ii) (and hence (ii*)) holds. Differentiating both sides of this inequality with respect to t at $t = 0^+$, we obtain

$$-\frac{1}{\|y\|}\,\mathrm{Re}\,\langle g(y), y\rangle \le -\|y\|\mu\frac{1-\|y\|^2}{2}. \tag{9.140}$$

This implies that $\omega^{\sharp}(s) \ge \mu$. Thus the decreasing function $\omega(s) \equiv a$ is an appropriate lower bound and the implication (ii) $\Rightarrow$ (i) follows from Theorem 9.2. The claimed estimate for the number μ is contained in Theorem 9.4 (i).

Now let us suppose again that inequality (ii) (hence, (ii*) and (9.140)) holds with some number $\mu > 0$. Setting in (9.140) $y = ru$, $u \in \partial\mathbb{B}$, $r \in (0,1)$ and letting r tend to zero (cf. the proof of Lemma 9.3), we get $\mathrm{Re}\,\langle g'(0)u, u\rangle \ge \frac{\mu}{2} > 0$. A direct calculation shows that

$$g'(0) = [(M_\tau)'(0)]^{-1}f'(\tau)(M_\tau)'(0) = Bf'(\tau)B^{-1}, \tag{9.141}$$

and so $\nu > 0$. Therefore, again by the Harnack inequality, we have

$$\mathrm{Re}\,\langle g(y), y\rangle \ge \mathrm{Re}\,\langle g'(0)y, y\rangle\frac{1-\|y\|}{1+\|y\|} \ge \nu\|y\|^2\frac{1-\|y\|}{1+\|y\|}. \tag{9.142}$$

On the other hand,

$$\frac{\partial ln\|G(t)y\|}{\partial t} = \frac{1}{2}\frac{\partial ln\|G(t)y\|^2}{\partial t} = \frac{1}{\|G(t)y\|^2}\,\mathrm{Re}\left\langle\frac{\partial G(t)y}{\partial t}, G(t)y\right\rangle$$

$$= -\frac{1}{\|G(t)y\|^2}\,\mathrm{Re}\,\langle g(G(t)y), G(t)y\rangle. \tag{9.143}$$

Also, it follows from the Schwarz Lemma that $\|G(t)y\| \le \|y\|$. Thus we have

$$\frac{\partial ln\|G(t)y\|}{\partial t} \le -\nu\frac{1-\|y\|}{1+\|y\|}. \tag{9.144}$$

Integrating this inequality, we obtain the following estimate:

$$ln\|G(t)y\| - ln\|y\| \le -\nu\frac{1-\|y\|}{1+\|y\|}t, \tag{9.145}$$

which coincides with (iii*).

Finally, if condition (iii*) holds, then differentiating it with respect to t at $t = 0^+$, we get

$$\mathrm{Re}\langle g(y), y\rangle \geq \frac{\nu}{(1 + \|y\|)^2} \geq \frac{\nu}{4} > 0. \tag{9.146}$$

Thus $\omega^\sharp(0) > 0$ and, in view of Theorem 9.4, the result follows. $\qquad \square$

Remark 9.9 *The above Corollary asserts that an exponential rate of convergence in the sense of the "distance" $d_\tau(\cdot)$ is equivalent to the same rate of convergence in the norm of $\mathcal{H}$. We remark in passing that when $\tau = 0$, estimate (iii) can be extended to an arbitrary Banach space (see [Poreda (1987)] and [Suffridge (1977)]). Recall also that the original topology and the topology generated by the hyperbolic metric on $\mathbb{B}$ are locally equivalent. Although global equivalence does not hold, we will prove that an exponential rate of convergence in the norm coincides with a global rate of exponential convergence with respect to the hyperbolic metric.*

Corollary 9.6 *Let $\{F(t)\}_{t\geq 0}$ be the flow generated by $f \in \mathcal{G}\mathrm{Hol}(\mathbb{B})$ with $f(\tau) = 0$, $\tau \in \mathbb{B}$. Then the flow $\{F(t)\}_{t\geq 0}$ has an exponential rate of convergence if and only if the following estimate of convergence in the hyperbolic metric holds:*

> *There exists a number $\eta > 0$ such that*
> $$\rho(F(t)x, \tau) \leq \rho(x, \tau) \cdot e^{-\eta \Lambda(x) t},$$
> *where $\Lambda(x) = e^{-2\rho(x,\tau)}$, $x \in \mathbb{B}$ and $t \geq 0$.* $\qquad$ (9.147)

Moreover, the maximal value of number η for which this inequality holds is $\nu = \inf_{\|x\|=1} \mathrm{Re}\,\langle Bf'(\tau)B^{-1}x, x\rangle$, where $B = P_\tau + \sqrt{1 - \|\tau\|^2}\,(I - P_\tau)$ if $\tau \neq 0$ and $B = I$ if $\tau = 0$.

Proof. Let us suppose that $\{F(t)\}_{t\geq 0}$ has an exponential rate of convergence. By Theorem 9.4, Corollary 9.5, Lemma 9.3 and formula (9.135) we have

$$\|M_\tau(F(t)x)\| \leq \|M_\tau(x)\| \cdot e^{-\nu \frac{1 - \|M_\tau(x)\|}{1 + \|M_\tau(x)\|} t}, \qquad x \in \mathbb{B}, \tag{9.148}$$

where the number ν is defined in (9.136).

Therefore one can write the estimate

$$\rho(F(t)x, \tau) = \rho\big(0, \|M_\tau(F(t)x)\|\big) \leq \rho(0, \|y\|)\, e^{-\nu \frac{1 - \|M_\tau(x)\|}{1 + \|M_\tau(x)\|} t},$$

$$\tag{9.149}$$

which coincides with (9.147) for $\eta = \nu$. Suppose now that inequality (9.147) holds with some number $\eta \geq 0$. Again, by using the flow $\{G(t)\}_{t\geq0}$ defined by (9.127) we can rewrite (9.147) as

$$\rho(G(t)y, 0) \leq \rho(y, 0) \cdot e^{-\eta\Lambda(y)t}, \quad \text{where} \quad \Lambda(y) = e^{-2\rho(y,0)}. \qquad (9.150)$$

Differentiating this inequality with respect to t at $t = 0^+$, we obtain

$$\mathrm{Re}\,\langle g(y), y^* \rangle \geq \eta \,\frac{\rho(y, 0)\Lambda(y)}{\|y\|}. \qquad (9.151)$$

Setting now $y = ru$, $u \in \partial\mathbb{B}$, $r \in (9, 1)$, and letting r tend to zero (cf. the proofs of Lemma 9.3 and Corollary 9.5), we get $\nu \geq \eta$, as claimed. $\qquad\square$

9.3.2 *Boundary sink point. Continuous version of the Julia–Wolff–Carathéodory theorem*

As a matter of fact, if $\mathcal{S} = \{F(t)\}_{t\geq0}$ converges to a boundary sink point $\tau \in \partial\mathbb{B}$ with a rate of convergence of exponential type,

$$d_\tau(F(t)x) \leq \exp\big(-t\omega(d_\tau(x))\big) \cdot d_\tau(x), \qquad (9.152)$$

where $\omega \in \mathcal{M}(0, \infty)$ is a decreasing function, then this estimate can be improved as follows:

$$d_\tau(F(t)x) \leq \exp(-t\omega(0))d_\tau(x), \qquad (9.153)$$

where $\omega(0) := \lim_{s\to0^+} \omega(s)$. In other words, we claim that if the inequality

$$\omega^\sharp(s) \geq \omega(s) \qquad (9.154)$$

holds for a decreasing ω, then the stronger inequality

$$\omega^\sharp(s) \geq \omega(0) \qquad (9.155)$$

also holds. In particular, this fact holds for the function $\omega = \omega_b$. This implies, in turn, that ω_b is actually constant: $\omega_b(s) = \omega_b(0) = \beta$ for all $s \in (0, \infty)$ and is equal to the so-called angular derivative of f (if it exists) at the point $\tau \in \partial\mathbb{B}$. Moreover, this number β gives the best rate of exponential convergence of $\mathcal{S} = \{G(t)\}_{t\geq0}$.

These facts can be proved by using a continuous analog of the classical Julia–Wolff–Carathéodory Theorem. We will need some additional notions, well-known in the finite-dimensional case (see, for example, [Cowen and MacCluer (1995)] and [Rudin (1980)]).

Definition 9.7   A curve $\Lambda : [0, 1) \mapsto \mathbb{B}$ is said to be asymptotically normal at a point $\tau \in \partial\mathbb{B}$ if

(i) $\qquad \lim_{s \to 1^-} \dfrac{\|\Lambda(s) - \lambda(s)\|^2}{1 - \|\lambda(s)\|^2} = 0;$

(ii) $\qquad \dfrac{\|\lambda(s) - \tau\|}{1 - \|\lambda(s)\|} \leq M < \infty, \quad 0 \leq s \leq 1,$

where $\lambda(s)$ is the orthogonal projection of $\Lambda(s)$ onto the complex line through 0 and τ:

$$\lambda(s) = \langle \Lambda(s), \tau \rangle \tau. \tag{9.156}$$

Definition 9.8   Let h be a holomorphic function on $\mathbb{B}$ with values in the complex plane $\mathbb{C}$. We say that h has a restricted limit L at $\tau \in \partial\mathbb{B}$ if h has the limit L along every curve which is **asymptotically normal** at τ.

Definition 9.9   Let $f : \mathbb{B} \mapsto \mathcal{H}$ be a holomorphic mapping on $\mathbb{B}$ and let $\tau \in \partial\mathbb{B}$. We say that f has a finite **angular derivative** at τ if for some element $y \in \mathcal{H}$, the function $h : \mathbb{B} \mapsto \mathcal{H}$, defined by

$$h(x) = \frac{\langle y - f(x), \tau \rangle}{1 - \langle x, \tau \rangle}, \tag{9.157}$$

has a finite restricted limit at τ. We denote this limit $\angle f'(\tau)$.

Theorem 9.5   *Let $f \in \mathrm{Hol}(\mathbb{B}, \mathcal{H})$ be the generator of a semigroup $\mathcal{S} = \{F(t)\}_{t \geq 0}$ of holomorphic self-mappings of $\mathbb{B}$. Suppose that f has no null-point in $\mathbb{B}$ and that $\tau \in \partial\mathbb{B}$ is the boundary sink point for $\mathcal{S}$. Then the following are equivalent:*

(1) the asymptotic behavior of $\mathcal{S}$ at τ is of exponential type;
(2) there is a positive number γ such that

$$d_\tau(F(t)x) \leq d^{-t\gamma} d_\tau(x), \quad x \in \mathbb{B} \text{ and } t \geq 0. \tag{9.158}$$

Moreover, if the angular derivative $\beta = \angle f'(\tau)$ of f at τ exists, then

(a) β is positive real number with $\beta = 2 \inf \{ \mathrm{Re} \langle f(x), x^* \rangle, \ x \in \mathbb{B} \};$
(b) the maximal γ which satisfies condition (ii) is exactly β.

To prove Theorem 9.5 we will need the following assertion [Elin *et al.* (2002)].

Lemma 9.5   *Let $F \in \mathrm{Hol}(\mathbb{B})$ be a holomorphic self-mapping of $\mathbb{B}$ with no fixed point in $\mathbb{B}$, and let τ be its boundary sink point. Then the curve*

$\Lambda : [0, 1) \mapsto \mathbb{B}$ *defined by*

$$\Lambda(s) = F(s\tau) \tag{9.159}$$

is asymptotically normal at τ.

Proof. Since τ is a sink point of F, it follows by Julia's lemma that there is a number $0 < \delta(F) \leq 1$ such that

$$d_\tau(F(x)) \leq \delta(F) \cdot d_\tau(x), \quad x \in \mathbb{B}. \tag{9.160}$$

For $0 < s < 1$ we have, by (9.156),

$$\begin{aligned}
\frac{1 - \|\Lambda(s)\|}{1 - s} \frac{1 + s}{1 + \|\Lambda(s)\|} &\leq \frac{1 - \|\lambda(s)\|}{1 - s} \frac{1 + s}{1 + \|\lambda(s)\|} \\
&\leq \frac{\|\tau - \lambda\|^2}{1 - \|\lambda(s)\|^2} \frac{1 - s^2}{(1 - s)^2} \leq \frac{\|\tau - \lambda\|^2}{1 - \|\Lambda(s)\|^2} \frac{1 - s^2}{(1 - s)^2} \\
&= \frac{d_\tau(F(s\tau))}{d_\tau(s\tau)} \leq \delta(F).
\end{aligned} \tag{9.161}$$

Hence,

$$\limsup_{s \to 1^-} \frac{1 - \|\Lambda(s)\|}{1 - s} \leq \limsup_{s \to 1^-} \frac{1 - \|\lambda(s)\|}{1 - s} \leq \delta(F). \tag{9.162}$$

On the other hand, the Julia–Wolff–Carathéodory theorem asserts that

$$\delta(F) = \liminf_{x \to \tau} \frac{1 - \|F(x)\|}{1 - \|x\|} = \angle F'(\tau). \tag{9.163}$$

Thus, by (9.161)–(9.163), we get

$$\begin{aligned}
\lim_{s \to 1^-} \frac{1 - \|\Lambda(s)\|}{1 - s} &= \lim_{s \to 1^-} \frac{1 - \|\lambda(s)\|}{1 - s} \\
&= \lim_{s \to 1^-} \frac{\|\tau - \lambda(s)\|}{1 - s} = \delta(F).
\end{aligned} \tag{9.164}$$

The last equality implies that

$$\lim_{s \to 1^-} \frac{\|\tau - \lambda(s)\|}{1 - \|\lambda(s)\|} = 1, \tag{9.165}$$

which proves condition (ii) of Definition 9.7.

To prove condition (i), we calculate as follows:

$$\lim_{s \to 1^-} \frac{\|\Lambda(s) - \lambda(s)\|^2}{1 - \|\lambda(s)\|^2} = \lim_{s \to 1^-} \frac{\|\Lambda(s)\|^2 + \|\lambda(s)\|^2 - 2\mathrm{Re}\langle \Lambda(s), \lambda(s) \rangle}{1 - \|\lambda(s)\|^2}$$

$$= \lim_{s \to 1^-} \frac{\|\Lambda(s)\|^2 + \|\lambda(s)\|^2 - 2\mathrm{Re}\overline{\langle \Lambda(s), \tau \rangle}\langle \Lambda(s), \tau \rangle}{1 - \|\lambda(s)\|^2}$$

$$= \lim_{s \to 1^-} \frac{\|\Lambda(s)\|^2 - \|\lambda(s)\|^2}{1 - \|\lambda(s)\|^2} = 1 - \lim_{s \to 1^-} \frac{1 - \|\Lambda(s)\|^2}{1 - \|\lambda(s)\|^2} \cdot \frac{1 - s}{1 - s}$$

$$= 1 - \frac{\delta(F)}{\delta(F)} = 0, \tag{9.166}$$

by (9.164). This proves condition (i), and we are done. $\qquad\square$

Proof of Theorem 9.5 Let condition (i) of the theorem hold, *i.e.*, for some decreasing function $\omega \in \mathcal{M}(0, \infty)$,

$$d_\tau(F(t)x) \le e^{-t\omega(d_\tau(x))} d_\tau(x) \tag{9.167}$$

or, explicitly,

$$\frac{|1 - \langle F(t)x, \tau \rangle|^2}{1 - \|F(t)x\|^2} \le e^{-2t\omega(d_\tau(x))} \frac{|1 - \langle x, \tau \rangle|^2}{1 - \|x\|^2} . \tag{9.168}$$

This is equivalent to the inequality

$$\frac{|1 - \langle F(t)x, \tau \rangle|^2}{1 - \langle x, \tau \rangle|^2} \le e^{-2t\omega(d_\tau(x))} \frac{1 - \|F(t)x\|^2}{1 - \|x\|^2} . \tag{9.169}$$

Once again it follows from the Julia–Wolff–Carathéodory theorem that for a fixed $t \ge 0$,

$$\delta(F(t)) := \liminf_{x \to \tau} \frac{1 - \|F(t)x\|}{1 - \|x\|}$$

$$= \angle [F(t)]'(\tau) := \lim_{x \to \tau} \frac{1 - \langle F(t)x, \tau \rangle}{1 - \langle x, \tau \rangle} , \tag{9.170}$$

where that last limit is taken along an asymptotically normal curve at τ.

Let us denote

$$\omega(0) = \lim_{s \to 1^-} \omega(d_\tau(s\tau)). \tag{9.171}$$

Thus, setting $x = s\tau$ in (9.169) and letting s tend to 1^-, we get

$$\delta^2(F(t)) \le e^{-2t\omega(0)} \delta(F(t)), \tag{9.172}$$

or

$$\delta(F(t)) \leq e^{-t\gamma}, \tag{9.173}$$

where we set $\gamma = 2\omega(0)$.

Now by using Julia's lemma, we obtain the implication (i)$\Rightarrow$(ii). The converse implication can be established by differentiating the inequality in (ii) at $t = 0^+$. Namely, we get

$$\mathrm{Re}\,\langle f(x), x^* \rangle \geq \frac{\gamma}{2} > 0. \tag{9.174}$$

So, one can set $\omega(s) \equiv \frac{\gamma}{2}$, and the asymptotic behavior of $\mathcal{S}$ at τ is seen to be of exponential type.

To prove the second part of the theorem, we first observe that f is a generator if and only if the equation

$$x + tf(x) = y \tag{9.175}$$

is solvable for all $t \geq 0$ and $y \in \mathbb{B}$ [Reich and Shoikhet (1996)].

The solution $x = \mathcal{J}_t(y) = (I + tf)^{-1}$ is the (nonlinear) resolvent of f. It has the following properties:

(1) for each $t \geq 0$, $\mathcal{J}_t : \mathbb{B} \mapsto \mathbb{B}$ is a holomorphic self-mapping of $\mathbb{B}$;
(2) for each $x \in \mathbb{B}$, the function $\mathcal{J}_t : \mathbb{R}^+ \mapsto \mathbb{B}$ is continuous and the semigroup $\mathcal{S} = \{F(t)\}_{t \geq 0}$ generated by f can be represented by the exponential formula

$$\lim_{n \to \infty} \left[\mathcal{J}_{\frac{t}{n}} \right]^n (x) = F(t)x, \quad x \in \mathbb{B}, \tag{9.176}$$

where the limit is taken with respect to the locally uniform topology of $\mathbb{B}$;
(3) if f has no null point in $\mathbb{B}$ and $\tau \in \partial\mathbb{B}$ is the sink point of $\mathcal{S}$, then for each $t > 0$, τ is also the sink point of $\mathcal{J}_t$, and moreover, the following approximations hold:

$$\lim_{t \to \infty} \mathcal{J}_t(x) = \tau, \quad x \in \mathbb{B}, \tag{9.177}$$

and

$$\lim_{t \to \infty} f(\mathcal{J}_t(x)) = 0, \quad x \in \mathbb{B}. \tag{9.178}$$

Now let us suppose that $\beta = \angle f'(\tau)$ exists (finitely). Then it follows by properties (1) and (3) above and Lemma 9.5 that for each $t > 0$, the curve

$\Lambda_t(s) := \mathcal{J}_t(s\tau) : [0,1) \mapsto \mathbb{B}$ is an asymptotically normal curve at $\tau \in \partial\mathbb{B}$. In addition, by equation (9.175) we have the identity

$$\Lambda_t(s) + tf(\Lambda_t(s)) = s\tau \tag{9.179}$$

for all $s \in [0,1)$ and $t \geq 0$.

Denote the angular derivative $\angle[\mathcal{J}_t]'(\tau)$ of $\mathcal{J}_t : \mathbb{B} \mapsto \mathbb{B}$ at the point τ by c_t. Again, by the Julia–Wolff–Carathéodory theorem,

$$c_t = \lim_{s \to 1^-} \frac{1 - \langle \mathcal{J}_t(s\tau, \tau\rangle}{1 - s}. \tag{9.180}$$

By (9.179) and (9.180) we get $\lim\limits_{s \to 1^-} f(\Lambda_t(s)) = 0$ and

$$\beta = \lim_{s \to 1^-} \frac{\langle f(\Lambda_t(s)), \tau\rangle}{\langle \Lambda_t(s), \tau\rangle - 1} = \lim_{s \to 1^-} \frac{1}{t} \cdot \frac{\langle (s\tau - \Lambda_t(s))\rangle}{\langle \Lambda_t(s), \tau\rangle - 1}$$

$$= \lim_{s \to 1^-} \frac{1}{t} \cdot \frac{-s + \langle \Lambda_t(s)), \tau\rangle}{1 - \langle \Lambda_t(s), \tau\rangle} = \lim_{s \to 1^-} \frac{1}{t} \cdot \left(\frac{\langle \Lambda_t(s)), \tau\rangle - 1}{1 - \langle \Lambda_t(s), \tau\rangle} + \frac{1 - s}{1 - \langle \Lambda_t(s), \tau\rangle} \right)$$

$$= \frac{1}{t}\left(-1 + \frac{1}{c_t}\right). \tag{9.181}$$

Thus we obtain that β is a non-negative real number and

$$c_t = \frac{1}{1 + t\beta}. \tag{9.182}$$

Hence,

$$d_\tau(\mathcal{J}_t(x)) \leq \frac{1}{1 + t\beta}\, d_\tau(x) \tag{9.183}$$

by Julia's lemma.

Applying now the exponential formula (see property (2) above) we get

$$d_\tau(F(t)x) \leq e^{-t\beta} d_\tau(x). \tag{9.184}$$

To conclude the proof of Theorem 9.5, it remains to be shown that if condition (ii), or, equivalently, inequality (9.174) holds for some $\gamma > 0$, then $\gamma \leq \beta$.

Indeed, setting $x = s\tau$ in (9.174) we get

$$\operatorname{Re}\langle f(s\tau), (s\tau)^*\rangle = \operatorname{Re}\left\langle f(s\tau), \frac{s\tau}{1 - s^2} - \frac{\tau}{1 - s} \right\rangle$$

$$= \frac{\operatorname{Re}\langle f(s\tau), \tau\rangle}{s - 1} \cdot \frac{1}{1 + s} \geq \gamma/2. \tag{9.185}$$

Letting now s tend to 1^-, we get

$$\frac{\mathrm{Re}\angle f'(\tau)}{2} = \frac{\beta}{2} \geq \frac{\gamma}{2}, \tag{9.186}$$

i.e., $\beta \geq \gamma$. This completes the proof of Theorem 9.5. $\square$

For the one-dimensional case, that is, when $\mathbb{B} = \Delta$, the open unit disk in the complex plane $\mathbb{C}$, the situation can be described as follows.

Theorem 9.6 ([Elin and Shoikhet (2001)]) *A mapping $f \in \mathcal{G}\mathrm{Hol}(\Delta)$ has no null point in Δ if and only if for some $\tau \in \partial\Delta$ the angular derivative*

$$\angle f'(\tau) = \beta \tag{9.187}$$

exists (finitely) with $\mathrm{Re}\beta \geq 0$.

Moreover, if $\mathcal{S} = \{F(t)\}_{t\geq 0}$ *is the flow generated by f, then*

$$\frac{|F(t)z - \tau|^2}{1 - |F(t)z|^2} \leq \exp\left(-t\mathrm{Re}\beta\right)\frac{|z - \tau|^2}{1 - |z|^2}, \tag{9.188}$$

i.e., the point τ is unique and the (globally) attractive sink point of $\mathcal{S}$.

This assertion is an infinitesimal version of the Julia–Wolff–Carathéodory theorem.

We remark in passing that more information on the asymptotic behavior of holomorphic and ρ-nonexpansive mappings and semigroups in the Hilbert ball can be found, for instance, in [Reich (1985); (1991); (1992)], [Reich and Shafrir (1987); (1990)] and [Reich and Shoikhet (1997b)].

9.4   Admissible Lower and Upper Bounds and Rates of Convergence

In this section we intend to examine the influence of certain estimates involving the generator on the asymptotic behavior of the semigroup it generates. As in the previous section, we will compare the solutions of the Cauchy problem (9.1) with the solutions of certain one-dimensional Cauchy problems, namely,

$$\begin{cases} \dfrac{\partial\beta}{\partial t} + \beta\omega(\beta) = 0 \\[2mm] \beta(0, s) = s \in [0, 1). \end{cases} \tag{9.189}$$

We begin with the following auxiliary assertions [Elin *et al.* (2004)].

Lemma 9.6 *Let ω be a continuous positive function on $[0,1)$. Then for all $s \in [0,1)$, the solution $\beta(t,s)$ of the Cauchy problem (9.189) is defined for all $t \geq 0$ and converges to 0 as $t \to +\infty$. In addition, if $m(s)$ and $M(s)$ are the minimum and maximum, respectively, of the function ω on the interval $[0,s]$, then $\beta(t,s)$ satisfies the following estimate:*

$$se^{-M(s)t} \leq \beta(t,s) \leq se^{-m(s)t}. \tag{9.190}$$

Proof. First, rewriting the differential equation of (9.189) in the form $\frac{\partial \beta}{\partial t} = -\beta\omega(\beta)$, we note that the solution $\beta(t,s)$ of (9.189) is decreasing with respect to t (when it is defined) and positive. Second, since

$$\int_s^{\beta(t,s)} \frac{dx}{x\omega(x)} = -t, \tag{9.191}$$

the convergence of the solution $\beta(t,s)$ to a certain limit $s_0 \geq 0$ as $t \to \infty$ is equivalent to the divergence of the integral

$$\int_s^{s_0} \frac{dx}{x\omega(x)}. \tag{9.192}$$

But this holds if and only if $s_0 = 0$.

Finally, since

$$\ln\beta(t,s) - \ln s = -\int_0^t \omega(\beta(r,s))dr, \tag{9.193}$$

the monotonicity of $\beta(\cdot,s)$ implies the last statement of the lemma. $\qquad\square$

Remark 9.10 *For a given $0 \leq t < \infty$, consider the two monotone sequences defined as follows:*

$$M_1 := M(s) = \max\{\omega(x) : x \in [0,s]\},$$
$$m_1 := m(s) = \min\{\omega(x) : x \in [0,s]\}, \tag{9.194}$$

and

$$M_{n+1} := \max\left\{\omega(x), \; x \in \left[se^{-M_n t}, s\right]\right\},$$
$$m_{n+1} := \min\left\{\omega(x), \; x \in \left[se^{-M_n t}, s\right]\right\}. \tag{9.195}$$

Note that the sequence $\{M_n\}_{n=1}^\infty$ is decreasing while the sequence $\{m_n\}_{n=1}^\infty$ is increasing. Hence the limits $A = \lim M_n$ and $B = \lim m_n$ exist. Iterating *the proof of the last statement of the preceding lemma, we obtain the validity of the estimates*

$$se^{-tA} \leq \beta(t, s) \leq se^{-tB}. \tag{9.196}$$

Furthermore, if we have more information about ω, then the latter inequalities can be rewritten in a more precise form, namely, if ω is increasing, then $A = \omega(s)$ and so $\beta(t, s) \geq s \exp(-t\omega(s))$; if ω is decreasing, then $B = \omega(s)$ and $\beta(t, s) \leq s \exp(-t\omega(s))$.

Lemma 9.7   *Let ω_1, and ω_2 be two continuous positive functions on $[0, 1)$ such that $\omega_1 \leq \omega_2$. Let β_1 and β_2 be the solutions of the following Cauchy problems:*

$$\begin{cases} \dfrac{\partial \beta_1}{\partial t} + \beta_1 \omega_1(\beta_1) = 0 \\ \beta_1(0, s) = s \end{cases} \quad \text{and} \quad \begin{cases} \dfrac{\partial \beta_2}{\partial t} + \beta_2 \omega_2(\beta_2) = 0 \\ \beta_2(0, s) = s, \end{cases}$$

$$\tag{9.197}$$

where $s \in [0, 1)$.

Let $m : [0, \infty) \mapsto [0, 1)$ be a differentiable nonnegative function such that $m(0) = s$ and

$$-m\omega_2(m) \leq \frac{dm}{dt} \leq -m\omega_1(m). \tag{9.198}$$

Then $\beta_2(t, s) \leq m(t) \leq \beta_1(t, s)$.

Proof.   The assertion is evidently true when $s = 0$. Let $t_0 > 0$ be small enough (so that $m(t) > 0$ for all $0 \leq t \leq t_0$). We have

$$\int_s^{m(t)} \frac{dy}{y\omega_1(y)} \leq -t \leq \int_s^{m(t)} \frac{dx}{x\omega_2(x)}. \tag{9.199}$$

On the other hand, the definitions of β_1 and β_2 imply that

$$-t = \int_s^{\beta_1(t,s)} \frac{dy}{y\omega_1(y)} = \int_s^{\beta_2(t,s)} \frac{dx}{x\omega_2(x)}. \tag{9.200}$$

From this we deduce that $\displaystyle\int_{\beta_1(t,s)}^{m(t)} \frac{dy}{y\omega_1(y)} \leq 0$ and $\displaystyle\int_{m(t)}^{\beta_2(t,s)} \frac{dx}{x\omega_2(x)} \leq 0$ for t small enough. Since the integrands are positive, we conclude that $\beta_2(t, s) \leq m((t) \leq \beta_1(t, s)$.

Now we will show that this inequality holds for all $t \geq 0$. Assume that there is $t > 0$ such that $m(t) < \beta_2(t, s)$. Let t_0 be the infimum of all such

t's. It is clear that $\beta_2(t_0, s) = m(t_0) = s_0$. Repeating our arguments, we get $\beta_2(t + t_0, s) \leq m(t + t_0)$ for t small enough. But this contradicts the choice of t_0. In a similar way one can also show that $\beta_1(t, s) \geq m(t)$ for all $t \geq 0$. $\qquad\square$

By N_τ we denote the class of all semi-complete vector–fields on $\mathbb{B}$ vanishing at the point $\tau \in \mathbb{B}$, *i.e.*,

$$N_\tau = \{f \in \mathcal{G}\mathrm{Hol}(\mathbb{B}) : f(\tau) = 0\}. \tag{9.201}$$

Lemma 9.8 *A mapping $f \in \mathrm{Hol}(\mathbb{B}, H)$ belongs to N_τ for some $\tau \in \mathbb{B}$ if and only if*

$$\mathrm{Re}\, \frac{\langle f(x), x \rangle}{1 - \|x\|^2} \geq \mathrm{Re}\, \frac{\langle f(x), \tau \rangle}{1 - \langle x, \tau \rangle}, \quad x \in \mathbb{B}. \tag{9.202}$$

Proof. First we note that $f \in \mathcal{G}\mathrm{Hol}(\mathbb{B})$ if and only if it is ρ-monotone, *i.e.*,

$$\mathrm{Re}\left[\frac{\langle f(x), x \rangle}{1 - \|x\|^2} + \frac{\langle f(y), y \rangle}{1 - \|y\|^2}\right] \geq \mathrm{Re}\, \frac{\langle f(x), y \rangle + \langle f(y), x \rangle}{1 - \langle x, y \rangle}. \tag{9.203}$$

If now $f(\tau) = 0$, $\tau \in \mathbb{B}$, then setting $y = \tau$ in (9.203) we get inequality (9.202).

Conversely, let (9.202) hold for some $\tau \in \mathbb{B}$. Denote

$$\chi^+ = \frac{x}{1 - \|x\|^2} - \frac{\tau}{1 - \langle \tau, x \rangle}. \tag{9.204}$$

Then (9.202) can be rewritten as

$$\mathrm{Re}\langle f(x), x^+ \rangle \geq 0, \quad x \in \mathbb{B}. \tag{9.205}$$

Now it can be shown that for each $k > 1 - \|\tau\|^2$, and for each pair of points $x \in \partial E_\tau(k)$ and $y \in \overline{E_\tau(k)}$, the following inequality holds:

$$\mathrm{Re}\langle x, x^+ \rangle \geq \mathrm{Re}\langle y, x^+ \rangle, \tag{9.206}$$

where

$$E_\tau(k) = \left\{x \in \mathbb{B} : \frac{|1 - \langle x, \tau \rangle|^2}{1 - \|x\|^2} < k\right\}, \quad k > 1 - \|\tau\|^2, \tag{9.207}$$

are ρ-balls in $\mathbb{B}$. Thus x^+ is a support functional of the convex set $\overline{E_\tau(k)}$ at the point $x \in \partial E_\tau(k)$. Now condition (9.205) implies that $f \in \mathcal{G}\mathrm{Hol}(E_\tau(k))$. Letting k tend to infinity, we also see that $f \in \mathcal{G}\mathrm{Hol}(\mathbb{B})$, because $\bigcup_{k > 1 - \|\tau\|^2} E_\tau(k) = \mathbb{B}$. This concludes the proof. $\qquad\square$

Proposition 9.1 *Let $f \in \mathrm{Hol}(\mathbb{B}, H)$ and let $u \ (= u(t, x))$ be a local solution of the Cauchy problem (9.1). Then for any two given continuous functions ω_ℓ and ω_u on $[0, q)$ such that ω_ℓ is decreasing and nonnegative, and ω_u is increasing, the following assertions are equivalent:*

(a) for some $\tau \in \mathbb{B}$ and for each $x \in \mathbb{B}$, there is a number $T = T(x) > 0$ such that

$$\|M_{-\tau}(x)\| \exp\big(-t\omega_u(\|M_{-\tau}(x)\|)\big) \leq \|M_{-\tau}(u(t, x))\|$$
$$\leq \|M_{-\tau}(x)\| \exp(-t\omega_\ell(\|M_{-\tau}(x)\|)) \qquad (9.208)$$

whenever $t \in [0, T)$;
(b) for some $\tau \in \mathbb{B}$ and for each $x \in \mathbb{B}\backslash\{\tau\}$,

$$\omega_\ell(\|M_{-\tau}(x)\|) \leq \frac{\sigma(\tau, x)}{\|M_{-\tau}(x)\|^2} \, \mathrm{Re}\, \langle f(x), x^\dagger \rangle \leq \omega_u(\|M_{-\tau}(x)\|),$$
$$(9.209)$$

where $x^\dagger = \frac{1}{1-\|x\|^2}\, x - \frac{1}{1-\langle \tau, x \rangle}\, \tau$.

When these assertions hold, the point τ in (a) and (b) is one and the same, and moreover, f is a strongly semi-complete vector field with $f(\tau) = 0$. In addition, $u \ (= u(t, x))$ is globally defined on $\mathbb{R}^+ \times \mathbb{B}$ and condition (a) holds for all $t \in \mathbb{R}^+$.

Proof.   Denote

$$\gamma_u(t, s) = se^{-t\omega_u(s)}, \quad \gamma_\ell(t, s) = se^{-t\omega_\ell(s)} \qquad (9.210)$$

and

$$m_\tau(t, x) = \|M_{-\tau}(u(t, x))\|. \qquad (9.211)$$

Then (a) can be written as

$$\gamma_u(t, \|M_{-\tau}(u(t, x))\|) \leq m_\tau(t, x) \leq \gamma_\ell(t, \|M_{-\tau}(u(t, x))\|), \quad t \in [0, T).$$
$$(9.212)$$

Since $\gamma_u(0, \|M_{-\tau}(u(t, x))\|) = m_\tau(0, x) = \gamma_\ell(t, \|M_{-\tau}(u(t, x))\|)$, these inequalities imply that

$$\frac{\partial \gamma_u(t, \|M_{-\tau}(x)\|)}{\partial t} \leq \frac{\partial m_\tau(t, x)}{\partial t} \leq \frac{\partial \gamma_\ell(t, \|M_{-\tau}(x)\|)}{\partial t} \qquad (9.213)$$

at the point $t = 0$, or equivalently,

$$-\omega_u(\|M_{-\tau}(x)\|) \cdot \|M_{-\tau}(x)\| \leq \frac{-\sigma(\tau, x)}{\|M_{-\tau}(x)\|} \operatorname{Re} \langle f(x), x^\dagger \rangle$$

$$\leq -\omega_\ell(\|M_{-\tau}(x)\|) \cdot \|M_{-\tau}(x)\|. \quad (9.214)$$

This yields the implication (a)$\Rightarrow$(b). In the other direction, let (b) hold. Then it follows by Lemma 9.8 that f is semi-complete with $f(\tau) = 0$. Hence the solution of the Cauchy problem $\begin{cases} \dfrac{du}{\partial t} + f(u) = 0, \\ u(0) = x \end{cases}$ is globally defined. As above, denote

$$m(t) = m_\tau(t, x) = \|M_{-\tau}(u(t, x))\| \quad (9.215)$$

and $s = \|M_{-\tau}(x)\|$. Then (b) can be written in the form

$$-m\omega_u(m) \leq \frac{\partial m}{\partial t} \leq -m\omega_\ell(m), \quad (9.216)$$

which coincides with (9.198). As a consequence of Lemma 9.7, we have

$$\beta_u(t, s) \leq m(t) \leq \beta_\ell(t, s), \quad (9.217)$$

where β_u and β_ℓ are the solutions of the Cauchy problems

$$\begin{cases} \dfrac{\partial \beta_u}{\partial t} + \beta_u \omega_u(\beta_u) = 0 \\ \beta_u(0, s) = s \end{cases} \quad \text{and} \quad \begin{cases} \dfrac{\partial \beta_\ell}{\partial t} + \beta_\ell \omega_\ell(\beta_\ell) = 0 \\ \beta_\ell(0, s) = s. \end{cases}$$

$$(9.218)$$

This implies the required inequality (a) for all $t \geq 0$ (see Remark 9.10). $\square$

Definition 9.10 Those functions ω_ℓ and ω_u which satisfy the assumptions of Proposition 9.1 and condition (b) of this proposition will be called admissible lower and upper bounds for $f \in \operatorname{Hol}(\mathbb{B}, H)$ (with respect to the point $\tau \in \mathbb{B}$).

Thus the existence of an admissible lower bound for f implies that f is a strongly semi-complete vector field and the flow generated by f is exponentially norm convergent to its stationary point τ, uniformly on each subset strictly inside $\mathbb{B}$.

Example 9.8 The mapping $f(z) = (z_1, z_2)$ is obviously a strongly semi-complete vector field on the unit ball in $\mathbb{C}^2$ equipped with any norm.

Clearly, for small $a \in \mathbb{C}$ the mapping

$$f_a(z) = (z_1 - a z_2^2, z_2) \tag{9.219}$$

is still a generator.

For which $a \in \mathbb{C}$ is the mapping f_a a generator in the domain

$$D_p = \{z \in \mathbb{C}^2 : |z_1|^p + |z_2|^p < 1\}, \quad p \geq 1? \tag{9.220}$$

A related calculation is given in [Suffridge (1977)]. The answer is the following:

$$|a| \leq \frac{1}{4^{\frac{1}{p}}} (p+1)^{1+\frac{1}{p}} (p-1)^{\frac{1}{p}-1}. \tag{9.221}$$

In the Hilbert ball $\mathbb{B} = D_2$ we observe that f_a satisfies the estimate

$$\|z\|^2(1 - \lambda\|z\|) \leq \operatorname{Re}\langle f_a(z), z\rangle \leq \|z\|^2(1 + \lambda\|z\|), \tag{9.222}$$

where $\lambda = |a|\dfrac{2}{3\sqrt{3}} \in [0,1]$. Those inequalities coincide exactly with condition (b) of Proposition 9.1 for the admissible lower and upper bounds defined by

$$\omega_\ell(s) = 1 - \lambda s \quad \text{and} \quad \omega_u(s) = 1 + \lambda s. \tag{9.223}$$

Hence by this proposition we obtain the following estimate for the semigroup $\{u(t, \cdot)\}$ generated by f_a:

$$\|z\| e^{-t(1+\lambda\|z\|)} \leq \|u(t, z)\| \leq \|z\| e^{-t(1-\lambda\|z\|)}. \tag{9.224}$$

Remark 9.11   *If ρ is the Poincaré metric on $\mathbb{B}$, then using the known (see [Goebel and Reich (1984)]) properties of ρ,*

$$\rho(x, y) = \rho(0, M_{-y}(x)) \tag{9.225}$$

and

$$\rho(0, kx) \leq k\rho(0, x), \quad 0 \leq k \leq 1, \tag{9.226}$$

we see that condition (a) in Proposition 9.1 implies the inequality

$$\rho(u(t, x), \tau) \leq e^{-t\omega(\|M_{-\tau}(x)\|)} \rho(x, \tau). \tag{9.227}$$

In other words, the exponential norm convergence implies the same rate of convergence in the hyperbolic metric.

Remark 9.12 *Generally speaking, the converse is not clear, because we know only local estimates: For each $x \in \mathbb{B}$ there are a neighborhood U of x and numbers $M = M(x, U) \geq 1$ and $m = m(x, U) \leq 1$ such that*

$$m\|x - y\| \leq \rho(x, y) \leq M\|x - y\| \tag{9.228}$$

for all $y \in U$ (see [Goebel and Reich (1984)]). Therefore the following question arises: Are there an admissible lower bound ω_ℓ and an admissible upper bound ω_u for $f \in N_\tau$ such that condition (b) of Proposition 9.1 is equivalent to the same or to a similar estimate of convergence in the metric ρ? We answer this question by establishing the following assertion.

Proposition 9.2 *Let $f \in \mathrm{Hol}(\mathbb{B}, H)$ and let $u = u(t, x)$ be the local solution of the Cauchy problem (9.1). Let ω_u and ω_ℓ be continuous positive functions defined on $[0, 1)$ such that*

$$\begin{cases} \omega_u(s) & \text{is increasing} \\[2mm] \dfrac{\omega_\ell(s)s}{(1 - s^2)\operatorname{arctanh} s} & \text{is decreasing.} \end{cases} \tag{9.229}$$

Then the following assertions are equivalent:

(a) ω_u is an admissible upper bound and ω_ℓ is an admissible lower bound (with respect to some point $\tau \in \mathbb{B}$) for f.

(b) For some $\tau \in \mathbb{B}$ and for each $x \in \mathbb{B}$, there is $T = T(x)$ such that

$$\rho(\tau, x) \exp\left(-t\omega_u(\|M_{-\tau}(x)\|) \frac{\|M_{-\tau}(x)\|}{\rho(\tau, x)\sigma(\tau, x)}\right) \leq \rho(\tau, u(t, x))$$

$$\leq \rho(\tau, x) \exp\left(-t\omega_\ell(\|M_{-\tau}(x)\|) \frac{\|M_{-\tau}(x)\|}{\rho(\tau, x)\sigma(\tau, x)}\right), \tag{9.230}$$

whenever $t \in [0, T)$. The point τ in (a) and (b) is one and the same. Moreover, for each $x \in \mathbb{B}$, the solution $u(\cdot, x)$ is well defined globally and (b) holds for all $t \geq 0$.

Proof. It is clear that condition (a) coincides with condition (b) of Proposition 9.1. As above, consider the function

$$m(t) = \|M_{-\tau}(u(t, x))\|. \tag{9.231}$$

If (a) holds, then it implies that

$$-m\omega_u(m) \leq \frac{dm}{dt} \leq -m\omega_\ell(m). \tag{9.232}$$

By Lemma 9.7, we obtain

$$\beta_u(t) \leq m(t) \leq \beta_\ell(t), \tag{9.233}$$

where β_u and β_ℓ are the solutions of the following Cauchy problems:

$$\begin{cases} \dfrac{\partial \beta_u}{\partial t} + \beta_u \omega_u(\beta_u) = 0 \\ \beta_u(0) = s = \|M_{-\tau}(x)\| \end{cases} \quad \text{and} \quad \begin{cases} \dfrac{\partial \beta_\ell}{\partial t} + \beta_\ell \omega_\ell(\beta_\ell) = 0 \\ \beta_e(0) = s = \|M_{-\tau}(x)\|. \end{cases} \tag{9.234}$$

This implies that

$$\ln\left(\frac{\operatorname{arctanh}\beta_u}{\operatorname{arctanh}s}\right) = -\int_0^t \frac{\beta_u(r)\omega_u(\beta_u(r))dr}{(1 - \beta_u^2(r))\operatorname{arctanh}\beta_u(r)} \tag{9.235}$$

and the same equality holds for β_ℓ and ω_ℓ. Now by using (9.233) and (9.227), we get condition (b). The implication (b)$\Rightarrow$(a) follows from the comparison of the derivatives in inequality (b) with respect to t at the point $t = 0^+$. $\qquad\square$

Now we will show that there are "universal" admissible lower and upper bounds $\omega_\wedge$ and $\omega^\wedge$.

Proposition 9.3   *For $f \in \operatorname{Hol}(\mathbb{B}, H)$ and any point $\tau \in \mathbb{B}$, the following assertions are equivalent:*

(i) f has an admissible lower bound ω_ℓ and an admissible upper bound ω_u with respect to the point $\tau \in \mathbb{B}$.

(ii) $f \in N_\tau$ and there are numbers $a \geq 0$ and $b \geq a$ such that

$$\|M_{-\tau}(x)\| \exp\left(-tb\frac{1 + \|M_{-\tau}(x)\|}{1 - \|M_{-\tau}(x)\|}\right) \leq \|M_{-\tau}(u(t, x))\|$$

$$\leq \|M_{-\tau}(x)\| \exp\left(-ta\frac{1 - \|M_{-\tau}(x)\|}{1 + \|M_{-\tau}(x)\|}\right), \tag{9.236}$$

where $u(t, x)$ is the solution of the Cauchy problem (9.1);

(iii) $f \in N_\tau$ and there are numbers $a \geq 0$ and $b \geq a$ such that

$$\rho(\tau, x) \exp\left(-\tau b \frac{\|M_{-\tau}(x)\|}{(1 - \|M_{-\tau}(x)\|)^2 \rho(\tau, x)}\right) \leq \rho(\tau, u(t, x))$$

$$\leq \rho(\tau, x) \exp\left(-ta \frac{\|M_{-\tau}(x)\|}{(1 + \|M_{-\tau}(x)\|)^2 \rho(\tau, x)}\right). \tag{9.237}$$

The numbers a and b in (ii) and (iii) can be chosen to be the same. In particular, if (i) holds, then one can put $a = \omega_\ell(0)$ and $b = \omega_u(0)$.

Proof. We intend to show that (i) is equivalent to

(i′) The functions

$$\omega_\wedge(s) = a\,\frac{1-s}{1+s} \qquad and \qquad \omega^\wedge(s) = b\,\frac{1+s}{1-s} \qquad (9.238)$$

(with some numbers a and b) are admissible lower and upper bounds.

In fact, all we need to prove is that (i)$\Rightarrow$(i′), since the reverse implication is evident.

Set $v(t,x) = M_{-\tau}(u(t, M_\tau(x)))$ and $\varphi(x) = -\,\frac{\partial v(0^+, x)}{\partial t}$. If $f \in \mathrm{Hol}(\mathbb{B}, H)$ has admissible lower and upper bounds ω_ℓ and ω_u, then by Proposition 9.1, $v(t,x)$ is well defined for all $t \geq 0$ and

$$\|x\| \exp(-t\omega_u(\|x\|)) \leq \|v(t,x)\| \leq \|x\| \exp(-t\omega_\ell(\|x\|)). \qquad (9.239)$$

Again by Proposition 9.1 we see that

$$\omega_\ell(\|x\|) \leq \frac{1}{\|x\|^2} \operatorname{Re} \langle \varphi(x), x \rangle \leq \omega_u(\|x\|). \qquad (9.240)$$

This implies that $\varphi(0) = 0$ and that

$$\omega_\ell(0) \leq \frac{1}{\|x\|^2} \operatorname{Re} \langle \varphi'(0)x, x \rangle \leq \omega_u(0). \qquad (9.241)$$

Now it follows by Proposition 9.1 that we also have

$$a\,\frac{1 - \|x\|}{1 + \|x\|} \leq \frac{1}{\|x\|^2} \operatorname{Re} \langle \varphi(x), x \rangle \leq b\,\frac{1 + \|x\|}{1 - \|x\|} \qquad (9.242)$$

when $0 \leq a \leq \omega_\ell(0)$ and $b \geq \omega_u(0)$. This means that the functions

$$\omega_\wedge(s) = a\,\frac{1-s}{1+s} \qquad and \qquad \omega^\wedge(s) = b\,\frac{1+s}{1-s} \qquad (9.243)$$

are also admissible lower and upper bounds for φ (and consequently for f). It is easy to verify that the function $\omega^\wedge$ is increasing and that the function $\omega_\wedge(s)s/(1 - s^2)\operatorname{arctanh} s$ is decreasing.

Finally, (i′) is equivalent to condition (ii) by Proposition 9.1 and to (iii) by Proposition 9.2. $\qquad\qquad\square$

Remark 9.13 *Actually, conditions (i)–(iii) of this proposition are equivalent to the following one:*

$f \in N_\tau$ and there are numbers $0 \le a \le b$ such that

$$a\|Bx\|^2 \le \operatorname{Re}\langle Bf'(\tau)x, \ Bx\rangle \le b\|Bx\|^2 \tag{9.244}$$

for all $x \in H$, where B is the linear operator defined by $B = P_\tau + \sqrt{1 - \|\tau\|^2}\,(I - P_\tau)$, $\tau \in \mathbb{B}\backslash\{0\}$, $P_\tau = \tau \frac{\langle \cdot, \tau \rangle}{\|\tau\|^2}$ and $B = I$ when $\tau = 0$. Indeed, by direct calculations one can show that $\varphi'(0) = Bf'(\tau)B^{-1}$ and (9.242) becomes (9.244)). In addition, by Proposition 9.1 and substituting the admissible lower and upper bounds (9.243), we conclude that the above conditions of Proposition 9.3 including (9.244) are equivalent to the inequality

$$a\left(\frac{\|M_{-\tau}(x)\|}{1 + \|M_{-\tau}(x)\|}\right)^2 \le \operatorname{Re}\langle f(x), x^\dagger\rangle \le b\left(\frac{\|M_{-\tau}(x)\|}{1 - \|M_{-\tau}(x)\|}\right)^2 \tag{9.245}$$

(compare with (i′)).

Chapter 10

Geometry of Domains in Banach Spaces

10.1 Biholomorphic Mappings in Banach Spaces and Generators on Biholomorphically Equivalent Domains

Let X and Y be two Banach spaces over the field of complex numbers C, and let $\mathcal{D} \subset X$ and $\Omega \subset Y$ be domains (open connected subsets) in X and Y, respectively.

Definition 10.1 A mapping $f \in \mathrm{Hol}(\mathcal{D}, \Omega)$ is said to be univalent on $\mathcal{D}$ if for each pair of distinct points x_1 and x_2 in $\mathcal{D}$ we have $f(x_1) \neq f(x_2)$.

In this case one can define the inverse mapping $f^{-1} : f(\mathcal{D}) \mapsto \mathcal{D}$. It is well known (see, for example, [Hervé (1963a)]) that if $X = \mathbb{C}^n$ and $Y = \mathbb{C}^m$ are finite dimensional complex spaces, then $f : \mathcal{D} \subset \mathbb{C}^n \mapsto \mathbb{C}^m$, $f \in \mathrm{Hol}(\mathcal{D}, \mathbb{C}^m)$, is univalent if and only if $n = m$ and f^{-1} is also holomorphic on $\Omega = f(\mathcal{D})$, $i.e.$, $f^{-1} \in \mathrm{Hol}(\Omega, \mathcal{D})$. However, this fact is no longer true in the infinite dimensional case (see counterexamples in [Abts (1980)] and [Suffridge (1972)]). Therefore, in the general case we give the following definition.

Definition 10.2 A univalent mapping $f \in \mathrm{Hol}(\mathcal{D}, \Omega)$, $\Omega = f(\mathcal{D})$, is said to be biholomorphic if $f^{-1} : \Omega \mapsto \mathcal{D}$ belongs to $\mathrm{Hol}(\Omega, \mathcal{D})$.

It is also known (see, for example, [Franzoni and Vesentini (1980)] and [Khatskevich and Shoikhet (1994a)] that if $f \in \mathrm{Hol}(\mathcal{D}, \Omega)$ is biholomorphic, then for each point $x \in \mathcal{D}$, the Fréchet derivative $A = f'(x)$ is a linear isomorphism between X and Y In this situation we will say that $\mathcal{D}$ and Ω are biholomorphically equivalent. As we have already mentioned, in this case X and Y must be linearly isomorphic Banach spaces. Generally speaking, the converse is not true. That is, even if X and Y are isomorphic Banach spaces and $f \in \mathrm{Hol}(\mathcal{D}, Y)$ has at each point x in $\mathcal{D}$ a continuously invertible Fréchet derivative, the mapping f needs not to be univalent on $\mathcal{D}$. Nevertheless, in this case the mapping f is biholomorphic on a neighborhood of each $x \in \mathcal{D}$ by the inverse function theorem. In this situation we will say that $f \in \mathrm{Hol}(\mathcal{D}, \Omega)$ is **locally biholomorphic**. The set of all univalent mappings from a domain $\mathcal{D} \subset X$ into X will be denoted by $\mathrm{Univ}(\mathcal{D})$. For the special case when $\mathcal{D}$ is the open unit ball of X, the subset of $\mathrm{Univ}(\mathcal{D})$ normalized by the conditions

$$f(0) = 0 \quad \text{and} \quad f'(0) = I \tag{10.1}$$

will be denoted by $S(\mathcal{D})$. This notation conforms to the one used in the classical one-dimensional case, when $\mathcal{D} = \Delta = \{z \in C : |z| < 1\}$. In this case we simply write

$$S\,(=S(\Delta)) = \{f \in \mathrm{Univ}(\Delta) : f(0) = 0 \quad \text{and} \quad f'(0) = 1\}. \tag{10.2}$$

That is, S consists of all the mappings $f \in \mathrm{Univ}(\Delta)$ such that f has the following Taylor series at the origin:

$$f(z) = z + \sum_{k=2}^{\infty} a_k z^k . \tag{10.3}$$

The following simple assertion is the key to our subsequent considerations.

Lemma 10.1 *Let $\mathcal{D}$ and Ω be two domains in a complex Banach space X such that $\Omega = f(\mathcal{D})$ for some biholomorphic mapping $f : \mathcal{D} \mapsto \Omega$. Then the classes $\mathcal{G}(\Omega)$ and $\mathcal{G}(\mathcal{D})$ of generators (semi-complete vector fields) on Ω and $\mathcal{D}$, respectively, are linearly isomorphic, i.e., there is a linear invertible operator T from the space $\mathrm{Hol}(\Omega, X)$ onto the space $\mathrm{Hol}(\mathcal{D}, X)$ which takes the set $\mathcal{G}(\Omega)$ onto the set $\mathcal{G}(\mathcal{D})$ (i.e., $\mathcal{G}(\mathcal{D}) = T(\mathcal{G}(\Omega))$). Moreover, such an isomorphism $T : \mathcal{G}(\Omega) \mapsto \mathcal{G}(\mathcal{D})$ can be given by the formulae*

$$T(\varphi)(\cdot) = [f'(\cdot)]^{-1}\varphi(f(\cdot)) \tag{10.4}$$

and

$$T^{-1}(g)(\cdot) = [f'(f^{-1}(\cdot))]g(f^{-1}(\cdot)), \tag{10.5}$$

where $\varphi \in \mathcal{G}(\Omega)$ and $g \in \mathcal{G}(\mathcal{D})$.

Proof. Let $\varphi \in \mathcal{G}(\Omega)$ and let $\{S_\varphi(t)\}_{t \geq 0}$ be the semigroup of holomorphic self-mappings on Ω generated by φ. Then it is clear that the family $\{G(t)\}_{t \geq 0}$ defined by

$$G(t) = f^{-1} \circ S(t) \circ f \tag{10.6}$$

is a semigroup of holomorphic self-mappings on $\mathcal{D}$. Since the X-valued function $S(\cdot)(x) : \mathbb{R}^+ \mapsto \Omega$ is differentiable for all $t \in [0, \infty)$, so is the function $G(\cdot)(x) : \mathbb{R}^+ \mapsto \mathcal{D}$. Thus, for each $x \in \mathcal{D}$, the strong limit

$$g(x) = \lim_{t \to 0^+} \frac{1}{t}\left(x - G(t)(x)\right) \qquad (10.7)$$

exists and $g \in \mathcal{G}(\mathcal{D})$ is a generator on $\mathcal{D}$. In other words, for each $\varphi \in \mathcal{G}(\Omega)$, the mapping $g = T(\varphi)$ belongs to $\mathcal{G}(\mathcal{D})$, where $T : \mathrm{Hol}(\Omega, X) \mapsto \mathrm{Hol}(\mathcal{D}, X)$ is defined by (10.4).

It is clear that T is an invertible linear operator and T^{-1} is given by (10.5). Changing the roles of φ and g, we see, by repeating the above considerations, that T^{-1} takes $\mathcal{G}(\mathcal{D})$ onto $\mathcal{G}(\Omega)$. Lemma 10.1 is proved. $\square$

Remark 10.1 *It is clear that if $\varphi \in \mathrm{aut}(\Omega)$, then $g = T\varphi \in \mathrm{aut}(\mathcal{D})$ and conversely. If, in particular, $\mathcal{D} = \Omega$ and $f \in \mathrm{Aut}(\mathcal{D})$, then the sets $\mathcal{G}(\mathcal{D})$ of all semi-complete vector fields and $\mathrm{aut}(\mathcal{D})$ of all complete vector fields are invariant under the operator $T : \mathrm{Hol}(\mathcal{D}, X) \mapsto \mathrm{Hol}(\mathcal{D}, X)$ defined by (10.4).*

Remark 10.2 *Let $\mathcal{D}$ and Ω be two domains in X and let $f : \mathcal{D} \mapsto \Omega$ be a biholomorphism of $\mathcal{D}$ onto Ω. Define the operator $T : \mathrm{Hol}(\Omega, X) \mapsto \mathrm{Hol}(\mathcal{D}, X)$ by (10.4). If $\varphi \in \mathrm{Hol}(\Omega, X)$ has a null point b in Ω, then so does $g = T\varphi$ in $\mathcal{D}$ and $a = f^{-1}(b)$ is a null point of g in $\mathcal{D}$. In addition, by direct calculations, we get*

$$g'(a) = [f'(a)]^{-1} \circ \varphi'(b) \circ f'(a). \qquad (10.8)$$

Thus, the linear operators $g'(a) : X \mapsto X$ and $\varphi'(b) : X \mapsto X$ are similar, hence they have the same spectrum. In particular, if $\varphi \in \mathcal{G}(\Omega)$ has a null point $b \in \Omega$, which is either quasi-regular, regular or strictly regular, then $a = f^{-1}(b)$ is also a quasi-regular, regular or strictly regular null point of g, respectively.

Using Lemma 10.1 and Remarks 10.1 and 10.2, one can present different parametric representations of semi-complete and complete vector fields. These representations are useful, inter alia, in finding geometric characterizations of biholomorphic mappings in Hilbert and Banach spaces.

In the next section we will turn to a general description of spirallike and starlike mappings defined on the unit ball in Banach space.

10.2 Starlike, Convex, and Spirallike Mappings

Definition 10.3 A set M in X is called **starshaped** with respect to the point w_0 if given any $w \in M$, the point $w_t = (1 - t)w_0 + tw$ also belongs to M for every t with $0 \le t \le 1$. That is, if M contains w, then it also contains the entire line segment joining w to w_0.

Definition 10.4 If $\mathcal{D}$ is a domain in X, then a biholomorphic mapping $f \in \mathrm{Hol}(\mathcal{D}, X)$ is said to be a **starlike mapping** on $\mathcal{D}$ with respect to the point $w_0 \in \overline{\Omega}$, the closure of the image $\Omega = f(\mathcal{D})$ of $\mathcal{D}$, if $\overline{\Omega}$ is a starshaped set with respect to the point w_0.

- If $w_0 \in \Omega$, then we say that f is **starlike with respect to an interior point**.
- If $w_0 \in \partial\Omega$, the boundary of Ω, then we say that f is **starlike with respect to a boundary point**.

Definition 10.5 If $\mathcal{D}$ is a domain in X, then a biholomorphic mapping $f \in \mathrm{Univ}(\mathcal{D})$ is said to be a convex mapping on $\mathcal{D}$ if its image $f(\mathcal{D}) = \Omega$ is a convex domain in X. It is clear that a biholomorphic mapping $f \in \mathrm{Univ}(\mathcal{D})$ is convex if and only if it is starlike with respect to any point $w_0 \in \overline{\Omega}$, the closure of the image $\Omega = f(\mathcal{D})$.

- A starlike mapping on $\mathcal{D}$ with respect to the origin is simply said to be, as in the classical case, starlike. In other words, a biholomorphic mapping $f \in \mathrm{Hol}(\mathcal{D}, X)$ is said to be starlike on $\mathcal{D}$ if the closure $\overline{\Omega}$ of the image $\Omega = f(\mathcal{D})$ of $\mathcal{D}$ is a starshaped set with respect to the origin.

In these definitions the origin is in $\overline{\Omega}$. If, in particular, the origin belongs to Ω, then the mapping f has a null point τ in $\mathcal{D}$. The set of all biholomorphic mappings on $\mathcal{D}$ which are starlike on $\mathcal{D}$ will be denoted by $\mathrm{Star}(\mathcal{D})$.

If there is indeed a point $\tau \in \mathcal{D}$ such that

$$f(\tau) = 0, \tag{10.9}$$

then we will write $f \in S_\tau^*(\mathcal{D})$. Of course, in this case such a point τ is unique because

$$S_\tau^*(\mathcal{D}) \subset \mathrm{Star}(\mathcal{D}) \subset \mathrm{Univ}(\mathcal{D}). \tag{10.10}$$

Again, in the one-dimensional case, when $X = \mathbb{C}$, we will simply write S^* to denote the family of all biholomorphic (univalent) starlike functions

f on the unit disk Δ normalized by the conditions $f(0) = 0$ and $f'(0) = 1$. That is,

$$S^* = S \cap S_0^*(\Delta). \tag{10.11}$$

In the sequel the spectrum of a linear operator A will be denoted by $\sigma(A)$.

Definition 10.6 A set M in X is said to be **spiral-shaped** (with respect to the origin) if there is a bounded linear operator $A : X \to X$ and a positive ε such that $\mathrm{Re}\lambda \geq \varepsilon > 0$ for all $\lambda \in \sigma(A)$ and such that for each $w \in M$ and $t \geq 0$, the point $e^{-tA}w$ also belongs to M.

Definition 10.7 If $\mathcal{D}$ is a domain in X, then a biholomorphic mapping $f \in \mathrm{Hol}(\mathcal{D}, X)$ is said to be a **spirallike mapping** on $\mathcal{D}$ if the closure $\overline{\Omega}$ of its image $\Omega = f(\mathcal{D})$ is a spiral-shaped set.

- Once again, if the origin belongs to Ω, then, as in the classical case, we will say that f is a **spirallike mapping on $\mathcal{D}$ with respect to an interior point**.
- Otherwise, if the origin belongs to $\partial\Omega$ the boundary of Ω, then we say that f is **spirallike with respect to a boundary point**.

The set of all biholomorphic mappings on $\mathcal{D}$ which are spirallike (with respect to the origin) on $\mathcal{D}$ will be denoted by $Spiral(\mathcal{D})$. If, in addition, there is a point $\tau \in \mathcal{D}$ such that

$$f(\tau) = 0, \tag{10.12}$$

then we will write $f \in Sp_\tau(\mathcal{D})$. Note also that if in Definition 10.6 the operator $A = I$, then Ω is actually starlike, *i.e.* $\mathrm{Star}(\mathcal{D}) \subset \mathrm{Spiral}(\mathcal{D})$. Consequently, $S_\tau^*(\mathcal{D}) \subset Sp_\tau(\mathcal{D})$.

10.2.1 *Starlike functions on the unit disk*

The concept of univalent starlike functions was first introduced by Alexander [Alexander (1915)] in 1915. In 1921, Nevanlinna [Nevalinna (1921)] made a more detailed study of this class. In particular, the following characterization of the class $S^* = S \cap S_0^*(\Delta)$ is due to him.

Theorem 10.1 *Let f be a univalent holomorphic mapping on the unit disk Δ such that*

$$f(0) = 0. \tag{10.13}$$

Then f is a starlike function on Δ if and only if

$$\operatorname{Re}\left[\frac{zf'(z)}{f(z)}\right] > 0, \quad z \in \Delta. \tag{10.14}$$

Intuitively, this result follows (as does most of the work on starlike functions on the unit disk) from the identity

$$\frac{\partial}{\partial\theta}\arg f(re^{i\theta}) = \operatorname{Re}\left\{\frac{re^{i\theta}f'(re^{i\theta})}{f(re^{i\theta})}\right\} \tag{10.15}$$

which is valid whenever the function f is holomorphic on Δ and not equal to zero at $z = re^{i\theta}$, $r > 0$, $\theta \in [0, 2\pi]$. Note also that if $f \in \operatorname{Hol}(\Delta, \mathbb{C})$ is locally biholomorphic, *i.e.*, $f'(z) \neq 0$ everywhere, and satisfies (10.14), then it is necessarily univalent.

Furthermore, condition (10.14) leads to the study of other interesting subclasses of $\operatorname{Univ}(\Delta)$. In particular, in 1936 Robertson [Robertson (1936)] introduced the class $S^*(\lambda)$ of starlike functions of order λ:

$$S^*(\lambda) = \left\{f \in S^* : \operatorname{Re}\left[\frac{zf'(z)}{f(z)}\right] > \lambda \geq 0, \ z \in \Delta\right\}. \tag{10.16}$$

In 1978 Wald [Wald (1978)] characterized starlike functions with respect to another center. Using our notions, his result can be reformulated in the following way.

Theorem 10.2   *Suppose that $f \in \operatorname{Hol}(\Delta, C)$ is either of the form*

$$f(z) = z + \sum_{k=2}^{\infty} a_k z^k \tag{10.17}$$

or of the form $f(z) = \sum_{k=1}^{\infty} b_k z^k$ with $f'(\tau) = 1$ for some $\tau \in \Delta$. Then the function $g(z) = f(z) - f(\tau)$ belongs to $S_\tau^*(\Delta)$ if and only if $\operatorname{Re} q(z) > 0$, where

$$q(z) = \frac{(z-\tau)(1-z\bar{\tau})f'(z)}{f(z)-f(\tau)} = \frac{(z-\tau)(1-z\bar{\tau})g'(z)}{g(z)},$$
$$z \neq \tau, \ q(\tau) = 1 - |\tau|^2. \tag{10.18}$$

We will see below that condition (10.18) can easily be obtained by using another approach in more general settings. Different applications of (10.18) are presented in Wald's thesis [Wald (1978)] (see also [Goodman (1983)]).

10.2.2 *Convex and close-to-convex functions on the unit disk*

Historically, the notion of a convex function arose earlier than the notion of a starlike function. This seems quite natural, since convexity has played a crucial role in the development of analysis and geometry.

In 1913 Study [Study (1913)] described univalent functions on a closed disk, the image of which is a convex set. However, his condition employs the second derivative of the function. On the other hand, it is clear that a domain Ω is convex if and only if it is starlike with respect to each one of its points. Using this fact, Sufferidge ([Suffridge (1973)] and [Suffridge (1977)]) gave another characterization of convex functions which also holds for higher dimensions (see Section 10.5).

Here we quote another classical result due to Alexander [Alexander (1915)] which provides an analytic connection between convex and starlike functions.

Theorem 10.3 *Suppose that f is a locally biholomorphic function on the disk $\Delta_r = \{z : |z| < r\}$. Then f is convex on Δ_r if and only if the function $f(z) = zf'(z)$ is starlike on Δ_r.*

As we mentioned above, in the study of the functions of the form

$$f(z) = \sum_{k=1}^{\infty} a_k z^k \tag{10.19}$$

that are holomorphic and univalent on the unit disk Δ, certain subclasses, the members of which share some simple geometric property, arise rather naturally. Further developments in the classical theory have often had analytic generalizations and extensions. In 1952 Kaplan [Kaplan (1952)] defined the class of close-to-convex functions as those functions of the form (10.19) with $a_1 = 1$ such that

$$\mathrm{Re}\left[\frac{f'(z)}{\varphi'(z)}\right] > 0, \quad z \in \Delta, \tag{10.20}$$

for some univalent convex function φ on Δ.

Since φ is convex if and only if $h(z) = z\varphi'(z)$ is starlike, (10.20) can be rewritten in the form

$$\mathrm{Re}\left[\frac{zf'(z)}{h(z)}\right] > 0, \tag{10.21}$$

where h is starlike on Δ. If, in particular, f is starlike itself, then we can take $h = f$, and (10.21) holds bacause

$$\mathrm{Re}\left[\frac{zf'(z)}{f(z)}\right] > 0 \qquad (10.22)$$

for $z \in \Delta$. Note also that condition (10.21) is often used as the definition of a close-to-convex function on Δ, and that it is employed as a basic condition in higher dimensional generalizations (see Section 10.5).

10.2.3   *Spirallike functions on the unit disk*

It seems that the first occurrence of the class $\mathrm{Spiral}(\Delta) = \mathrm{Sp}_0(\Delta)$ arose when condition (10.14) was modified analytically by inserting the factor $e^{i\theta}$:

$$\mathrm{Re}\left[e^{i\theta}\frac{zf'(z)}{f(z)}\right] > 0, \quad z \in \Delta. \qquad (10.23)$$

(See Montel [Montel (1933)] and [Špaček (1933)].) Actually, this definition is compatible with our Definition 10.7.

Proposition 10.1   *Let $f \in \mathrm{Hol}(\Delta, C)$ have the form*

$$f(z) = z + \sum_{k=2}^{\infty} a_k a^k \qquad (10.24)$$

(i.e., $f(0) = 0$ and $f'(0) = 1$). Then $f \in \mathrm{Sp}_0(\Delta)$ if and only if condition (10.23) holds for some $\theta \in (-\pi/2, \pi/2)$.

Note also that for a fixed $\theta \in (-\pi/2, \pi/2)$, a function f satisfying (10.23) is called θ-spirallike.

10.3   Higher-Dimensional Extensions and the Dynamical Approach

Until 1970 the literature on geometric properties of biholomorphic mappings in higher dimensional space ($\mathbb{C}^n$, Hilbert spaces and Banach spaces) is rather limited (see [Cartan (1933)], [Matsuno (1955)] and [Suffridge (1970)]). Cartan [Cartan (1933)] was the first mathematician who suggested the study of starlike and convex mappings in several complex variables despite the fact that many properties of univalent functions on the unit disk (*e.g.*, the Riemann mapping theorem, Carathéodory's theorem on kernel convergence and the Bieberbach–de Branges theorem) fail in the higher dimensional case.

In 1970 Suffridge [Suffridge (1970)] established, inter alia, a necessary and sufficient condition for starlikeness which generalizes Theorem 10.1 to higher dimensions. He used the principle of subordination and one-dimensional ideas due to Robertson [Robertson (1961)]. Furthermore, he also used a similar approach to describe starlike and spirallike domains in Banach spaces (see [Suffridge (1973)], [Suffridge (1977)], [Heath and Suffridge (1979)], [Gong (1999)], [Pfaltzgraff and Suffridge (1975)]).

Pfaltzgraff and Suffridge [Pfaltzgraff and Suffridge (1975)] gave a characterization of close-to-starlike holomorphic mappings which in the one-dimensional case coincides with a characterization of close-to-convex functions.

Roughly speaking, the idea in these considerations is the following one. If $f \in S$ (*i.e.*, f is a univalent holomorphic mapping on the unit disk Δ with $f(0) = 0$ and $f'(0) = 1$), then the condition of starlikeness (10.14) can be rewritten in the form

$$f(z) = f'(z)g(z), \tag{10.25}$$

where $g \in \mathrm{Hol}(\Delta, C)$ has the form

$$g(z) = zp(z) \tag{10.26}$$

with

$$\mathrm{Re}\, p(z) > 0, \quad z \in \Delta. \tag{10.27}$$

Conditions (10.26) and (10.27) are equivalent to

$$g(0) = 0 \tag{10.28}$$

and

$$\mathrm{Re}\, g(z)\bar{z} > 0, \quad z \in \Delta, \ z \neq 0. \tag{10.29}$$

The latter condition can easily be generalized to the case of Hilbert and Banach spaces. Namely, let X be a complex Banach space and let X^* be the dual of X. By $\langle x, x^* \rangle$ we denote the action of a linear functional x^* in X^* on an element x of X.

Recall that the mapping $J : X \to 2^{X^*}$ defined by

$$J(x) = \left\{ x^* \in X^* : \langle x, x^* \rangle = \|x\|^2 = \|x^*\|^2 \right\}, \quad x \in X, \tag{10.30}$$

is called the (normalized) duality mapping.

Let $\mathcal{D}$ be the open unit ball in X. We now define two families of holomorphic mappings on $\mathcal{D}$. Let N denote the subset of those mappings $g \in \mathrm{Hol}(\mathcal{D}, X)$ which satisfy

$$g(0) = 0 \quad \text{and} \quad \mathrm{Re}\langle g(x), x^* \rangle > 0 \tag{10.31}$$

for all $0 \neq x \in \mathcal{D}$ and $x^* \in J(x)$. Let

$$M = \{g \in N : g'(0) = I\}, \tag{10.32}$$

where I denotes the identity operator on X. The following three assertions are due to Suffridge [Suffridge (1973)] and [Suffridge (1977)].

Theorem 10.4   *Let f be a locally biholomorphic mapping on $\mathcal{D}$ (i.e., $f'(x)$ is a bounded linear operator with a bounded inverse for each $x \in \mathcal{D}$) with $f(0) = 0$. Then f is starlike if and only if the following condition holds:*

$$f(x) = f'(x)[g(x)], \quad x \in \mathcal{D}, \tag{10.33}$$

for some $g \in M$.

Note that condition (10.33) and the inclusion $g \in N$ imply the inclusion $g \in M$.

Theorem 10.5   *Let f be a locally biholomorphic mapping on $\mathcal{D}$ and set*

$$f(x) - f(y) = f'(x)[w(x, y)] \tag{10.34}$$

for all $x, y \in \mathcal{D}$. Then f is a convex mapping on $\mathcal{D}$ if and only if $\mathrm{Re}\langle w(x, y), x^* \rangle > 0$ whenever $\|y\| < \|x\|$ and $x^* \in J(x)$.

As it turns out, the class N is also useful in the characterization of spirallike mappings.

Theorem 10.6   *Let A be a bounded linear operator which is strongly accretive, i.e., there is $\varepsilon > 0$ such that*

$$\mathrm{Re}\langle Ax, x^* \rangle \geq \varepsilon \|x\|^2 \tag{10.35}$$

for all $x \in X$ and $x^* \in J(x)$.

Suppose that $f \in \mathrm{Hol}(\mathcal{D}, X)$ is a locally biholomorphic mapping on $\mathcal{D}$ which satisfies the conditions

$$f(0) = 0 \tag{10.36}$$

and

$$f'(0) = I. \tag{10.37}$$

Then f is spirallike (relative to A) if and only if it satisfies the equation

$$Af(x) = f'(x)[g(x)] \tag{10.38}$$

with some $g \in N$.

Remark 10.3 *Actually, Theorem 10.4 follows from Theorem 10.6. Indeed, if f is a locally biholomorphic mapping, then $f'(0) = B$ is invertible. Setting $A = I$, one can consider the starlike mapping $\tilde{f} = B^{-1}f$ which satisfies conditions (10.36) and (10.37) and equation (10.33) with $g \in N$. As we have mentioned above, g must belong to M. Hence f also satisfies (10.33) with the same g.*

Remark 10.4 *In the above remark we used the auxiliary mapping $\tilde{f}$ because of the normalizing condition (10.37) of Theorem 10.6. This condition is essential in the proof of Theorem 10.6 in [Suffridge (1973)] and [Suffridge (1977)]. The crucial point in that proof is that condition (10.37), $f'(0) = I$, implies $g'(0) = A$.*

In fact, these conditions are needed because of the assumption (10.35) that A is strongly accretive. This assumption was used by Sufiridge [Suffridge (1977)] in his definition of spirallike mappings. Our definition 10.7 is more general, since the spectrum of each strongly accretive operator lies strictly in the right half–plane. This enables us, in particular, to avoid the normalization (10.37). Note also that in contrast with (10.35), our requirement of A is independent of any equivalent norm on X. Since we are interested in the geometric properties of Ω, our definition seems to be more natural.

Remark 10.5 *It seems that for $\mathbb{C}^n$, Hilbert and Banach spaces the papers [Pfaltzgraff (1975)] and [Pfaltzgraff and Suffridge (1975)] (see also [Gurganus (1975)]) were the first ones where the ideas of applying dynamical systems appeared. (In the one-dimensional case such ideas were employed earlier by Löwner, Kufarev [Golusin (1969); Goodman (1983)], Robertson [Robertson (1961)] and Brickman [Brickman (1973)].) Pfaltzgraff, for example, applied generalized Löwner differential equations to characterize subordination chains in $\mathbb{C}^n$ and univalent mappings on the unit ball. He also suggested this approach for Banach spaces.*

Now we will turn to a general description of spirallike and starlike mappings defined on the unit ball in Banach space (see [Aharonov *et al.* (2003); Elin *et al.* (2001)] for the one-dimensional case and compare [Elin *et al.* (2000)] for a more general situation).

Theorem 10.7　*Let $\mathcal{D}$ be the open unit ball in a Banach space X and let $f \in \mathrm{Hol}(\mathcal{D}, X)$ be e biholomorphic mapping on $\mathcal{D}$. Then f is spirallike on $\mathcal{D}$ if and only if there exist a linear operator A such that its spectrum lies strictly inside the right half-plane and a real number m, $m \leq 0$, such that the following condition holds:*

$$\mathrm{Re}\,\langle [f'(z)^{-1} A f(z), z^* \rangle \geq m(1 - \|z\|^2) \tag{10.39}$$

for all $z \in \mathcal{D}$ and $z^* \in J(z)$. If $A = I$ in (10.39), then f is actually starlike. In addition, if f has a null point in $\mathcal{D}$, then it is spirallike (starlike) with respect to an interior point.

Proof.　If f is spirallike, then there exists a linear operator A with its spectrum strictly in the right-half plane such that for each $y \in \Omega = f(\mathcal{D})$ and $t \geq 0$, the element $e^{-tA}y$ is in Ω. This means that the vector field $\varphi = A$ is semi-complete on Ω. Hence, by Lemma 10.1, the vector field

$$g = T(A) = [f'(\cdot)]^{-1} A f(\cdot) \tag{10.40}$$

is semi-complete on $\mathcal{D}$.　Therefore (10.39) is a consequence of formula (3.5.19) in [Elin *et al.* (2004)].　Conversely, if (10.39) holds, then, again by the above-mentioned formula, the vector field g defined by (10.40) is semi-complete on $\mathcal{D}$. Therefore $A = T^{-1}(g)$ is semi-complete on $\Omega = f(\mathcal{D})$. That is, for each $y \in \Omega$, the curve $\{e^{-tA}y\}_{t\geq 0} \subset \Omega$ is the solution of the Cauchy problem

$$\begin{cases} \dfrac{du}{dt} + Au = 0 \\[2mm] u(0) = y. \end{cases} \tag{10.41}$$

Now it follows from the assumption on A and Definition 10.7 that Ω is spiral-shaped.

Note also that if f has a null point $a \in \mathcal{D}$, then so does g, and $f(a) = g(a) = 0 \in \Omega$ is the limit point of $e^{-tA}y$ for each $y \in \Omega$. Hence in this case f is spirallike with respect to an interior point. Otherwise, that is, if $0 \in \partial\Omega$, the mapping f is spirallike with respect to a boundary point.

In addition, it is clear that if $A = I$, then the curve $\{e^{-t}y\}_{t\geq 0}$ is a straight line. In other words, Ω is star-shaped, and f is starlike.　　□

Remark 10.6 *As we saw in the proof of Theorem 10.7, in order to determine when $f \in \mathrm{Hol}(\mathcal{D}, X)$ satisfying (10.39) is spirallike (starlike) with respect to an interior point, we need to know whether f has an interior null point. If it does not, then there is a sequence $\{x_n\} \subset \mathcal{D}$ such that $x_n \to y \in \partial \mathcal{D}$ and*

$$f(x_n) \to 0, \quad n \to \infty. \tag{10.42}$$

Indeed, it is sufficient to set $x_n = f(e^{-nA}z)$, $n = 0, 1, 2, \ldots$ for any $z \in f(\mathcal{D})$.

Generally speaking, the above question is equivalent to the existence of a null point of the semi-complete vector field g defined by (10.40). Moreover, one can understand the condition of spirallikeness (starlikeness) via the differential equation

$$f'(z)g(z) = Af(z), \tag{10.43}$$

where g is a strongly semi-complete vector field.

In addition, it turns out that if f is a solution of (10.43) such that $f'(z)$ is invertible for all $z \in \mathcal{D}$, then f is actually univalent, hence biholomorphic.

Indeed, since g is strongly semi-complete, then by definition, it has an interior null point $\tau \in \mathcal{D}$, which is locally uniformly attractive for the semigroup $\{S_g(t)\}_{t \geq 0}$ generated by g. Furthermore, there is a neighborhood $U \subset \mathcal{D}$ of the point τ such that f is biholomorphic on U. Denote $V = f(U)$. Suppose by way of contradiction that f is not univalent on $\mathcal{D}$, that is, there are two distinct points x_1 and x_2 in $\mathcal{D}$ such that $f(x_1) = f(x_2) = y \in f(\mathcal{D})$. But then $\{S_g(t)x_1\}$ and $\{S_g(t)x_2\}$ both converge to τ as t tends to infinity, and therefore there is $t_0 \geq 0$ such that both $S_g(t_0)x_1 = u_1$ and $S_g(t_0)x_2 = u_2$ belong to U.

Since $u(t) = e^{-tA}$ solves the Cauchy problem (10.41), we have

$$S_g(t_0) = f^{-1} \circ e^{-tA} \circ f. \tag{10.44}$$

Hence

$$y_i = f(u_i) = e^{-t_0 A}y \in V, \quad i = 1, 2. \tag{10.45}$$

This implies that $y_1 = y_2$ and consequently, $u_1 = S_g(t_0)x_1 = u_2 = S_g(t_0)x_2$ which is impossible by the univalence of $S_g(t_0)$. Finally, note that (10.43) and the chain rule also imply the equality

$$A = f'(\tau) \circ g'(\tau) \circ [f'(\tau)]^{-1}, \tag{10.46}$$

which means that the spectrum of A lies strictly in the right half–plane. So we have proved the following assertion.

Proposition 10.2   *Let $\mathcal{D}$ be a domain in X and let $f \in \mathrm{Hol}(\mathcal{D}, X)$ be locally biholomorphic on $\mathcal{D}$. Then f is spirallike if and only if there are a strongly semi-complete vector field g on $\mathcal{D}$ and a linear operator $A : X \to X$ such that f satisfies the differential equation (10.43). If, in addition, $\mathcal{D}$ is a hyperbolic domain endowed with a metric ρ which induces the norm topology, then there is a point $\tau \in \mathcal{D}$ such that $f(\tau) = 0$ and for each ρ-ball B_r centered at τ, $B_r = \{x \in \mathcal{D} : \rho(x, \tau) < r\}$, the image $f(B_r) = \Omega_r$ is spiral-shaped (star-shaped, when $A = I$).*

Remark 10.7   *The last assertion follows from the fact that each ρ-ball B_r centered at τ is invariant under the semigroup $\{S(t)_{t\geq 0}$ generated by g. As a matter of fact, we will see below that in the above situation there is an equivalent norm of X and a ball B in this norm centered at τ such that $f(B)$ is spiral-shaped (star-shaped).*

Corollary 10.1   *Let $\mathcal{D}$ be the open unit ball in X and let $f \in \mathrm{Hol}(\mathcal{D}, X)$ be a locally biholomorphic mapping on $\mathcal{D}$ which satisfies the condition*

$$\mathrm{Re}\,\langle [f'(z)]^{-1} A f(z), z^* \rangle \geq \alpha(\|z\|) \|z\|, \quad z \in \mathcal{D}, \quad z^* \in J(z) \quad (10.47)$$

for some linear operator $A : X \to X$ and some real continuous function $\alpha : [0, 1] \to \mathbb{R}$ such that $\alpha(1) > 0$. Then f is a univalent spirallike (starlike) mapping on $\mathcal{D}$.

Remark 10.8   *Condition (10.47) is a sufficient condition for f to be spirallike (or starlike) but it is not necessary. It would be nice, of course, to find a condition which is both necessary and sufficient for f to be spirallike (starlike). However, as we have already mentioned, to tackle this problem we need to find a condition which recognizes whether $f \in \mathrm{Hol}(\mathcal{D}, X)$ (or equivalently, $g = [f'(\cdot)]^{-1} A f(\cdot) \in \mathcal{G}(\mathcal{D})$) has an interior null point in $\mathcal{D}$. Generally speaking, these problems seem to be quite complicated. Moreover, there are examples of semi-complete vector fields on the closed unit balls $\overline{\mathcal{D}}$ of Banach spaces which have no mull point in $\overline{\mathcal{D}}$ [Khatskevich et al. (1995a)]. Nevertheless, these problems can be solved completely in the case of the Hilbert ball [Elin et al. (2004)].*

Recall that for a bounded linear operator $A : X \to X$ we denote by $\sigma(A)$ the spectrum of A. The number

$$\kappa_+(A) = \max\{\mathrm{Re}\lambda : \lambda \in \sigma(A)\} \quad (10.48)$$

is called the upper exponential index of A. it is well known (see [Daletskii and Krein (1970)]) that

$$\kappa_+(A) = \lim_{t \to \infty} \frac{\log \|e^{tA}\|}{t} \tag{10.49}$$

and for each $\omega > \kappa_+$ there is a positive number $N = N(\omega)$ such that

$$\|e^{tA}\| \le N e^{\omega t}, \quad t \ge 0. \tag{10.50}$$

The number

$$\kappa_-(A) = \min\{\operatorname{Re}\lambda : \lambda \in \sigma(A)\} = \lim_{t \to \infty} \frac{\log \|e^{-tA}\|}{-t} \tag{10.51}$$

is called the lower exponential index of A.

We also have that for each $\omega < \kappa_-$, there is a positive number $m = m(\omega)$ such that $\|e^{-tA}\| \le m e^{-\omega t}$.

If now $\mathcal{D}$ is a bounded domain in X and $g \in \mathcal{G}(\mathcal{D})$ has a null point $\tau \in \mathcal{D}$, then in a neighborhood of τ the mapping g can be represented by the Taylor series

$$g(x) = A(x - \tau) + \sum_{k=k}^{\infty} P_k(x - \tau), \tag{10.52}$$

where $A = g'(\tau)$, $l \ge 2$, and P_k are homogeneous forms of order k. Thus g is a strongly semi-complete vector field if and only if $\kappa_-(A) > 0$.

Proposition 10.3 *Let $\mathcal{D}$ be a bounded domain in X and let $g \in \mathcal{G}(\mathcal{D})$ admit the representation (10.52) in a neighborhood of $\tau \in \mathcal{D}$. Suppose that $A = g'(\tau)$ satisfies the condition*

$$0 < \kappa_+(A) < l\kappa_-(A). \tag{10.53}$$

Then for a given invertible operator $B \in L(X)$ such that $BA = AB$, the differential equation

$$Af(x) = f'(x)g(x) \tag{10.54}$$

has a unique solution $f \in \operatorname{Hol}(\mathcal{D}, X)$ which satisfies the initial conditions

$$\begin{cases} f(\tau) = 0 \\ f'(\tau) = B. \end{cases} \tag{10.55}$$

If $\{S(t)\}_{t\geq 0}$ is the semigroup generated by g, then this solution can be represented by the formula

$$f(x) = \lim_{t\to\infty} Be^{At}(S(t)x - \tau), \tag{10.56}$$

where the limit in (10.56) is taken with respect to the topology of locally uniform convergence over $\mathcal{D}$.

Proof.   We assume for simplicity that $\tau = 0$.

Step 1.   First we note that there is a norm $\|\cdot\|_A$ equivalent to the original norm of X and a ball $\mathbf{B}$ in this norm centered at $\tau = 0$ such that $g \in \mathcal{G}(\mathbf{B})$, i.e., $\mathbf{B}$ is invariant under the semigroup $\{S(t)\}_{t\geq 0}$ generated by g. Indeed, for a given $\mu \in (0, \kappa_-(A))$, there is $m > 0$ such that $\|e^{-tA}\| \leq me^{-\mu t}$. Setting

$$\|x\|_A = \sup_{t>0} \|e^{(\mu I - A)t}x\| \tag{10.57}$$

we get $\|x\| \leq \|x\|_A \leq m\|x\|$ and

$$\|e^{-As}x\|_A \leq e^{-\mu s}\|x\|_A, \quad s \geq 0. \tag{10.58}$$

The last inequality implies in turn that for each $x^* \in X^*$ such that $\mathrm{Re}\langle x, x^*\rangle = \|x\|_A^2 = \|x^*\|_A^2$ (i.e., $x^* \in J_A(x)$),

$$\mathrm{Re}\langle Ax, x^*\rangle \geq \mu\|x\|_A^2 > 0, \quad x \neq 0. \tag{10.59}$$

Now choose any ball $B_r = \{x \in X : \|x\|_A < r\} \subset \mathcal{D}$ centered at the origin with radius $r > 0$ and represent $g \in \mathcal{G}(\mathcal{D})$ by using the Taylor series (10.52) for $\tau = 0$. We have

$$g(x) = Ax + \sum_{k=l}^{\infty} P_k(x), \quad l \geq 2, \tag{10.60}$$

where P_k are homogeneous polynomials of order $k \geq l$. If

$$M = \sup_{\|x\|\leq r} \|g(x) - Ax\|_A, \tag{10.61}$$

then it follows by the Schwarz Lemma that

$$\|g(x) - Ax\|_A \leq \frac{M}{r^l}\|x\|_A^l \tag{10.62}$$

for all $x \in B_r$. Therefore we get

$$\mathrm{Re}\langle g(x), x^* \rangle \geq \mu \|x\|_A^2 - \frac{M}{r^l} \|x\|_A^{l+1}$$

$$= \|x\|_A^2 \left(\mu - \frac{M}{r^l} \|x\|_A^{l-1} \right) > 0 \tag{10.63}$$

for all $x \in \mathbf{B} := \left\{ x \in X : \|x\|_A < \min\left\{ \left(\frac{\mu r^l}{M}\right)^{\frac{1}{l-1}}, r \right\} \right\}$, $x \neq 0$ and $x^* \in J_A(x)$.

Hence g is a semi-complete vector field on $\mathbf{B}$.

Step 2. Let now $h : \mathbf{B} \to X$ be any bounded holomorphic mapping on $\mathbf{B}$ such that

$$h(0) = 0 \quad \text{and} \quad \mathcal{D}^n h(0) = 0 \quad \text{for} \quad 1 \leq n \leq l - 1. \tag{10.64}$$

we claim that

$$\lim_{t \to \infty} e^{tA} h(S(t)x) = 0 \tag{10.65}$$

for all $x \in \mathbf{B}$. Indeed, without any loss of generality we may assume that the radius of $\mathbf{B}$ is 1. As above, (10.64) and the Schwarz Lemma imply that

$$\|h(x)\|_A \leq M(h)\|x\|_A^l, \quad x \in \mathbf{B}, \tag{10.66}$$

where $M(h) = \sup_{x \in \mathbf{B}} \|h(x)\|_A$. In addition, it follows by Proposition 3.6.4 in [Elin *et al.* (2004)] that for each $x \in \mathbf{B}$,

$$\|S(t)x\|_A \leq e^{-\mu t} \frac{\|x\|_A}{(1 - \|x\|_A)^2} \left(1 - \|S(t)x\|_A \right)^2$$

$$\leq e^{-\mu t} \frac{\|x\|_A}{(1 - \|x\|_A)^2}. \tag{10.67}$$

Now using (10.66), (10.67) and assumption (10.53), we can choose $\mu \in (0, \kappa_-)$ and $\omega > \kappa_+$ such that $\omega < l\mu$. Then we get

$$\left\| e^{tA} h(S(t)x) \right\|_A \leq N(\omega) e^{\omega t} \|h(S(t)x)\|_A$$

$$\leq N(\omega) \cdot M(h) e^{\omega t} \|S(t)x\|_A^l$$

$$\leq N(\omega) M(h) e^{(\omega - l\mu)t} \frac{\|x\|^l}{(1 - \|x\|)^{2l}} \to 0 \tag{10.68}$$

when t tends to infinity. The claim is proved. $\qquad\square$

Step 3. Now we show that the limit in (10.56) exists for all $s \in \mathcal{D}$. First, let us restrict ourselves to **B**. Consider the mapping $u : \mathbb{R}^+ \times \mathbf{B} \mapsto X$ defined as follows:

$$u(t, x) =: e^{At} S(t) x, \quad t \geq 0, \quad x \in \mathbf{B}. \tag{10.69}$$

By calculations and the definition of $S(t)$, we get

$$\frac{\partial u(t, z)}{\partial t} = A e^{At} S(t) x - e^{At} g(S(t)x) = -e^{At} h(S(t)x), \tag{10.70}$$

where $h(x) = g(x) - Ax$ obviously satisfies (10.64). So we get from (10.70) and (10.68) that for each $x \in \mathbf{B}$ and $t_1, t_2 \in \mathbb{R}^+$,

$$\left\| u(t_1, x) - (t_2, x) \right\|_A \leq N(\omega) M(h) \int_{t_1}^{t_2} e^{(\omega - l\mu)dt} \cdot \frac{\|x\|^2}{(1 - \|x\|)^2}. \tag{10.71}$$

Thus the limit in (10.56) exists pointwise. Denote this limit by f. Let now $\widetilde{\mathcal{D}}$ be any ball strictly inside $\mathcal{D}$. It follows from (10.53) that $S(t) \to 0$ uniformly on $\widetilde{\mathcal{D}}$. Hence there is a positive p such that $y = S(p)x \in \mathbf{B}$ for each $x \in \widetilde{\mathcal{D}}$. Furthermore, it follows from the semigroup property that for $x \in \widetilde{\mathcal{D}}$,

$$\lim_{t \to \infty} B e^{A(p+t)} S(p + t) x = \lim_{t \to \infty} B e^{Ap} e^{At} S(t) S(p) x$$
$$= \lim_{t \to \infty} B e^{Ap} e^{At} S(t) y = e^{Ap} f(y) = e^{Ap} f(S(p)x). \tag{10.72}$$

This concludes Step 3.

Step 4. Next we will show the solvability of equation (10.54). In fact, we claim that the mapping $f : \mathcal{D} \mapsto X$ defined by (10.56) is a solution of this equation with the initial data $f(0) = 0$ and $f'(0) = B$. Indeed, substituting $f = B \lim_{t \to \infty} e^{At} S(t)$ in the right-hand side of (10.54), we have by equation (5.2) in [Reich and Shoikhet (1996)],

$$f'(x)g(x) = B \lim_{t \to \infty} e^{tA} \left[\frac{\partial S(t)x}{\partial x} g(x) \right] = B \lim_{t \to \infty} e^{tA} g(S(t)x)$$
$$= B \lim_{t \to \infty} \left[e^{tA} g'(0) S(t) x + h(S(t)x) \right], \tag{10.73}$$

where $h = g - A$. Again, by Step 2, we conclude that

$$\lim_{t \to \infty} e^{At} h(S(t)x) = 0 \tag{10.74}$$

and

$$f'(x)g(x) = BA \left[\lim_{t \to \infty} e^{At} S(t) x \right] = A f(x). \tag{10.75}$$

Step 5. Finally, it remains to show that if equation (10.54) has another solution which satisfies the same initial conditions, then it must coincide with the mapping f defined by (10.56). In fact, if $\tilde{f} : \mathcal{D} \mapsto X$ satisfies (10.54), $\tilde{f}(0) = 0$ and $\tilde{f}'(0) = B$, then we obtain

$$A\tilde{f}(S(t)x) = \tilde{f}'(S(t)x)g(S(t)x) = -\tilde{f}'(S(t)x)\frac{\partial S(t)x}{\partial t} \qquad (10.76)$$

for all $x \in \mathcal{D}$ and $t \geq 0$. If we denote $G(t,x) = \tilde{f}(S(t)x)$, $t \geq 0$, then this equality means that

$$AG(t,x) = -\frac{\partial G(t,x)}{\partial t} . \qquad (10.77)$$

In addition,

$$G(0,x) = \tilde{f}(x). \qquad (10.78)$$

Therefore, solving (10.77) with the initial condition (10.78), we get

$$G(t,x) = e^{-tA}\tilde{f}(x), \qquad (10.79)$$

or

$$\tilde{f}(x) = e^{tA}G(t,x) = e^{tA}\tilde{f}(S(t)s) = e^{tA}\big[BS(t)x + h(S(t)x)\big], \quad x \in \mathcal{D}, \ t \geq 0, \qquad (10.80)$$

where $h = \tilde{f} - B$. Letting t tend to infinity and using Step 2, we get $\tilde{f}(x) = f(x)$. Proposition 10.3 is proved.

Corollary 10.2 *Let $\mathcal{D}$ be a bounded domain in X and let $g \in \mathcal{G}(\mathcal{D})$ satisfy the conditions*

$$g(\tau) = 0 \qquad (10.81)$$

and

$$g'(\tau) = I \qquad (10.82)$$

for some $\tau \in \mathcal{D}$. Then for a given invertible operator $B : X \mapsto X$, the differential equation

$$f(x) = f'(x)g(x) \qquad (10.83)$$

has a unique solution which satisfies the initial conditions

$$f(\tau) = 0 \quad \text{and} \quad f'(\tau) = B. \qquad (10.84)$$

If $\{S(t)\}_{t\geq 0}$ is the semigroup generated by g, then this solution can be represented as

$$f = B \lim_{t \to \infty} e^t (S(t) - \tau). \qquad (10.85)$$

Moreover, f is biholomorphic on $\mathcal{D}$ and $f(\mathcal{D})$ is a star-shaped domain.

Corollary 10.3   *Let $\mathcal{D}$ be a bounded domain in X and let $f : \mathcal{D} \mapsto X$ be a starlike mapping on $\mathcal{D}$ with respect to an interior point τ in $\mathcal{D}$. Then there is a ball $B \subset \mathcal{D}$ centered at τ such that $f(B)$ is star-shaped. Moreover, if ρ is a metric on $\mathcal{D}$ assigned to $\mathcal{D}$ by a Schwarz–Pick system, then the image of each ρ-ball B centered at τ is a star-shaped domain.*

Proposition 10.4   *A bounded domain $\mathcal{D}$ in X is biholomorphically equivalent to a star-shaped domain if and only if there exists $g \in \mathcal{G}(\mathcal{D})$ such that for some $\tau \in \mathcal{D}$, $g(\tau) = 0$ and $g'(\tau) = I$.*

10.4   Distortion Theorems for Starlike Mappings on the Unit Ball

A celebrated theorem of Köbe asserts that the image of any univalent function on the unit disk of the form $z + \sum_{k=2}^{\infty} a_k z^k$ covers the disk of radius $\frac{1}{4}$ centered at the origin. This fact is no longer true in higher dimensions (see [Cartan (1933)]). Nevertheless, Barnard, Fitzgerald and Gong showed that a similar assertion can be established for starlike mappings on the unit ball of a Banach space normalized at the origin (see [Barnard *et al.* (1991)]). An improved result was obtained in [Chuaqui (1995)] for the so-called strongly starlike mappings. Actually, covering estimates (as well as growth estimates) turn out to depend on lower and upper admissible bounds for semi-complete vector fields related to starlike mappings.

Let $\mathcal{D}$ be (as above) a domain in a Banach space X and let $f \in \mathrm{Hol}(\mathcal{D}, X)$. We recall that f is starlike on $\mathcal{D}$ (with respect to an interior point τ), i.e., $f \in S_\tau^*(\mathcal{D})$, if and only if it satisfies the differential equation

$$f(x) = f'(x)g(x) \qquad (10.86)$$

for some strongly semi-complete vector field g on $\mathcal{D}$ such that

$$g(\tau) = 0 \qquad (10.87)$$

and

$$g'(\tau) = I. \tag{10.88}$$

We will call such a mapping g the characteristic mapping of f, and the evolution equation

$$\begin{cases} \dfrac{\partial u(t,x)}{\partial t} + g(u(t,x)) = 0 \\ u(0,x) = x \in \mathcal{D} \end{cases} \tag{10.89}$$

will be called the characteristic equation for f defined by (10.86). Recall also that if $\mathcal{D}$ is bounded, then f can be found by the formula

$$f(x) = A \lim_{t \to \infty} e^t [u(t,x) - \tau], \tag{10.90}$$

where $A = f'(\tau)$ and there is a ball $B \subset \mathcal{D}$ centered at τ such that f is starlike on B. Set $f_1(x) = A^{-1}f(x)$, $x \in \mathcal{D}$. Then f_1 also satisfies (10.86) and is starlike on both $\mathcal{D}$ and B. So, one can try to establish growth and covering estimates of both f and f_1 on B. If, in addition, $\mathcal{D}$ is a symmetric domain, then it can be realized as the unit ball in a JB^*-triple system, and we are able to study the behavior of these mappings on the whole of $\mathcal{D}$ by using similar tools to the ones we employed in a Hilbert space. Indeed, first we note that in this case the domain $\mathcal{D}$ is homogeneous, *i.e.*, for each $x \in \mathcal{D}$ there is an automorphism $M_{-\tau} \in \mathrm{Aut}(\mathcal{D})$ such that $M_{-\tau}(\tau) = 0$. The mapping $\tilde{f} = f_1 \circ M_{-\tau}$ is also starlike, since $\tilde{f}(\mathcal{D}) = f_1(\mathcal{D}) = A^{-1}f(\mathcal{D})$.

At the same time, Lemma 10.1 implies that the mapping

$$\tilde{g} = \left([M_{-\tau}(\cdot)]'\right)^{-1}[g(M_\tau(\cdot))] \tag{10.91}$$

belongs to $\mathcal{G}(\mathcal{D})$ and is normalized by

$$\tilde{g}(0) = 0 \tag{10.92}$$

and

$$\tilde{g}'(0) = I. \tag{10.93}$$

Hence, by the chain rule and (10.86), we get

$$\begin{aligned} \tilde{f}(x) &= f(M_{-\tau}(x)) = f'(M_{-\tau}(x))[g(M_{-\tau}(x))] \\ &= \tilde{f}'(x)[[M_{-\tau}(x)]']^{-1}[g(M_{-\tau}(x))] = \tilde{f}'(x)\tilde{g}(x). \end{aligned} \tag{10.94}$$

Thus $\tilde{g}$ is the characteristic mapping for $\tilde{f}$ and

$$v(t, x) = M_{-\tau}\big(u(t, M_\tau(x))\big) \tag{10.95}$$

is the solution of the characteristic equation for $\tilde{f}$.

Let now $\omega = (\omega_1, \omega_2)$ be a pair of positive real functions on $[0, 1)$ such that the Cauchy problems

$$\begin{cases} \dfrac{\partial \beta_i(t, s)}{\partial t} + \beta_i(t, s)\omega_i(\beta_i(t, s)) = 0 \\ \beta_i(0, s) = s \in [0, 1), \quad i = 1, 2, \end{cases} \tag{10.96}$$

have unique solutions $\beta_i(t, s) \in [0, 1)$ for all $t \geq 0$, $s \in [0, 1)$. Assume also that ω_1 is an increasing function while ω_2 is decreasing.

We will say that f is ω-starlike on $\mathcal{D}$ if ω_1 and ω_2 are admissible lower and upper bounds for $\tilde{g}$, i.e., for each $x \in \mathcal{D}$ and $x^* \in J(x)$,

$$\omega_2(\|x\|)\|x\|^2 \geq \mathrm{Re}\langle \tilde{g}(x), x^* \rangle \geq \omega_1(\|x\|)\|x\|^2. \tag{10.97}$$

Since $\tilde{g}'(0) = I$, the existence of $\omega = (\omega_1, \omega_2)$ is again provided by the universal lower and upper bounds.

$$\omega_\wedge(\|x\|) = \frac{1 - \|x\|}{1 + \|x\|}, \quad \omega^\wedge \|x\| = \frac{1 + \|x\|}{1 - \|x\|}. \tag{10.98}$$

However, even for the one-dimensional case there are many examples when ω_1 and ω_2 can be improved and may be chosen more precisely (see examples below).

Let now γ_i be functions defined on $(0, 1)$ such that

$$\frac{\gamma_i'(s)}{\gamma_i(s)} = \frac{1}{s\omega_i(s)}, \quad i = 1, 2, \tag{10.99}$$

i.e., $ln\gamma_i(s)$ is a primitive function for $\frac{\omega_i(0)}{s\omega_i(s)}$, $i = 1, 2$. Observe that, since the Cauchy problem (10.96) is globally solvable for $t \in [0, \infty)$, and its solution $\beta_i(t, s)$ goes to zero as t tends to ∞, it follows that the integrals

$$\int_0^s \frac{dt}{t\omega_i(t)} \tag{10.100}$$

are divergent and tend to ∞. Hence γ_i can be continuously extended to $[0, 1)$ by $\gamma_i(0) = 0$. In order that γ_i be defined uniquely, we set, for example, $\gamma_i(1/2) = 1$. Then it is an increasing function on $(0, 1)$ by (10.99).

Now (10.96) and (10.99) imply that

$$\gamma_i(\beta_i(t,s)) = e^{-t}\gamma_i(s), \quad i = 1, 2. \tag{10.101}$$

Recalling (10.95) and arguing as in the Hilbert space case, we obtain the inequalities

$$\beta_2(t, \|x\|) \leq \|v(t,x)\| \leq \beta_1(t, \|x\|) \tag{10.102}$$

which, in turn, will be used in the proof of the following assertion, where $\omega = (\omega_1, \omega_2)$ [Elin *et al.* (2004)].

Proposition 10.5 *Let X be a complex Banach space such that its open unit ball $\mathcal{D}$ is homogeneous and let $f \in \mathrm{Hol}(\mathcal{D}, X)$ be ω-starlike on $\mathcal{D}$, with $f(\tau) = 0$, $\tau \in \mathcal{D}$. Let γ_i, $i = 1, 2$, be the solutions of (10.99) defined above. Then the following estimates hold:*

$$\|A^{-1}\|^{-1}k(\|x\|)\gamma_2(\|x\|) \leq \|f(M_{-\tau}(x))\| \leq K(\|x\|)\gamma_1(\|x\|)\|A\|, \tag{10.103}$$

where $A = f'(\tau)$,

$$K(p) := \sup_{s \in [0,p]} \frac{s}{\gamma_1(s)}, \quad 0 < p \leq 1, \tag{10.104}$$

$$k(p) := \sup_{s \in [0,p]} \frac{s}{\gamma_2(s)}, \quad 0 < p \leq 1, \tag{10.105}$$

and $M_{-\tau}$ is an automorphism of $\mathcal{D}$ which takes τ into 0. In particular,

$$f(\mathcal{D}) \supset \mathcal{D}_r, \tag{10.106}$$

where $\mathcal{D}_r$ is the ball centered at the origin of radius

$$r = \|A^{-1}\|^{-1} \lim_{s \to 1^-} \gamma_2(s)k(s). \tag{10.107}$$

Proof. Since γ_1 and γ_2 are increasing functions, we have by (10.101) and (10.102) that

$$\gamma_1(\|v(t,x)\|) \leq \gamma_1(\beta_1(t, \|x\|)) = e^{-t}\gamma_1(\|x\|) \tag{10.108}$$

and

$$\gamma_2(\|v(t,x)\|) \geq \gamma_2(\beta_2(t, \|x\|)) = e^{-t}\gamma_2(\|x\|). \tag{10.109}$$

These inequalities and definitions (10.104) and (10.105) imply that

$$e^{-t}\gamma_2(\|x\|) \cdot k(\|x\|) \le \|v(t,x)\| \le e^{-t}\gamma_1(\|x\|) \cdot K(\|x\|) \qquad (10.110)$$

because $\|v(t,x)\| \le \|x\|$ by the Schwarz Lemma. Using now the fact that $\tilde{f}(x) = A^{-1}f(M_{-\tau}(x)) = \lim_{t\to\infty} e^t v(t,x)$, we get (10.103). Since $\tilde{f}(\mathcal{D}) = A^{-1}f(\mathcal{D})$ we also obtain (10.106) from the left-hand inequality of (10.103)⬚

Remark 10.9   *Observe that it follows by (10.92), (10.93) and (10.97) that $\omega_1(0) \le 1$ while $\omega_2(0) \ge 1$. If the limits $\lim_{s\to 0^+} \gamma_i'(s)$, $i = 1, 2$, exist and are positive, then $K(p)$ in (10.104) is finite and $k(p)$ in (10.105) is positive. In this case $w_i(0)$, $i = 1, 2$, must actually equal 1. Indeed, it follows by (10.99) that*

$$\gamma_i'(0^+) = \lim_{s\to 0^+} \frac{\gamma_i(s)}{s} = \lim_{s\to 0^+} \gamma_i'(s)$$

$$\lim_{s\to 0^+} \frac{\gamma_i(s)}{s} \cdot \lim_{s\to 0^+} \omega_i(s) = \gamma_i'(0^+) \cdot \omega_i(0), \qquad i = 1, 2. \qquad (10.111)$$

Hence $\omega_i(0) = 1$, $i = 1, 2$. In other words, in this case we must find functions w_i such that $w_i(0) = 1$, $i = 1, 2$.

Example 10.1   For each starlike mapping $f \in \mathrm{Hol}(\mathcal{D}, X)$, lower and upper admissible bounds $\omega_i(s)$, $i = 1, 2$, can be chosen as follows:

$$\omega_1(s) = \frac{1 - cs}{1 + cs} \quad \text{and} \quad \omega_2(s) = \frac{1 + cs}{1 - cs} \qquad (10.112)$$

for some $c \in [0, 1]$. If $0 \le c < 1$, then f is said to be strongly starlike. We have

$$\gamma_1(s) = \frac{cs}{(1 - cs)^2}, \quad \gamma_2(s) = \frac{cs}{(1 + cs)^2} \qquad (10.113)$$

and

$$K(p) = k(p) = \frac{1}{c}. \qquad (10.114)$$

So, we get the growth and covering estimates

$$\frac{\|x\|}{(1 + c\|x\|)^2} \le \|f(M_{-\tau}(x))\| \le \frac{\|x\|}{(1 - c\|x\|)^2}. \qquad (10.115)$$

In particular,

$$f(\mathcal{D}) \supset \mathcal{D}_{\frac{1}{(1+c)^2}}. \qquad (10.116)$$

Corollary 10.4 *Let X be a complex Banach space such that its open unit ball is a homogeneous domain. Then the image of every starlike mapping (with respect to an interior point) covers the ball of radius $\frac{1}{4}$ centered at the origin.*

Proof. Set $c = 1$ in (10.116).

Note that the characteristic functions of many classical examples of starlike mappings have lower and upper bounds of the form

$$\omega_{1,2}(s) = 1 \mp \lambda s. \tag{10.117}$$

In this case, solving the equations

$$\frac{\gamma'_{1,2}(s)}{\gamma_{1,2}(s)} = \frac{1}{s(1 \mp \lambda s)} \tag{10.118}$$

we get $\gamma_{1,2}(s) = \frac{s}{1 \mp \lambda s}$ and $K(p) = k(p) = 1$. Thus we obtain the improved estimates

$$\frac{\|x\|}{1 + \lambda\|x\|} \le \|f(x)\| \le \frac{\|x\|}{1 - \lambda\|x\|}. \tag{10.119}$$

$\square$

Now let us consider some different situations.

Example 10.2 Let $\mathcal{D} = \Delta$ be the open unit disk in $\mathbb{C}$, and let

$$f(z) = \frac{z}{1 - z}. \tag{10.120}$$

It is easy to see that f is a convex, hence a starlike function, $f(0) = 0$, and that its characteristic function is $g(z) = z - z^2$ (recall that $g(z) = f(z)/f'(z)$). In this case $\omega_i(s)$, $i = 1, 2$, can be chosen as $\omega_1(s) = 1 - s$ and $\omega_2(s) = 1 + s$.

Consequently, $\lambda = 1$ and

$$\frac{|z|}{1 + |z|} \le |f(z)| \le \frac{|z|}{1 - |z|}. \tag{10.121}$$

In particular, $f(\Delta) \supset \Delta_{\frac{1}{2}}$.

Example 10.3 Let $\mathcal{D}_2$ be the open unit ball in $\mathbb{C}^2$, $a \in \mathbb{C}$, and let $g_a : \mathcal{D}_2 \mapsto \mathbb{C}^2$ be defined by

$$g_a(z_1, z_2) := \left(z_1 - az_2^2, z_2\right). \tag{10.122}$$

We already know that this mapping is a generator if $\|a\| \leq \frac{3\sqrt{3}}{2}$ and that $\omega_{1,2}$ are of the form (10.117) with $\lambda = |a|\frac{2}{3\sqrt{3}}$. If now f_a is the starlike mapping corresponding to g_a, then we see that it satisfies the estimates (10.119).

Example 10.4 In a similar way as in Example 10.3 one can consider a more general case:

$$g_{a,b}(z_1, z_2) := \left(z_1 - az_2^2, \ z_2 - bz_1^2\right). \tag{10.123}$$

Now the question is, for which a and b the mapping $g_{a,b}$ is a semi-complete vector field on

$$\mathcal{D}_p = \left\{(z_1, z_2) \in \mathbb{C}^2 : |z_1|^p + |z_2|^p < 1\right\}. \tag{10.124}$$

We will give an answer for the cases $p = 1$ and $p = 3$ (for $p \neq 1,3$ the calculations are not so simple).

1. $\underline{p = 1}$. In this case a support functional z^* at $z \in \mathbb{C}^2$ is given by

$$\langle w, z^* \rangle = \left(w_1 \frac{|z_1|}{z_1} + w_2 \frac{|z_2|}{z_2}\right) (|z_1| + |z_2|) \tag{10.125}$$

and

$$\langle g_{a,b}(z), z^* \rangle = \left(|z_1| + |z_2| - az_2^2 \frac{|z_1|}{z_1} - bz_1^2 \frac{|z_2|}{z_2}\right) (|z_1| + |z_2|). \tag{10.126}$$

After some computations we get

$$\|z\|^2(1 - \lambda\|z\|) \leq \mathrm{Re}\langle g_{a,b}(z), z^* \rangle \leq \|z\|^2(1 + \lambda\|z\|), \tag{10.127}$$

where

$$\|z\| = |z_1| + |z_2|, \quad \lambda = \frac{|a|\,|b|}{|a| + |b|}. \tag{10.128}$$

Thus $g_{a,b}$ is a semi-complete vector field if $\frac{1}{|a|} + \frac{1}{|b|} \geq 1$. (In fact, this condition is also necessary.) Therefore the starlike mapping corresponding to $g_{a,b}$ satisfies the estimates (10.119). In particular,

$$\frac{\|z\|}{1 + \|z\|} \leq \|f_{2,2}(z)\| \leq \frac{\|z\|}{1 - \|z\|} \tag{10.129}$$

and

$$\frac{2\|z\|}{2 + \|z\|} \leq \|f_{1,1}(z)\| \leq \frac{2\|z\|}{2 - \|z\|}. \tag{10.130}$$

2. $\underline{p = 3}$. Now the support functional at $z \in \mathbb{C}^2$ is given by

$$\langle w, z^* \rangle = \left(w_1 \frac{|z_1|^3}{z_1} + w_2 \frac{|z_2|^3}{z_2} \right) \Big/ \left(|z_1|^3 + |z_2|^3 \right)^{\frac{1}{3}}. \quad (10.131)$$

Consequently,

$$\langle g_{a,b}(z), z^* \rangle = \frac{|z_1|^3 + |z_2|^3 - a z_2^2 \frac{|z_1|^3}{z_1} - b z_1^2 \frac{|z_2|^3}{z_2}}{\left(|z_1|^3 + |z_2|^3 \right)^{\frac{1}{3}}}. \quad (10.132)$$

It is easy to observe that

$$\|z\|^2 \left(1 - \frac{|a| + |b|}{2^{\frac{4}{3}}} \|z\| \right) \le \mathrm{Re}\langle g_{a,b}(z), z^* \rangle \le \|z\|^2 \left(1 + \frac{|a| + |b|}{2^{\frac{4}{3}}} \|z\| \right),$$

$$(10.133)$$

where $\|z\| = \left(|z_1|^3 + |z_2|^3 \right)^{\frac{1}{3}}$. Therefore the mapping $g_{a,b}$ is a semi-complete vector field if $|a| + |b| < 2^{\frac{4}{3}}$ (Again, this condition is also necessary.) The corresponding starlike mapping $f_{a,b}$ satisfies the estimate (10.119) with $\lambda = \frac{|a| + |b|}{2^{\frac{4}{3}}}$.

Example 10.5 Consider the mapping $f : \mathcal{D}_2 \mapsto \mathbb{C}^2$ defined by

$$f_1(z) := \frac{z_1 - z_2}{z_1 e^{z_1} - z_2 e^{z_2}} z_1 e^{z_1} \quad (10.134)$$

$$f_2(z) := \frac{z_1 - z_2}{z_1 e^{z_1} - z_2 e^{z_2}} z_2 e^{z_2}. \quad (10.135)$$

Calculations show that f is univalent on $\mathcal{D}_2$ and

$$g(z_1, z_2) := [f'(z_1, z_2)]^{-1} f(z_1, z_2) = (z_1(1 + z_2), z_2(1 + z_1))$$

$$(10.136)$$

(*cf.* [Chuaqui (1995)]).

Since $g : \mathcal{D}_2 \mapsto \mathbb{C}^2$ is a semi-complete vector field, it follows that f is starlike. In addition, in this case lower and upper bounds ω_1 and ω_2 can be chosen as in (10.117) with $\lambda = \frac{1}{\sqrt{2}}$. Hence

$$\frac{\|z\|}{1 + \frac{1}{\sqrt{2}} \|z\|} \le \|f(z)\| \le \frac{\|z\|}{1 - \frac{1}{\sqrt{2}} \|z\|} \quad (10.137)$$

and $f(\mathcal{D}_2)$ covers an open ball of radius $\frac{\sqrt{2}}{\sqrt{2}+1}$.

10.5 Differential Equations for Starlike and Spirallike Mappings in $\mathcal{H} = \mathbb{C}^n$

In what follows we consider the complex Euclidean space $\mathcal{H} = \mathbb{C}^n$. In this case we write

$$\langle z, w \rangle = \sum_{k=1}^{n} z_k \overline{w_k} \qquad (10.138)$$

for vectors $z = (z_1, \ldots, z_n)$ and $w = (w_1, \ldots, w_n)$ in $\mathcal{H}$. Thus

$$\mathbb{B} = \left\{ z \in \mathcal{H} : \|z\| = \sqrt{|z_1|^2 + \cdots + |z_n|^2} < 1 \right\} \qquad (10.139)$$

is the open unit ball in $\mathcal{H}$.

For a linear operator $\Gamma \in L(\mathcal{H}, \mathcal{H})$ we will write $\mathrm{Re}\,\Gamma > 0$ if $\mathrm{Re}\,\lambda > 0$ for all eigenvalues λ of Γ. Note that this is equivalent to $\mathrm{Re}\langle \Gamma x, x \rangle \geq \varepsilon \|x\|^2$, $\varepsilon > 0$, $x \in \mathcal{H}$.

As above we denote by I the identity operator on $\mathcal{H}$.

Recall that a univalent mapping $h \in \mathrm{Hol}(\mathbb{B}, \mathcal{H})$ is said to be spirallike on $\mathbb{B}$ if there exists a bounded linear operator $\Gamma : \mathcal{H} \mapsto \mathcal{H}$ with $\mathrm{Re}\,\Gamma > 0$ such that for each $z \in B$ and $t \geq 0$,

$$e^{-t\Gamma} h(z) \in h(\mathbb{B}). \qquad (10.140)$$

If this inclusion holds with $\Gamma = I$, the mapping h is called starlike.

Furthermore, if $h(z) = 0$ for some point $z \in B$, the mapping h is said to be spirallike (starlike) with respect to an interior point. Otherwise, $0 \in \partial h(\mathbb{B})$ and the mapping h is said to be spirallike (starlike) with respect to a boundary point.

In this section we study a system of partial differential equations in $\mathbb{B} \subset \mathbb{C}^n$ connected to the classes of spirallike and starlike mappings on B with respect to a boundary point.

It was shown in [Elin *et al.* (2000)] that such mappings are solutions of the differential equation

$$\Gamma h(z) = h'(z) g(z), \qquad (10.141)$$

where $\Gamma \in L(\mathcal{H}, \mathcal{H})$ with $\mathrm{Re}\,\Gamma > 0$ and f is an (infinitesimal) generator on $\mathbb{B}$ without null point. By $h'(z)$ we denote the linear operator on $\mathcal{H} = \mathbb{C}^n$

defined by the Jacobi matrix:

$$h'(z) = \begin{pmatrix} \frac{\partial h_1(z)}{\partial z_1} & \cdots & \frac{\partial h_1(z)}{\partial z_n} \\ \vdots & \vdots & \vdots \\ \frac{\partial h_n(z)}{\partial z_1} & \cdots & \frac{\partial h_n(z)}{\partial z_n} \end{pmatrix}, \tag{10.142}$$

at the point $z \in \mathbb{B}$.

Here we concentrate on generators of one-dimensional type (o.d.t.). In this case the above equation can be written in the following form:

$$\Gamma h(z) = (1 - \langle z, \tau \rangle) p(z) h'(z)(z - \tau), \tag{10.143}$$

where $\tau \in \partial \mathbb{B}$, $p \in \mathrm{Hol}(\mathbb{B}, \mathbb{C})$ with $\mathrm{Re}\, p(z) > 0$, and $\Gamma \in L(\mathcal{H})$ is a bounded linear operator with $\mathrm{Re}\,\Gamma > 0$.

In particular, we will discuss in the sequel the existence, uniqueness and univalence of solutions of (10.143).

By $C : \mathbb{B} \mapsto \mathcal{H}$ we denote the Cayley transform of the unit ball:

$$C(z) = \frac{1}{1 - \langle z, \tau \rangle} (z + \tau). \tag{10.144}$$

By Π we denote Siegel's domain in $\mathcal{H}$:

$$\Pi = \left\{ w \in \mathcal{H} : \mathrm{Re}\langle w, \tau \rangle + |\langle w, \tau \rangle|^2 > \|w\|^2 \right\}. \tag{10.145}$$

Sometimes it is convenient to use the orthogonal projections P and Q defined by

$$Pz = \langle z, \tau \rangle \tau \quad \text{and} \quad Qz = z - Pz, \tag{10.146}$$

and "partial coordinates":

$$z_1 = \langle z, \tau \rangle \quad \text{and} \quad z_2 = Qz. \tag{10.147}$$

So, we write $z = (z_1, z_2)$ for $z = z_1 \tau + z_2 \in \mathcal{H}$. Note that z_1 is a scalar while z_2 is a 'vector coordinate'. We have

$$C(z) = \left(\frac{1 + z_1}{1 - z_1}, \frac{1}{1 - z_1} z_2 \right) \tag{10.148}$$

and $\Pi = \left\{ w \in H : \mathrm{Re}\, w_1 > \|w_2\|^2 \text{ where } w_1 = \langle w, \tau \rangle, w_2 = Qw \right\}$.

We also note that for any mapping F holomorphic in a domain $\Omega \subset \mathcal{H}$,

$$\frac{\partial F(z)}{\partial z_1} = F'(z)\tau, \tag{10.149}$$

where $F'(z)$ denotes the Fréchet derivative of F at z.

The following more or less known fact will be useful in the sequel (*cf.* [Rudin (1980)]).

Lemma 10.2 *The Cayley transform maps $\mathbb{B}$ biholomorphically onto Π and its inverse mapping is defined by*

$$z = C^{-1}(w) = \frac{2}{1 + w_1} w - \tau. \tag{10.150}$$

Moreover,

$$(1 - z_1)C'(z)(z - \tau) = -2\tau \tag{10.151}$$

and

$$\frac{\partial C^{-1}(w)}{\partial w_1} = \frac{1}{1 + w_1}\tau - \frac{1}{(1 + w_1)^2}w. \tag{10.152}$$

Lemma 10.3 *A mapping $h \in \mathrm{Hol}(\mathbb{B}, \mathcal{H})$ is a solution of equation* (10.143) *if and only if the mapping $\tilde{h} \in \mathrm{Hol}(\Pi, \mathcal{H})$ defined by $\tilde{h}(w) = h \circ C^{-1}(w)$ is a solution of the differential equation*

$$\frac{\partial \tilde{h}(w)}{\partial w_1} + \tilde{p}(w)\Gamma\tilde{h}(w) = 0 \tag{10.153}$$

for some $\tilde{p} \in \mathrm{Hol}(\Pi, \mathbb{C})$ with $\mathrm{Re}\,\tilde{p}(w) > 0$.

The functions $p \in \mathrm{Hol}(\mathbb{B}, \mathbb{C})$ in (10.143) *and $\tilde{p} \in \mathrm{Hol}(\Pi, \mathbb{C})$ in* (10.153) *are connected by the formula*

$$\tilde{p}(C(z)) \cdot p(z) = \frac{1}{2}. \tag{10.154}$$

Proof. Verifying (10.143) (respectively, (10.153)) with $h = \tilde{h} \circ C$ (respectively, $\tilde{h} = h \circ C^{-1}$) and $p = \frac{1}{2\tilde{p} \circ C}$ (respectively, $\tilde{p} = \frac{1}{2p \circ C^{-1}}$), we get our assertion.

$\square$

Using now Lemma 10.2 one can study the solvability and properties of equation (10.153) instead of equation (10.143).

To proceed, we need the following lemma.

Lemma 10.4

(a) *([de Fabritiis (1994), Proposition 3.3]) For every function $b \in$ Hol$(\Pi, \mathbb{C})$ there exists $k \in$ Hol$(\Pi, \mathbb{C})$ such that*

$$\frac{\partial k(w)}{\partial w_1} = b(w). \tag{10.155}$$

(b) *For any two functions $k_1, k_2 \in$ Hol$(\Pi, \mathbb{C})$, both satisfying (10.155), the difference $\kappa(w) = k_1(w) - k_2(w)$ does not depend on w_1 and so is an entire function on $\mathcal{H}$, i.e.., $\kappa \in$ Hol$(\mathcal{H}, \mathbb{C})$.*

(c) *The difference $k(w_1, w_2) - k(1, w_2)$ does not depend on the choice of a function $k \in$ Hol$(\Pi, \mathbb{C})$ satisfying equation (10.155).*

Proof. Assertion (a) was proved in [de Fabritiis (1994)].

Assume now that k_1 and k_2 are two solutions of (10.155), i.e., $\frac{\partial k_i(w)}{\partial w_1} = b(w)$, $i = 1, 2$. Then for the function $\kappa(w) = k_1(w) - k_2(w)$ we have $\frac{\partial \kappa(w)}{\partial w_1} = 0$. So, this function is holomorphic in Π and does not depend on $w_1 = \langle w, \tau \rangle$. Hence, κ is holomorphic in $\mathcal{H}$. Indeed, for each $w_2 \in Q\mathcal{H}$, one can find $w_1 \in \mathbb{C}$ such that Re$w_1 > \|w_2\|^2$, i.e., $w = (w_1, w_2)$ belongs to Π. Since $\kappa \in$ Hol$(\Pi, \mathbb{C})$ and does not depend on w_1, it follows that κ belongs to Hol$(Q\mathcal{H}, \mathbb{C})$, hence $\kappa \in$ Hol$(\mathcal{H}, \mathbb{C})$, too.

To prove assertion (c), we can now write $k_1 = k_2 + \kappa$ with $\frac{\partial \kappa(w)}{\partial w_1} = 0$. Therefore,

$$\begin{aligned}
k_1(w_1, w_2) - k_1(1, w_2) &= \left[k_2(w_1, w_2) + \kappa(w_1, w_2)\right] - \left[k_2(1, w_2) + \kappa(1, w_2)\right] \\
&= \left[k_2(w_1, w_2) + k_2(1, w_2)\right] + \left[\kappa(w_1, w_2) - \kappa(1, w_2)\right] \\
&= \left[k_2(w_1, w_2) - k_2(1, w_2)\right]. \tag{10.156}
\end{aligned}$$

$\square$

Further, for any function $k \in$ Hol$(\mathcal{D}, \mathbb{C})$, $\mathcal{D} \subset \mathcal{H}$, and any bounded linear operator $\Gamma \in L(\mathcal{H})$, one can define a holomorphic operator–valued mapping $e^{\Gamma k(w)} \in$ Hol$(\mathcal{D}, L(\mathcal{H}))$ as follows:

$$e^{\Gamma k(w)} a = \sum_{n=0}^{\infty} \frac{k^n(w)}{n!} \Gamma^n a, \quad a \in \mathcal{H}. \tag{10.157}$$

Let now $k \in$ Hol$(\Pi, \mathbb{C})$ satisfy (10.155) with $b(w) = \tilde{p}(w)$ and Re$\Gamma > 0$ as in equations (10.143) and (10.153).

For $\tilde{h} \in$ Hol$(\Pi, \mathcal{H})$ we define $g \in$ Hol$(\Pi, \mathcal{H})$ by

$$g(w) = e^{\Gamma k(w)} \tilde{h}(w). \tag{10.158}$$

Lemma 10.5 *A mapping $\tilde{h} \in \mathrm{Hol}(\Pi, \mathcal{H})$ is a solution of equation (10.153) if and only if the mapping g defined by (10.158) does not depend on w_1 and, consequently, $g \in \mathrm{Hol}(\mathcal{H}, \mathcal{H})$.*

Proof. Differentiating the mapping $\tilde{h} \in \mathrm{Hol}(\Pi, \mathcal{H})$ defined by $\tilde{h}(w) = e^{-\Gamma k(w)} g(w)$, we have

$$
\begin{aligned}
\tilde{h}(w+a) - \tilde{h}(w) &= e^{-\Gamma k(w)} \big\{ e^{-\Gamma(k(w+a)-k(w))} g(w+a) - g(w) \big\} \\
&= e^{-\Gamma k(w)} \big\{ \big[e^{-\Gamma(k(w+a)-k(w))} - I \big] g(w+a) + [g(w+a) - g(w)] \big\}.
\end{aligned}
$$

$$(10.159)$$

Hence, we conclude that

$$
\tilde{h}'(w)a = e^{-\Gamma k(w)} \big\{ g'(w)a - (k'(w)a)\Gamma g(w) \big\}. \tag{10.160}
$$

Consequently,

$$
\tilde{h}'(w)\tau = e^{-\Gamma k(w)} \left\{ \frac{\partial g(w)}{\partial w_1} - \tilde{p}(w)\Gamma g(w) \right\}. \tag{10.161}
$$

Substituting this expression in (10.153), we obtain that

$$
\frac{\partial g(w)}{\partial w_1} = 0. \tag{10.162}
$$

As in the proof of Lemma 10.4 one can conclude also that $g \in \mathrm{Hol}(\mathcal{H}, \mathcal{H})$. Conversely, let g be defined by (10.158) and assume that $g'(w)\tau = 0$. Then

$$
\begin{aligned}
g(w+a) &- g(w) \\
&= e^{\Gamma k(w)} \big\{ \big[e^{\Gamma(k(w+a)-k(w))} - I \big] \tilde{h}(w+a) + [\tilde{h}(w+a) - \tilde{h}(w)] \big\}
\end{aligned}
$$

$$(10.163)$$

and

$$
g'(w)a = e^{\Gamma k(w)} \big\{ \tilde{h}'(w)a + (k'(w)a)\Gamma \tilde{h}(w) \big\}. \tag{10.164}
$$

By our assumption for $a = \tau$ we get the equality

$$
e^{\Gamma k(w)} \big(\tilde{h}'(w)\tau + \tilde{p}(w)\Gamma \tilde{h}(w) \big) = 0, \tag{10.165}
$$

which implies our assertion. $\qquad\square$

To formulate our next results we introduce the two following classes:

$$K := \left\{ k \in \mathrm{Hol}(\Pi, \mathbb{C}) : \frac{\partial k(w)}{\partial w_1} = \tilde{p}(w), \right.$$

$$\left. \text{where} \quad \tilde{p}(w) = \frac{1}{2p(C^{-1}(w))}, \ \mathrm{Re}\,\tilde{p}(w) > 0 \right\} \qquad (10.166)$$

and $\Phi := \left\{ \phi \in \mathrm{Hol}(QB, \mathcal{H}) : e^{\Gamma k(1,\cdot)}\phi \in \mathrm{Hol}(Q\mathcal{H}, \mathcal{H}) \ \text{for some} \ k \in K \right\}$.

Theorem 10.8

(a) *Let* $h \in \mathrm{Hol}(\mathbb{B}, \mathcal{H})$ *be a solution of* (10.143):

$$\Gamma h(z) = (1 - z_1)p(z)h'(z)(z - \tau). \qquad (10.167)$$

Then $h(0, \cdot) \in \Phi$.

(b) *Let* $\phi \in \Phi$. *Then equation* (10.143) *has a unique solution* $h \in \mathrm{Hol}(\mathbb{B}, \mathcal{H})$ *which satisfies the initial data*

$$h(0, \zeta) = \phi(\zeta). \qquad (10.168)$$

Moreover, this solution can be represented by the formula

$$h(z) = e^{-\Gamma k(C(z))}\left[e^{\Gamma k(1,\zeta)}\phi(\zeta) \right]\Big|_{\zeta = \frac{z_2}{1 - z_1}} \qquad (10.169)$$

and does not depend on the choice of $k \in K$.

(c) *If* $\phi \in \Phi$, *then the mapping* h, *defined by* (10.169), *is locally univalent on* $\mathbb{B}$ *if and only if,*

$$\det\!\big(\Gamma\phi(\zeta)\ \phi'(\zeta)\big) \neq 0. \qquad (10.170)$$

Proof. (a) Let h be a solution of (10.143). Then by Lemma 10.3 the mapping $\tilde{h} = h \circ C^{-1}$ is a solution of (10.153). It is clear that

$$h(0, \zeta) = \tilde{h}(1, \zeta). \qquad (10.171)$$

Define g by (10.158): $g(w) = e^{\Gamma k(w)}\tilde{h}(w)$, where $k \in K$. By Lemma 10.5, the mapping g does not depend on w_1, i.e., $g'(w)\tau = 0$. Hence, actually, $g \in \mathrm{Hol}(Q\mathcal{H}, \mathcal{H})$. Then

$$h(0, \zeta) = \tilde{h}(1, \zeta) = e^{-\Gamma k(1,\zeta)}g(\zeta) \in \Phi. \qquad (10.172)$$

(b) Let $\phi \in \Phi$, i.e., for some $k \in K$,

$$\phi(\zeta) = e^{-\Gamma k(1,\zeta)}g(\zeta), \qquad (10.173)$$

and $g \in \mathrm{Hol}(Q\mathcal{H}, \mathcal{H})$. By Lemma 10.5 the mapping $\tilde{h}$, where

$$\tilde{h}(w) = e^{-\Gamma k(w)} g(w_2), \quad w_2 = Qw, \tag{10.174}$$

is a solution of (10.153), and the mapping $h = \tilde{h} \circ C$ solves (10.143).

Moreover,

$$h(0, \zeta) = \tilde{h} \circ C(0, \zeta) = \tilde{h}(1, \zeta) = e^{-\Gamma k(1,\zeta)} g(\zeta) = \phi(\zeta). \tag{10.175}$$

Let $h_*(0, \zeta) = \phi(\zeta)$. As we have seen, for some $k_* \in K$ the mapping

$$g_*(w) = e^{\Gamma k_*(w)} h_*(C^{-1}(w)) \tag{10.176}$$

does not depend on w_1 and g_*, and, in fact, belongs to $\mathrm{Hol}(Q\mathcal{H}, \mathcal{H})$. Furthermore, by Lemma 10.4 we already know that

$$k_*(w_1, w_2) - k_*(1, w_2) = k(w_1, w_2) - k(1, w_2), \tag{10.177}$$

where $k \in K$. Then

$$
\begin{aligned}
h_*(w) &= e^{-\Gamma k_* * w)} g_*(w)\big|_{w=C(z)} \\
&= e^{-\Gamma(k_*(w_1,w_2) - k_*(1,w_2)) + \Gamma k_*(1,w_2)} g_*(w_2)\big|_{w=C(x)} \\
&= \left[e^{-\Gamma(k(w_1,w_2) - k(1,w_2))} \left(e^{\Gamma k_*(1,\zeta)} g_*(\zeta) \right) \right]\Big|_{w=C(z),\,\zeta=\frac{1}{1-z_1} z_2} \\
&= e^{-\Gamma k(C(z))} \left[e^{\Gamma k(1,\zeta)} \phi(\zeta) \right]\Big|_{\zeta=\frac{1}{1-z_1} z_2}. \tag{10.178}
\end{aligned}
$$

Consequently, a solution of (10.143) with (10.168) is unique and does not depend on the choice of a function $k \in K$.

(c) Note that representation (10.169) can be rewritten in the form $h = \tilde{h} \circ C$, where

$$\tilde{h}(w_1, \zeta) = e^{\Gamma(k(1,\zeta) - k(w_1,\zeta))} \phi(\zeta), \quad (w_1, \zeta) \in \Pi, \tag{10.179}$$

and C is as above, the Cayley transform of the unit ball $\mathbb{B}$. Since the Cayley transform is a biholomorphic mapping (see Lemma 10.2), it is enough to show that the mapping $\tilde{h}$ defined by (10.179) is locally univalent if and only if (10.170) holds. To do this, we just calculate the Jacobian $J(\tilde{h})$ of the mapping $\tilde{h}$:

$$J(\tilde{h}) = \det\left(\frac{d\tilde{h}(w_1, \zeta)}{d(w_1, \zeta)} \right) = \left(\frac{\partial \tilde{h}(w_1, \zeta)}{\partial w_1} \; \frac{\partial \tilde{h}(w_1, \zeta)}{\partial \zeta} \right), \tag{10.180}$$

where

$$\frac{\partial \tilde{h}(w_1, \zeta)}{\partial w_1} = e^{\Gamma(k(1,\zeta)-k(w_1,\zeta))}(-\tilde{p}(w_1,\zeta))\Gamma\phi(\zeta) \qquad (10.181)$$

and

$$\frac{\partial \tilde{h}(w_1, \zeta)}{\partial \zeta} = e^{\Gamma(k(1,\zeta)-k(w_1,\zeta))}\left(\frac{\partial(k(1,\zeta)-k(w_1,\zeta))}{\partial \zeta}\Gamma\phi(\zeta) + \frac{\partial\phi(\zeta)}{\partial \zeta}\right).$$
$$(10.182)$$

So,

$$J(\tilde{h}) = -\tilde{p}(w_1,\zeta)\det\left(e^{\Gamma(k(1,\zeta)-k(w_1,\zeta))}\right)\det\left(\Gamma\phi(\zeta)\frac{\partial\phi(\zeta)}{\partial\zeta}\right) \quad (10.183)$$

(here we understand the operator $e^{\Gamma(k(1,\zeta)-k(w_1,\zeta)}$ as a matrix). This formula implies our claim. $\qquad\square$

Example 10.6 Consider the function $p \in \mathrm{Hol}(\mathbb{B}, \mathbb{C})$ defined as follows:

$$p(z) = \frac{1 + z_1^2}{1 - z_1^2}. \qquad (10.184)$$

Then using (10.154), we find the function $\tilde{p}(w) = \frac{w_1}{1+w_1^2}$. So solutions of equation (10.143) can be found by formula (10.168):

$$h(z) = e^{-\frac{1}{2}\Gamma\log\left(\frac{1+w_1^2}{2}\right)}\phi(\zeta)\Big|_{w_1 = \frac{1+z_1}{1-z_1}, \zeta = \frac{1}{1-z_1}z_2}, \qquad (10.185)$$

where $\phi \in \Phi$.

Take the operator

$$\Gamma = \begin{pmatrix} \mu_1 & 0 & \dots & 0 \\ 0 & \mu_2 & \dots & 0 \\ \vdots & \vdots & \ddots & \vdots \\ 0 & 0 & \dots & \mu_n \end{pmatrix}, \qquad (10.186)$$

where $\mu_i \in \mathbb{C}$, $i = 1, \dots, n$, and suppose that the mapping ϕ satisfies the following condition:

$$Q\phi(\zeta) = -\phi_1(\zeta)\zeta, \quad \text{where} \quad \phi_1 = \langle \phi, \tau \rangle \in \mathrm{Hol}(Q\mathbb{B}, \mathbb{C}). \quad (10.187)$$

In this case, the coordinates of the solution h are seen to be

$$h_1(z) = \frac{(1 - z_1)^{\mu_1 - 1}}{(1 + z_1^2)^{\frac{\mu_1}{2}}} \cdot \phi_1\left(\frac{1}{1 - z_1}Qz\right)(1 - z_1), \qquad (10.188)$$

and

$$h_i(z) = -\frac{(1-z_1)^{\mu_i-1}}{(1+z_1^2)^{\frac{\mu_i}{2}}} \cdot \phi_1\left(\frac{1}{1-z_1}Qz\right)z_i, \quad i = 2,\ldots,n. \quad (10.189)$$

It is easy to verify (by using Theorem 10.8 (c)) that this mapping is locally univalent if and only if the function ϕ_1 does not vanish on $Q\mathbb{B}$. This fact will be also useful in the sequel.

From the point of view of Theorem 10.8, the question of the (global) univalence of solutions remains open. Below we will prove the existence of a globally univalent solution for the important case in which $\Gamma = \beta I$, where $\beta = \uparrow(f_\tau)'(\tau) := \lim_{r\to 1^-}((f_\tau)'(r\tau))$. We already know that β is a real nonnegative number. In other words, in the remaining part of this section we will deal with the differential equation

$$\beta h(z) = h'(z)f(z), \tag{10.190}$$

where

$$f(z) = (1 - \langle z,\tau\rangle)p(z)(z-\tau), \ \tau \in \partial\mathbb{B}, \ \operatorname{Re}p(z) \geq 0, \ z \in \mathbb{B}. \tag{10.191}$$

Observe also that a univalent solution h (if it exists) of this equation is a starlike mapping on $\mathbb{B}$ with respect to the boundary point τ, because $h(\tau) = \lim_{r\to 1^-} h(r\tau) = 0$ (see, for example, [Elin *et al.* (2000)]).

Recall that a curve $\Lambda : [0,1) \mapsto B$ is said to be asymptotically normal at a boundary point $\tau \in \partial\mathbb{B}$ if

$$\lim_{s\to 1^-} \frac{\|\Lambda(s) - \lambda(s)\tau\|^2}{1 - |\lambda(s)|^2} = 0 \tag{10.192}$$

and

$$\frac{|\lambda(s) - 1|}{1 - |\lambda(s)|} \leq M \leq \infty, \quad 0 \leq s < 1, \tag{10.193}$$

where $\lambda(s) = \langle \Lambda(s),\tau\rangle$ (see, for example, [Cowen and MacCluer (1995)] or [Rudin (1980)]).

A holomorphic function $\psi : \mathbb{B} \mapsto \mathbb{C}$ is said to be differentiable at a boundary point $\tau \in \partial\mathbb{B}$ if for some number $c \in \mathbb{C}$ and for every curve $\Lambda : [0,1) \mapsto \mathbb{B}$ asymptotically normal at τ, the limit

$$\lim_{z\to\tau, z\in\Lambda} \frac{\psi(z) - c}{\langle z - \tau,\tau\rangle} =: \angle\psi'(\tau) \tag{10.194}$$

exists finitely and does not depend on Λ. This limit is called the *angular derivative of the function* ψ at the point $\tau \in \partial\mathbb{B}$.

We say that an o.d.t. generator (10.191) with $\tau \in \partial\mathbb{B}$ is of a *strongly hyperbolic type*, if the angular derivative

$$\angle(f_\tau)'(\tau) := \lim_{z \to \tau} \frac{\langle f(z), \tau \rangle}{\langle z, \tau \rangle - 1} \left(= \angle \lim_{z \to \tau}(1 - \langle z, \tau \rangle)p(z) \right) = \beta \quad (10.195)$$

exists with $\beta > 0$, and the semigroup $\mathcal{S} = \{F_t\}_{t \geq 0}$ generated by f converges to τ nontangentially (see [Cowen and MacCluer (1995)]).

Theorem 10.9 *Let f be a generator of a strongly hyperbolic type with $\angle(f_\tau)'(\tau) = \beta$. Then the equation (10.190):*

$$\beta h(z) = h'(z)f(z) \quad (10.196)$$

has a globally univalent solution $h \in \mathrm{Hol}(\mathbb{B}, \mathbb{C}^n)$ which is a starlike mapping of one-dimensional type.

To prove this theorem we need the following lemma.

Lemma 10.6 *Let $f \in \mathcal{G}[\tau]$, $\tau \in \partial B$, be a generator of a strongly hyperbolic type with the angular derivative $\angle(f_\tau)'(\tau) = \beta$. Assume that there exists a real function $c(t) : \mathbb{R}^+ \mapsto \mathbb{R}^+$, $c(t) \neq 0$, and a sequence $t_k \in \mathbb{R}^+$, $k = 1, 2, \ldots$, $t_k \to \infty$, such that the locally uniform limit*

$$h(z) = \lim_{k \to \infty} c(t_k)(F_{t_k}(z) - \tau) \quad (10.197)$$

exists. Then h satisfies the equation

$$\beta h(z) = h'(z)f(z). \quad (10.198)$$

Proof. First we note that by the Weierstraß Theorem

$$h'(z) = \lim_{k \to \infty} c(t_k)\frac{\partial F_{t_k}(z)}{\partial z}. \quad (10.199)$$

On the other hand, it is well-known that F_t, $t \geq 0$, satisfies the differential equation

$$\frac{\partial F_{t_k}(z)}{\partial t} + \frac{\partial F_{t_k}(z)}{\partial z}f(z) = 0 \quad (10.200)$$

(see, for example, [Reich and Shoikhet (1996)]).

In addition, it follows from the Cauchy problem that

$$\frac{\partial F_{t_k}(z)}{\partial z}f(z) = f(F_t(z)). \quad (10.201)$$

Thus, for given $c(t)$ and t_k, we obtain

$$c(t_k)\frac{\partial F_{t_k}(z)}{\partial z}\,f(z) = c(t_k)\big(1 - \langle F_{t_k}(z),\tau\rangle\big)p\,(F_{t_k}(z))(F_{t_k}(z) - \tau).$$

$$(10.202)$$

Noting that

$$\lim_{k\to\infty}\big(1 - \langle F_{t_k}(z),\tau\rangle\big)p\,(F_{t_k}(z)) = \lim_{k\to\infty}\frac{\langle f(F_{t_k}(z)),\tau\rangle}{\langle F_{t_k}(z),\tau\rangle - 1} = \beta, \quad (10.203)$$

and letting k tend to infinity in (10.202), we get

$$\beta h(z) = h'(z)f(z), \qquad\qquad (10.204)$$

and we are done. $\qquad\qquad\qquad\qquad\qquad\qquad\qquad\qquad\qquad\qquad\square$

Proof of Theorem 10.9 Let $\mathcal{S} = \{F_t\}_{t\geq 0}$ be the semigroup generated by f. Define $h_t \in \mathrm{Hol}(\mathbb{B},\mathcal{H})$, $t \geq 0$, by

$$h_t(z) = \frac{F_t(z) - \tau}{\langle F_t(0) - \tau,\tau\rangle}, \quad t \geq 0. \qquad\qquad (10.205)$$

Consider $\langle h_t(z),\tau\rangle = \frac{\langle F_t(z)-\tau,\tau\rangle}{\langle F_t(0)-\tau,\tau\rangle}$. For any $t \geq 0$ and for all $z \in \mathbb{B}$, the numerator and the denominator of this fraction take values in the left half-plane $\mathbb{C}_- = \{\xi \in \mathbb{C} : \mathrm{Re}\,\xi < 0\}$ only. Hence, $\arg\langle h_t(z),\tau\rangle < \pi$, i.e., $\langle h_t(z),\tau\rangle$ is never a real negative number. Therefore, the family $\{h_t\}_{t\geq 0}$ is normal (see, for example, [Montel (1933)]). Therefore there exists a sequence $t_k \to \infty$ such that $\{h_{t_k}\}$ locally uniformly converges either to infinity or to a holomorphic mapping $h \in \mathrm{Hol}(\mathbb{B},\mathcal{H})$.

But it follows from Theorem 10.8 that $QF_t(0) = 0$, hence, $h_t(0) = \tau$ for all $t \geq 0$. Thus, for some sequence $\{t_k\}$, $t_k \to \infty$, there exists the locally uniform limit

$$h(z) = \lim_{k\to\infty} h_{t_k}(z) \in \mathrm{Hol}(\mathbb{B},\mathcal{H}). \qquad\qquad (10.206)$$

Setting $c(t) = \frac{1}{\langle F_t(0)-\tau,\tau\rangle}$, we obtain by Lemma 10.6 that the mapping h defined by (10.206) satisfies the differential equation

$$\beta h(z) = h'(z)f(z). \qquad\qquad (10.207)$$

Next we prove that the mapping h is locally univalent. Indeed, by Theorem 10.8,

$$Qh(z) = \lim_{k \to \infty} \frac{1 - \langle F_{t_k}(z), \tau \rangle}{(\langle F_{t_k}(0), \tau \rangle - 1)(1 - \langle z, \tau \rangle)} \, Qz = \frac{-\langle h(z), \tau \rangle}{(1 - \langle z, \tau \rangle)} \, Qz.$$

$$(10.208)$$

Hence the restriction $\phi = h\big|_{Q\mathbb{B}}$ (which, by Theorem 10.8 (a), belongs to Φ) must satisfy the same condition:

$$Q\phi(\zeta) = \frac{-\langle \phi(\zeta), \tau \rangle}{(1 - \langle \zeta, \tau \rangle)} \, Q\zeta = -\langle \phi(\zeta), \tau \rangle \zeta. \qquad (10.209)$$

Since in our assumptions $\Gamma = \beta I$, we have

$$\det\big(\Gamma \phi(\zeta) \, \phi'(\zeta)\big) = \beta \det\big(\phi(\zeta) \, \phi'(\zeta)\big) = \beta(-1)^{n-1} \cdot \langle \phi(\zeta), \tau \rangle^n. \qquad (10.210)$$

Thus, by Theorem 10.8 (c), we have to show that the function $\langle \phi(\zeta), \tau \rangle$ does not vanish in $\mathbb{B}$. Suppose that there exists a point $\zeta_0 \in Q\mathbb{B}$ such that $\langle \phi(\zeta_0), \tau \rangle = 0$. By g and g_k we denote the functions

$$g(\lambda) = \left\langle \phi\left(\lambda \frac{\zeta_0}{\|\zeta_0\|}\right), \tau \right\rangle \qquad (10.211)$$

and

$$g_k(\lambda) = \left\langle h_{t_k}\left(0, \lambda \frac{\zeta_0}{\|\zeta_0\|}\right), \tau \right\rangle = \frac{\left\langle F_{t_k}\left(0, \lambda \frac{\zeta_0}{\|\zeta_0\|}\right) - \tau, \tau \right\rangle}{\langle F_{t_k}(0) - \tau, \tau \rangle}, \qquad (10.212)$$

defined on the open unit disk Δ of the complex plane $\mathbb{C}$.

By the construction we see that all of the functions g and g_k are holomorphic on Δ and $g(\lambda_0) = 0$, where $\lambda_0 = \|\zeta_0\|$ and $g = \lim_{k \to \infty} g_k$. Since $g_k(0) = 1$, we have $g(0) = 1$, so g is not identically zero. Consequently, its null points are isolated. Hence for some circle $\Lambda = \{\lambda : |\lambda - \lambda_0| = \delta\}$,

$$\min_{\Lambda} |g(\lambda)| = \varepsilon > 0. \qquad (10.213)$$

Choosing k big enough, we get

$$\max_{\Lambda} |g(\lambda) - g_k(\lambda)| < \varepsilon. \qquad (10.214)$$

It follows from the Rouché Theorem that, for such k, the function g_k has at least one null point $\lambda_1 \in \Delta$. By the definition of g_k, we see that

$$\left\langle F_{t_k}\left(0, \lambda_1 \frac{\zeta_0}{\|\zeta_0\|}\right), \tau \right\rangle = 1. \tag{10.215}$$

But this equality is impossible because all the elements F_t, $t \geq 0$, of the semigroup S are self-mappings of $\mathbb{B}$. This contradiction proves (by using Theorem 10.8 (c)) that the mapping h must be locally univalent on $\mathbb{B}$.

Now using the multidimensional analog of the Rouché Theorem (see, for example, [Aizenberg and Yuzhakov (1983)]) and the fact that each F_t, $t > 0$ is globally univalent on $\mathbb{B}$ (see, for example, [Abate (1992)] and [Gurganus (1975)]) we are able to prove that h is, in fact, globally univalent.

Suppose that the mapping h takes the value w_0 at some point z_0. Because $h - w_1$ is locally univalent, it has isolated null points only.

Let $\{r_m\}$ be a sequence of positive numbers, $r_m \to 1^-$, such that on all the spheres $S_m := \{z : \|z\| = r_m\}$ the mapping $h - w_0$ is different from 0. By compactness arguments, for any fixed number m we have

$$\min_{S_m} \|h(z) - w_0\| = \varepsilon_m > 0. \tag{10.216}$$

At the same time, for some k big enough we have

$$\max_{S_m} \|h(z) - h_{t_k}(z)\| < \varepsilon_m. \tag{10.217}$$

By the multidimensional version of the Rouché Theorem (see, for example, [Aizenberg and Yuzhakov (1983)]), the mappings $h - w_0$ and $h_{t_k} - w_0$ have the same number of null points. The first of them vanishes at least at the point z_0, while the second one is globally univalent, hence cannot vanish at more than one point in $\mathbb{B}$. Therefore, $h - w_0$ has at most one null point inside the sphere S_m. Since m is arbitrary and $r_m \to 1$, this mapping is globally univalent.

Finally, we need to prove that h is of o.d.t.

To this end, it is sufficient to show that all the mapping h_t are of the form (10.143). Using Theorem 10.8 one calculates

$$h_t(z) = \frac{1 - \langle F_t(z), \tau \rangle}{\langle F_t(0) - \tau, \tau \rangle}\left(\frac{Qz}{1 - \langle z, \tau \rangle} - \tau\right)$$

$$= (1 - \langle z, \tau \rangle)\frac{1 - \langle F_t(z), \tau \rangle}{(\langle F_t(0), \tau \rangle - 1)(1 - \langle z, \tau \rangle)^2}(z - \tau). \tag{10.218}$$

By setting $p(z) = \frac{1 - \langle F_t(z), \tau \rangle}{(\langle F_t(0), \tau \rangle - 1)(1 - \langle x, \tau \rangle)^2}$, we conclude our proof. $\qquad \square$

Example 10.7 Consider equation (10.190) with

$$f(z) = \left(z_1^2 - 1, (1 + z_1)z_2, \ldots, (1 + z_1)z_n\right), \tag{10.219}$$

which is, actually, a system of partial differential equations:

$$2h_i(z) = (z_1^2 - 1)\frac{\partial h_i(z)}{\partial z_1} + \sum_{k=2}^{n}(1 + z_1)z_k \frac{\partial h_i(z)}{\partial z_k}, \quad i = 1, \ldots, n. \tag{10.220}$$

Solving the Cauchy problem which in this case has the form

$$\begin{cases} \dfrac{\partial u_1(t, z)}{\partial t} + \left(u_1^2(t, z) - 1\right) = 0 \\[2mm] \dfrac{\partial u_i(t, z)}{\partial t} + \left(1 + u_1(t, z)\right) u_i(t, z) = 0, \quad i = 2, \ldots, n \\[2mm] u_1(0, z) = z_1 \\[2mm] u_i(0, z) = z_i, \quad i = 2, \ldots, n \end{cases} \tag{10.221}$$

we find that

$$u(t, z) - \tau = \frac{2}{(1 - z_1) + e^{2t}(1 + z_1)}\,(z - \tau). \tag{10.222}$$

Thus, the mappings h_t defined in (10.205) is seen to be

$$h_t(z) = \frac{1 + e^{2t}}{(1 - z_1) + e^{2t}(1 + z_1)}\,(\tau - z) \tag{10.223}$$

and the limit

$$h(z) = \lim_{t \to \infty} h_t(z) = \frac{1}{1 + z_1}\,(\tau - z) \tag{10.224}$$

exists. It is easy to verify that the mapping $h(z)$ satisfies the system (10.220) which is equivalent to (10.190). Hence it is a univalent mapping with respect to a boundary point.

Furthermore, $h\big|_{Q\mathbb{B}}$ is of the form

$$h(\xi) = \tau - \xi, \;\; \xi = (0, z_2 \ldots, z_n). \tag{10.225}$$

Therefore, it follows from Theorem 10.8 that $h(z)$ defined by (10.224) is the unique solution of (10.190) (or (10.220)).

On the other hand, once the initial condition (10.225) is given, one can find h directly by formula (10.169) which is presented in Theorem 10.8.

Indeed, in our example $p(z) = \frac{1+z_1}{1-z_1}$. Hence one can find, by (10.154), that $\tilde{p}(w) = \frac{1}{2w_1}$ and $k(w) = \frac{1}{2}\log w_1$ (see Lemma 10.4). Substituting now $k(w)$, $\Gamma = 2I$, and $\phi(\zeta) = (1, -\zeta)$, $\zeta \in \mathbb{C}^{n-1}$, we get, by (10.169), the representation

$$h(z) = e^{-2 \cdot \frac{1}{2}\log w_1}(\tau - \xi)\Big|_{w_1 = \frac{1+z_1}{1-z_1}, \xi = \frac{1}{1-z_1}Q_z} =$$

$$\left(\frac{1-z_1}{1+z_1}, \frac{-z_2}{1+z_1}, \ldots, \frac{-z_n}{1+z_1}\right), \qquad (10.226)$$

which coincides with (10.224).

Bibliography

Abate, M. (1988a). Horospheres and iterates of holomorphic maps, *Math. Z.* **198**, pp. 225–238.

Abate, M. (1988b). Converging semigroups of holomorphic maps, *Atti Accad. Naz. Lincei Rend. Cl. Sci. Fis. Mat. Natur.* **82**, pp. 223–227.

Abate, M. (1989a). Common fixed points of commuting holomorphic maps, *Math. Ann.* **283**, pp. 645–655.

Abate, M. (1989b). Iteration Theory of Holomorphic Maps on Taut Manifolds, Mediterranean Press, Rende.

Abate, M. (1992) The infinitesimal generators of semigroups of holomorphic maps, *Ann. Mat. Pura Appl.* **161**, pp. 167–180.

Abate, M. (1998). The Julia–Wolff–Carathéodory theorem in polydisks, *J. Analyse Math.* **74**, pp. 275–306.

Abate, M and Vigué, J.-P. (1991). Common fixed points in hyperbolic Riemann surfaces and convex domains, *Proc. Amer. Math. Soc.* **112**, pp. 503–512.

Abts, D. (1980). On injective holomorphic Fredholm mappings of index 0 in complex Banach spaces, *Comment. Math. Univ. Carolinae* **21**, pp. 513–525.

Aharonov, D., Elin, M., Reich, S. and Shoikhet, D. (1999a). Parametric representations of semi-complete vector fields on the unit balls in $\mathbb{C}^n$ and Hilbert space, *Atti Accad. Naz. Lincei* **10**, pp. 229–253.

Aharonov, D., Elin, M. and Shoikhet, D. (2003). Spirallike functions with respect to a boundary point, *J. Math. Anal. Appl.* **280**, pp. 17–29.

Aharonov, D., Reich, S. and Shoikhet, D. (1999b). Flow invariance conditions for holomorphic mappings in Banach spaces, *Math. Proceedings of the Royal Irish Academy* **99A**, pp. 93–104.

Aizenberg, L., Reich, S. and Shoikhet, D. (1996). One-sided estimates for the existence of null points of holomorphic mappings in Banach spaces, *J. Math. Anal. Appl.* **203**, pp. 38–54.

Aizenberg, L. and Yuzhakov, A. (1983). Integral Representations and Residues in Multidimensional Complex Analysis, AMS, Providence, RI.

Alexander, J. W. (1915). Functions which map the interior of the unit circle upon simple regions, *Annals of Math.* **17**, pp. 12–22.

Aleksandrov, A. B. (1994). Function theory in the ball, in: Khenkin, G. M. and

Vitushkin, A. G. (eds.), Several Complex Variables II, Springer–Verlag, pp. 107–178.

Aleksandrov, I. A. (1976). Parametric Extensions in the Theory of Univalent Functions, Nauka, Moscow.

Arazy, J. (1987). An application of infinite dimensional holomorphy to the geometry of Banach spaces, *Lecture Notes in Math.* **1267**, pp. 122–150.

Barbu, V. (1976). Nonlinear Semigroups and Differential Equations in Banach Spaces, Noordhoff, Leyden.

Barnard, R. W., FitzGerald, C. H. and Gong, S. (1991). The growth and 1/4-theorems for starlike mappings in $\mathbb{C}^n$, *Pacific J. Math.* **150**, pp. 13–22.

Berkson, E., Kaufman, R. and Porta, H. (1974). Möbius transformations on the unit disk and one-parametric groups of isometries of H^p, *Trans. Amer. Math. Soc.* **199**, pp. 223–239.

Berkson E. and Porta, H. (1978). Semigroups of analytic functions and composition operators, *Michigan Math. J.* **25**, pp. 101–115.

Brézis, H. (1970). On a characterization of flow–invariant sets, *Communications on Pure and Applied Mathematics* **23**, pp. 261–263.

Brézis, H. (1973). Opérateurs Maximaux Monotones, North Holland, Amsterdam.

Brickman, L. (1973). Φ-like analytic functions, I, *Bull. Amer. Math. Soc.* **79**, pp. 555–558.

Browder, F. (1976). Nonlinear Operators and Nonlinear Equations of Evolution in Banach Spaces, Proc. Symp. Pure Math., Vol. 18, Amer. Math. Soc., Providence, R.I.

Budzyńska, M. (2004). Holomorphic retracts in domains with the locally uniformly linearly convex Kobayashi distance, *Contemporary Math.* **364**, pp. 27–34.

Burckel, R. B. (1981). Iterating analytic self-maps of discs, *Am. Math. Monthly* **88**, pp. 396–407.

Carathéodory, C. (1929). Uber die Winkelderivierten von beschränkten Analytischen Funktionen, *Sitzungsber. Preuss. Akad. Wiss. Berlin*, pp. 39–54.

Carathéodory, C. (1954). Theory of Functions of a Complex Variable, Chelsea, New York.

Cartan, H. (1933) Sur la possibilité d'étendre aux fonctions de plusieurs variables complexes la theorie des fonctions univalents, in [Montel (1933)], pp. 129–155.

Cartan, H. (1967). Calcul Différentiel, Hermann, Paris.

Cartan, H. (1986). Sur les retractions d'une variété, *C.R. Acad. Sci. Paris* **303**, pp. 715–716.

Chae, S. B. (1985). Holomorphy and Calculus in Normed Spaces, Marcel Dekker, New York and Basel.

Chen, G.-N. (1984). Iteration for holomorphic maps of the open unit ball and the generalized upper half-plane of $\mathbb{C}^n$, *J. Math. Anal. Appl.* **98**, pp. 305–313.

Chirka, E. (1985). Complex Analytic Sets, Nauka, Moscow.

Chu, C.-H. and Mellon, P. (1997). Iteration of compact holomorphic maps on a Hilbert ball, *Proc. Amer. Math. Soc.* **125**, pp. 1771–1777.

Chuaqui, M. (1995). Applications of subordination chains to starlike mappings in

$\mathbb{C}^n$, *Pacific J. Math.*, **168**, pp. 33–48.

Cowen, C. C. and MacCluer, B. D. (1995). Composition Operators on Spaces of Analytic Functions, CRC Press, Boca Raton, FL.

Daletskii, Yu. L. and Krein, M. G. (1970). Stability of Solutions of Differential Equations in a Banach Space, Nauka, Moscow.

Deimling, K. (1974). Zeros of accretive operators, *Manuscripta Math.* **13**, pp. 365–374.

Deimling, K. (1992). Multivalued Differential Equations, W. de Gruyter & Co., Berlin.

Denjoy, A. (1926). Sur l'itération des fonctions analytiques, *C. R. Acad. Sci. Paris* **182**, pp. 255–257.

Dineen, S. (1989). The Schwarz Lemma, Clarendon Press, Oxford.

Dineen, S., Timoney, R. M. and Vigué, J.-P. (1985). Pseudodistances invariantes sur les domaines d'un espace localement convexe, *Ann. Scuola Norm. Sup. Pisa* **12**, pp. 515–529.

Dixmier, P. J. (1969). Les C^*-Algèbres et Leurs Représentations, Paris, Gauthier–Villars.

Duc, Thai Do (1992). The fixed points of holomorphic maps on a convex domain, *Ann. Polon. Math.* **56**, pp. 143–148.

Dunford, N. (1943). Spectral theory I. Convergence to projections, *Trans. Amer. Math. Soc.* **54**, pp. 185–217.

Dunford, N. and Schwarz, J. T. (1958). Linear Operators, Interscience Publ., New York and London.

Duren, P. (1983). Univalent Functions, Springer, Berlin.

Earle, C. J. and Hamilton, R. S. (1970). A fixed point theorem for holomorphic mappings, *Proc. Symp. Pure Math.*, Vol. 16, Amer. Math. Soc., Providence, R.I., pp. 61–65.

Edelstein, M. (1962). On fixed and periodic points under contractive mappings, *J. London Math. Soc.* **37**, pp. 74–79.

Edelstein, M. (1964). On non-expansive mappings of Banach spaces, *Proc Cambridge Phil. Soc.* **60**, pp. 439–447.

Edelstein, M. (1972). The construction of an asymptotic center with a fixed point property, *Bull. Amer. Math. Soc.* **78**, pp. 206–208.

Elin, M., Harris, L. A., Reich, S. and Shoikhet, D. (2002a). Evolution equations and geometric function theory in J^*-algebras, *J. of Nonlinear and Convex Anal.* **3**, pp. 81-121.

Elin, M., Reich, S. and Shoikhet, D. (2000). Holomorphically accretive mappings and spiral-shaped functions of proper contractions, *Nonlinear Analysis Forum* **5**, pp. 149–161.

Elin, M., Reich, S. and Shoikhet, D. (2001). Dynamics of inequalities in geometric function theory, *J. of Inequal. & Appl.* **6**, pp. 651–664.

Elin, M., Reich, S. and Shoikhet, D. (2002). Asymptotic behavior of semigroups of ρ-nonexpansive and holomorphic mappings on the Hilbert ball, *Ann. Mat. Pura Appl.* **181**, pp. 501–526.

Elin, M., Reich, S. and Shoikhet, D. (2004). Complex dynamical systems and the geometry of domains in Banach spaces, *Dissertations Math.* **427**.

Elin M. and Shoikhet, D. (2001). Dynamic extension of the Julia–Wolff–Carathéodory Theorem, *Dynamic Systems and Applications* **10**, pp. 421–438.

de Fabritiis, C. (1994). On the linearization of a class of semigroups on the unit ball of $\mathbb{C}^n$, *Ann. Mat. Pura Appl.* **166**, pp. 363–379.

Fan, Ky (1978). Analytic functions of a proper contraction, *Math. Z.* **160**, pp. 275–290.

Fan, Ky (1979). Julia's lemma for operators, *Math. Ann.* **239**, pp. 241–245.

Fan, Ky (1982). Iterations of analytic functions of operators, *Math. Z.* **179**, pp. 293–298.

Fan, Ky (1983). Iterations of analytic functions of operators II, *Linear and Multilinear Algebra* **12**, pp. 295–304.

Fan, Ky (1986). The angular derivative of an operator-valued analytic function, *Pacific J. Math.* **121**, pp. 67–72.

Fan, Ky (1988). Inequalities for proper contractions and strictly dissipative operators, *Linear Algebra Appl.* **105**, pp. 237–248.

Fischer, G. (1976). Complex Analytic Geometry, Lecture Notes in Math., **538**, Springer, Berlin.

Forelli, F. (1964). The isometries of H^p, *Canad. J. Math.* **16**, pp. 721–728.

Franzoni, T. and Vesentini, E. (1980). Holomorphic Maps and Invariant Distances, North-Holland, Amsterdam.

Goebel, K. (1982a). Uniform convexity of Carathéodory's metric on the Hilbert ball and its consequences, *Symposia Mathematica* **26**, pp. 163–179.

Goebel, K. (1982b). Fixed points and invariant domains of holomorphic mappings of the Hilbert ball, *Nonlinear Analysis* **6**, pp. 1327–1334.

Goebel, K. and Kirk, W. A. (1990). Topics in Metric Fixed Point Theory, Cambridge University Press, Cambridge.

Goebel, K. and Reich, S. (1982). Iterating holomorphic self-mappings of the Hilbert ball, *Proc. Japan Acad.* **58**, pp. 349–352.

Goebel, K. and Reich, S. (1984). Uniform Convexity, Hyperbolic Geometry and Nonexpansive Mappings, Marcel Dekker, New York and Basel.

Goebel, K., Sekowski, T. and Stachura, A. (1980). Uniform convexity of the hyperbolic metric and fixed points of holomorphic mappings in the Hilbert ball, *Nonlinear Analysis* **4**, pp. 1011–1021.

Gohberg, I. and Markus, A. (1960). Characteristic properties of a pole of the resolvent of a linear closed operator. *Ucheny Zapiski Beltskogo Gosped.* **5**, pp. 71–76.

Golusin, G.M. (1969). Geometric Theory of Functions of a Complex Variable, Amer. Math. Soc., Providence, RI.

Gong, S. (1999). Convex and Starlike Mappings in Several Complex Variables, Science Press, Beijing–New York & Kluwer Acad. Publ., Dordrecht–Boston–London.

Goodman, A. W. (1983). Univalent Functions, Vols. I, II, Mariner Publ. Co., Tampa, FL.

Gurganus, K. R. (1975). Φ-like holomorphic functions in $\mathbb{C}^n$ and Banach space, *Trans. Amer. Math. Soc.* **205**, pp. 389–406.

Harris, L. A. (1969). Schwarz's lemma in normed linear spaces, *Proc. Nat. Acad. Sci. U.S.A.* **62**, pp. 1014–1017.

Harris, L. A. (1971a). A continuous form of Schwarz's lemma in normed linear spaces, *Pacific J. Math.* **38**, pp. 635–639.

Harris, L. A. (1971b). The numerical range of holomorphic functions in Banach spaces, *Amer. J. Math.* **93**, pp. 1005–1019.

Harris, L. A. (1974a). Bounded symmetric homogeneous domains in infinite-dimensional spaces, *Lecture Notes in Math.* **364**, pp. 13–40.

Harris, L. A. (1974b). The numerical range of functions and best approximation, *Proc. Camb. Phil. Soc.* **76**, pp. 133–141.

Harris, L. A. (1977). On the size of balls covered by analytic transformations, *Monatshefte für Mathematik* **83**, pp. 9–23.

Harris, L. A. (1979). Schwarz–Pick systems of pseudometrics for domains in normed linear spaces, *Advances in Holomorphy*, North Holland, Amsterdam, pp. 345–406.

Harris, L. A. (1981). A generalization of C^*-algebras, *Proc. London Math. Soc.* **42**, pp. 331–361.

Harris, L. A., Reich, S. and Shoikhet, D. (2000). Dissipative holomorphic functions, Bloch radii, and the Schwarz lemma, *J. Analyse Math.* **82**, pp. 221–232.

Harris, L. A. (2003). Fixed points of holomorphic mappings for domains in Banach spaces, *Abstract and Appl. Analysis* **2003**, pp. 261–274.

Harris, T. E. (1963). The Theory of Branching Processes, Springer, Berlin.

Hayden, T. L. and Suffridge, T. J. (1971). Biholomorphic maps in Hilbert space have a fixed point, *Pacific J. Math.* **38**, pp. 419–422.

Hayden, T. L. and Suffridge, T. J. (1976). Fixed points of holomorphic maps in Banach spaces, *Proc. Amer. Math. Soc.* **60**, pp. 95–105.

Heath, L. F. and Suffridge, T. J. (1979). Starlike, convex, close-to-convex, spiral-like and ϕ-like maps in a commutative Banach algebra with identity, *Trans. Amer. Math. Soc.* **250**, pp. 195–212.

Heins, M. H. (1941). On the iteration of functions which are analytic and single-valued in a given multiply-connected region, *Amer. J. Math.* **63**, pp. 461–480.

Helmke, U. and Moore, J. B. (1994). Optimization and Dynamical Systems, Springer, London.

Hervé, M. (1963a). Several Complex Variables: Local Theory, Tata Institute of Fundamental Research, Bombay and Oxford Univ. Press, London.

Hervé, M. (1989). Analyticity in Infinite Dimensional Spaces, Walter de Gruyter, Berlin.

Hille, E. and Phillips, R. (1957). Functional Analysis and Semigroups, Amer. Math. Soc., Providence, R.I.

Isidro, J. M. and Stacho, L. L. (1984). Holomorphic Automorphism Groups in Banach Spaces: An Elementary Introduction, North Holland, Amsterdam.

Isidro, J. M. and Vigué, J. P. (1984). The group of biholomorphic automorphisms of symmetric Siegel domains and its topology, *Ann. Scuola Norm. Sup. Pisa* **11**, pp. 343–351.

Jarnicki, M. and Pflug, P. (1993). Invariant Distances and Metrics in Complex Analysis, Walter de Gruyter, Berlin.

Jordan, D. W. and Smith, P (1987). Nonlinear Ordinary Differential Equations, Clarendon Press, Oxford.

Kakutani, S. (1941). A generalization of Brouwer's fixed point theorem, *Duke Math. J.* **8**, pp. 457–459.

Kapeluszny, J., Kuczumow, T. and Reich, S. (1999a). The Denjoy–Wolff theorem in the open unit ball of a strictly convex Banach space, *Advances in Math.* **143**, pp. 111–123.

Kapeluszny, J., Kuczumow, T. and Reich, S. (1999b). The Denjoy–Wolff theorem for condensing holomorphic mappings, *J. Functional Anal.* **167**, pp. 79–93.

Kaplan, W. (1952). Close-to-convex schlicht functions, *Michigan Math. J.* **1**, pp. 169–185.

Kato, T. (1966). Perturbation Theory for Linear Operators, Springer, Berlin.

Kato, T. (1967). Nonlinear semigroups and evolution equations, *J. Math. Soc. Japan* **19**, pp. 508–520.

Kaup, W. (1983).A Riemann mapping theorem for bounded symmetric domains in complex Banach spaces, *Math. Z.* **183**, pp. 503–529.

Kaup, W. and Vigué, J.-P. (1990). Symmetry and local conjugacy on complex manifolds, *Math. Ann.* **268**, pp. 329–340.

Kelley, J. L. (1975). General Topology, Springer, Berlin.

Khatskevich, V., Reich, S. and Shoikhet, D. (1995a). Fixed point theorems for holomorphic mappings and operator theory in indefinite metric spaces, *Integral Equations Operator Theory* **22**, pp. 305–316.

Khatskevich, V., Reich, S. and Shoikhet, D. (1998). Asymptotic behavior of solutions of evolution equations and construction of holomorphic retractions, *Math. Nachr.* **189**, pp. 171–178.

Khatskevich, V. and Shoikhet, D. (1984). Fixed points of analytic operators in a Banach space and applications, *Siberian Math. J.* **25**, 188–200.

Khatskevich, V. and Shoikhet, D. (1994a). Differentiable Operators and Nonlinear Equations, Birkhäuser, Basel.

Kirk, A. and Sims, B. (Eds.) (2001). Handbook of Metric Fixed Point Theory, Kluwer Academic Publishers, Dordrecht.

Koliha J. J. (1974). Power convergence and pseudoinverses of operators in Banach spaces, *J. Math. Anal. Appl.* **48**, pp. 446–469.

Krasnoselskii, M. A., Vainikko, G. M., Zabreiko, P. P., Ruticki, Ya. B. and Stecenko, V. Ya. (1969). Approximate Solution of Operator Equations, Nauka, Moscow.

Krasnoselskii, M. A. and Zabreiko, P. P. (1984). Geometrical Methods of Nonlinear Analysis, Springer, Berlin.

Krein, S. G. (1971). Linear Differential Equations in a Banach Space, Amer. Math. Soc., Providence, RI.

Kryczka, A. and Kuczumow, T. (1997). The Denjoy–Wolff-type theorem for compact k_{B_H}-nonexpansive maps in a Hilbert ball, *Ann. Univ. Mariae Curie-Sklodowska* **51**, pp. 179–183.

Kubota, Y. (1983). Iteration of holomorphic maps of the unit ball into itself, *Proc.*

Amer. Math. Soc., **88**, pp. 476–480.

Kuczumow, T. (1984). Common fixed points of commuting holomorphic mappings in Hilbert ball and polydisc, *Nonlinear Analysis* **8**, pp. 417–419.

Kuczumow, T. (1985). Nonexpansive retracts and fixed points of nonexpansive mappings in the Cartesian product of n Hilbert balls, *Nonlinear Anlysis* **9**, pp. 601–604.

Kuczumow, T., Reich, S. and Shoikhet, D. (2001a). The existence and non-existence of common fixed points for commuting families of holomorphic mappings, *Nonlinear Anal.* **43**, pp. 45–59.

Kuczumow, T., Reich, S. and Shoikhet, D. (2001b). Fixed point of holomorphic mappings: a metric approach, in: Handbook of Metric fixed Point Theory, Kluwer, Dordrecht, pp. 437–515.

Kuczumow, T. and Stachura, A. (1990). Iterates of holomorphic and k_D-nonexpansive mappings in convex domains in $\mathbb{C}^n$, *Advances in Math.* **81**, pp. 90–98.

Kufarev, P. P. (1943). On one-parameter families of analytic functions, *Mat. Sb.* **13**, pp. 87–118.

Kufarev, P. P. (1947). A theorem on solutions of a differential equation, *Uchen. Zap. Tomsk. Gos. Univ.* **5**, pp. 20–21.

Lebedev, N. A. (1975). The Square Principle in the Theory of Univalent Functions, Nauka, Moskow.

Lempert, L. (1982). Holomorphic retracts and intrinsic metrics in convex domains, *Anal. Math.* **8**, pp. 257–261.

Löwner, K. (1923). Untersuchungen über schlichte konforme Abbildungen des Einheitskreises, I, *Math. Ann.* **89**, pp. 103–121.

Lumer, G. and Phillips, R. S. (1961). Dissipative operators in a Banach space, *Pacific J. Math.* **11**, pp. 679–698.

Lyubich, Yu and Zemanek, J. (1994). Precompactness in the uniform ergodic theory, *Studia Mathematica* **112**, pp. 89–97.

Lyzzaik, A. (1984). On a conjecture of M.S. Robertson, *Proc. Amer. Math. Soc.* **91**, pp. 108–110.

MacCluer, B. D. (1983). Iterates of holomorphic self-maps of the unit ball in $\mathbb{C}^n$, *Michigan Math. J.* **30**, pp. 97–106.

Martin, R. H., Jr. (1973). Differential equations on closed subsets of a Banach space, *Trans. Amer. Math. Soc.*, **179**, pp. 399–414.

Matsuno, T. (1955). Star-like theorems and convex-like theorems in the complex vector space, *Sci. Rep. Tokyo Kyoiku Daigaku* **5**, pp. 88–95.

Mazet, P. (1992). Les points fixes d'une application holomorphe d'un domaine borné dans lui-même admettent une base de voisinages convexes stable, *C. R. Acad. Sci. Paris* **314**, pp. 197–199.

Mazet, P. and Vigué, J.-P. (1991). Points fixes d'une application holomorphe d'un domaine borné dans lui-même, *Acta Math.* **166**, pp. 1–26.

Mazet, P. and Vigué, J.-P. (1992). Convexité de la distance de Carathéodory et points fixes d'applications holomorphes, *Bull. Sci. Math.* **16**, pp. 285–305.

P. Mellon. (1996). Another look at results of Wolff and Julia type for J^*-algebras, *J. Math. Anal. Appl.* **198**, pp. 444–457.

Mercer, P. R. (1992). Proper maps, complex geodesics and iterates of holomorphic maps on convex domains in $\mathbb{C}^n$, *Contemporary Mathematics*, **137**, pp. 339–342.

Mercer, P. R. (1993). Complex geodesics and iterates of holomorphic maps on convex domains in $\mathbb{C}^n$, *Trans. Amer. Math. Soc.* **338**, pp. 201–211.

Mercer, P. R. (1997). Sharpened versions of the Schwarz lemma, *J. Math. Anal. Appl.* **205**, pp. 508–511.

Mercer, P. R. (1999). On a strengthened Schwarz–Pick inequality, *J. Math. Anal. Appl.* **234**, pp. 735–739.

Mercer, P. R. (2000). Another look at Julia's lemma, *Complex Variables Theory Appl,* **43**, pp. 129–138.

Montel, P. (1933). Leçons sur les Fonctions Univalentes ou Multivalentes, Gauthier–Villars, Paris.

Nevanlinna, R. (1921). Über die konforme Abbildung Sterngebieten, *Oeversikt av Finska-Vetenskaps Societeten Ferhandlinger* **63(A)**, No. 6.

Nussbaum, R. D. (1994). Finsler structures for the part metric and Hilbert's projective metric and applications, *Diff. and Integral Equations* **7**, pp. 1649–1707.

Pfaltzgraff, J. A. (1974). Subordination chains and univalence of holomorphic mappings in $\mathbb{C}^n$, *Math. Ann.* **210**, pp. 55–68.

Pfaltzgraff, J. A. (1975). Subordination chains and quasiconformal extension of holomorphic maps in $\mathbb{C}^n$, *Ann. Acad. Sci. Fen. Ser. A Math.* **1**, pp. 13–25.

Pfaltzgraff, J. A. and Suffridge, T. J. (1975). Close-to-starlike holomorphic functions of several variables, *Pacific J. Math.* **57**, pp. 271–279.

Poreda, T. (1987). On the univalent holomorphic maps of the unit polydisc in $\mathbb{C}^n$ which have the parametric representation I, II, *Ann. Univ. Mariae Curie-Sklodowska* **41**, pp. 105–113, pp. 115–121.

Potapov, V. P. (1960). The multiplicative structure of J-contractive matrix functions, *Amer. Math. Soc. Transl.* **2**,15, pp. 231–243.

Reich, S. (1975). Minimal displacement of points under weakly inward pseudo-lipschitzian mappings, *Atti Accad. Naz. Lincei* **59**, pp. 40–44.

Reich, S. (1976). On fixed point theorems obtained from existence theorems for differential equations, *J. Math. Anal. Appl.* **54**, pp. 26–36.

Reich, S. (1985). Averaged mappings in the Hilbert ball, *J. Math. Anal. Appl.* **109**, pp. 199–206.

Reich, S. (1991). The asymptotic behavior of a class of nonlinear semigroups in the Hilbert ball, *J. Math. Anal. Appl.* **157**, pp. 237–242.

Reich, S. (1992). Approximating fixed points of holomorphic mappings, *Math. Japon.* **37**, pp. 457–459.

Reich, S and Shafrir, I. (1987). The asymptotic behavior of firmly nonexpansive mappings, *Proc. Amer. Math. Soc.* **101**, pp. 246–250.

Reich, S. and Shafrir, I. (1990). Nonexpansive iterations in hyperbolic spaces, *Nonlinear Analysis* **15**, pp. 537–558.

Reich, S. and Shoikhet, D. (1996). Generation theory for semigroups of holomorphic mappings in Banach spaces, *Abstr. Appl. Anal.* **1**, pp. 1–44.

Reich, S. and Shoikhet, D. (1997a). Semigroups and generators on convex domains with the hyperbolic metric, *Atti. Accad. Naz. Lincei* **8**, pp. 231–250.

Reich, S. and Shoikhet, D. (1997b). The Denjoy–Wolff theorem, *Ann. Univ. Mariae Curie-Sklodowska* **51**, pp. 219–240.

Reich, S. and Shoikhet, D. (1998a). A characterization of holomorphic generators on the Cartesian product of Hilbert balls, *Taiwanese J. Math.* **2**, pp. 383–396.

Reich, S. and Shoikhet, D. (1998b). Metric domains, holomorphic mappings and nonlinear semigroups, *Abstr. Appl. Anal.* **3**, pp. 203–228.

Reich, S. and Shoikhet, D. (1998c). Averages of holomorphic mappings and holomorphic retractions on convex hyperbolic domains, *Studia Math.* **130**, pp. 231–244.

Reiffen, H. J. (1965). Die Carathéodorysche Distanz und ihre zugehörige Differentialmetrik, *Math. Ann.* **161**, pp. 315–324.

Robertson, M. S. (1936). On the theory of univalent functions, *Ann. of Math.* **37**, pp. 374–408.

Robertson, M. S. (1961). Applications of the subordination principle to univalent functions, *Pacific J. Math.* **11**, pp. 315–324.

Robertson, M. S. (1981). Univalent functions starlike with respect to a boundary point, *J. Math. Anal. Appl.* **81**, pp. 327–345.

Rudin, W. (1973). Functional Analysis, McGraw-Hill, New York.

Rudin, W. (1974). Real and Complex Analysis, McGraw-Hill, New York.

Rudin, W. (1978). The fixed point sets of some holomorphic maps, *Bull. Malaysian Math. Soc.* **1**, pp. 25–28.

Rudin, W. (1980). Function Theory on the Unit Ball in $\mathbb{C}^n$, Springer, Berlin.

Sevastyanov, B. A. (1971). Branching Processes, Nauka, Moscow.

Shabat, B. (1976). Introduntion to Complex Analysis, Part 2, Nauka, Moscow.

Shafrir, I. (1992a). Common fixed points of commuting holomorphic mappings in the product of n Hilbert balls, *Michigan Math. J.* **39**, pp. 281–287.

Shafrir, I. (1992b). Coaccretive operators and firmly nonexpansive mappings in the Hilbert ball, *Nonlinear Analysis* **18**, pp. 637–648.

Shapiro, J. H. (1993). Composition Operators and Classical Function Theory, Springer, Berlin.

Shoikhet, D. (1986). The invariance principle in the fixed point theory of analytic operators, preprint, Institute of Physics, Siberian Branch of the Academy of Sciences of the USSR, No. 33M.

Shoikhet, D. (1993). Some properties of Fredholm mappings in Banach analytic manifolds, *Integral Equations Operator Theory* **16**, pp. 430–451.

Silverman, H. and Silvia, E. M. (1990). Subclasses of univalent functions starlike with respect to a boundary point, *Houston J. Math.* **16**, pp. 289–299.

Sine, R. (1989). Behavior of iterates in the Poincaré metric, *Houston Journal of Mathematics* **15**, pp. 273–289.

Špaček, L. (1933). Přispěvek k teorii funkci prostych, *Časopis Pěst. Mat.*, **62**, pp. 12–19.

Stachura, A. (1985). Iterates of holomorphic self-maps of the unit ball in Hilbert spaces, *Proc. Amer. Math. Soc.* **93**, pp. 88–90.

Study, E. (1913). Konforme Abbildung Einfachzusammenhägender Bereiche,

B. G. Teubner, Leipzig and Berlin.

Suffridge, T. J. (1970). The principle of subordination applied to functions of several variables, *Pacific J. Math.* **33**, pp. 241–248.

Suffridge, T. J. (1972). A holomorphic function having a discontinuous inverse, *Proc. Amer. Math. Soc.* **31**, pp. 629–630.

Suffridge, T. J. (1973). Starlike and convex maps in Banach space, *Pacific J. Math.* **46**, pp. 575–589.

Suffridge, T. J. (1974). Common fixed points of commuting holomorphic maps of the hyperball, *Michigan Math. J.* **21**, pp. 309–314.

Suffridge, T. J. (1977). Starlikeness, convexity and other geometric properties of holomorphic maps in higher dimensions, *Complex Analysis (Proc. Conf. Univ. Kentucky, Lexington, KY, 1976)*, Lecture Notes in Math. **599**, pp. 146–159.

Thorp, E. and Whitley, R. (1967). The strong maximum modulus theorem for analytic functions into a Banach space, *Proc. Amer. Math. Soc.* **18**, pp. 640–646.

Trenogin, V. A. (1980). Functional Analysis, Nauka, Moscow.

Upmeier, H. (1986). Jordan Algebras in Analysis, Operator Theory and Quantum Mechanics, CBMS, Reg. Conf. Ser. in Math., Vol. **67**, Amer. Math. Soc., Providence, RI.

Vesentini, E. (1983). Su un teorema di Wolff e Denjoy, *Rend. Sem. Mat. Fis. Milano* **53**, pp. 17–25.

Vesentini, E. (1985). Iterates of holomorphic mappings, *Uspekhi Mat. Nauk* **40**, pp. 13–16.

Vesentini, E. (1987a). Semigroups of holomorphic isometries, *Advances in Math.* **65**, pp. 272–306.

Vesentini, E. (1987b). Holomorphic families of holomorphic isometries, *Lecture Notes in Mathematics* **1227**, pp. 290–302.

Vesentini, E. (1991). Krein spaces and holomorphic isometries of Cartan domains, *Geometry and Complex Variables*, Marcel Dekker, New York, pp. 409–413.

Vesentini, E. (1992). Holomorphic isometries of Cartan domains of type four, *Atti. Accad. Naz. Lincei* **3**, pp. 287–294.

Vesentini, E. (1994a). Semigroups of holomorphic isometries, in *Complex Potential Theory*, Kluwer, Dordrecht, pp. 475–548.

Vesentini, E. (1994b). Semigroups of linear contractions for an indefinite metric, *Atti. Accad. Naz. Lincei Mem.* **2**, pp. 53–83.

Vesentini, E. (1996a). Conservative operators, in *Partial Differential Equations and Applications*, Marcel Dekker, New York, pp. 303–311.

Vigué, J.-P. (1986). Sur les points fixes d'applications holomorphes, *C.R. Acad. Sci. Paris* **303**, pp. 927–930.

Vigué, J.-P. (1991a). Fixed points of holomorphic mappings in a bounded convex domain in $\mathbb{C}^n$, *Proc. Symp. Pure Math.* **52**, pp. 579–582.

Wald, J. K. (1978) On starlike functions, Ph. D. Thesis, Univ. of Delaware, Newark, Delaware.

Webb, J. R. L. (1996). Zeros of weakly inward accretive mappings via A-proper maps, in: Kartsatos, A. G. (ed.), Theory and Applications of Nonlinear

Operators of Accretive and Monotone Type, Marcel Dekker, New York, pp. 289–297.

Wlodarczyk, K. (1985). Iterations of holomorphic maps of infinite dimensional homogeneous domains, *Monatsh. Math.* **99**, pp. 153–160.

Wlodarczyk, K. (1986). Studies of iterations of holomorphic maps in J^*-algebras and complex Hilbert spaces, *Quart. J. Math. Oxford* **37**, pp. 245–256.

Wlodarczyk, K. (1987). Julia's lemma and Wolff's theorem for J^*-algebras, *Proc. Amer. Math. Soc.* **99**, pp. 472–476.

Wlodarczyk, K. (1995). Fixed points and invariant domains of expansive holomorphic maps in complex Banach spaces, *Advances in Mathematics* **110**, pp. 247–254.

Wolff, J. (1926a). Sur l'itération des fonctions holomorphes dans une region, et dont les valeurs appartiennent à cette région, *C. R. Acad. Sci. Paris* **182**, pp. 42–43.

Wolff, J. (1926b). Sur l'itération des fonctions bornées, *C. R. Acad. Sci. Paris* **182**, pp. 200–201.

Wolff, J. (1926c). Sur une généralisation d'un théorème de Schwarz, *C. R. Acad. Sci. Paris* **182**, pp. 918–920.

Yang, P. (1978). Holomorpic curves and boundary regularity of biholomorpic maps of pseudoconvex domains, preprint.

Yosida, K. (1974). Functional Analysis, Springer, Berlin, Fourth Edition.

Index